AF771423

Oleg Okun and Giorgio Valentini (Eds.)

Applications of Supervised and Unsupervised Ensemble Methods

Studies in Computational Intelligence, Volume 245

Editor-in-Chief
Prof. Janusz Kacprzyk
Systems Research Institute
Polish Academy of Sciences
ul. Newelska 6
01-447 Warsaw
Poland
E-mail: kacprzyk@ibspan.waw.pl

Further volumes of this series can be found on our homepage: springer.com

Vol. 223. Zbigniew W. Ras and Agnieszka Dardzinska (Eds.)
Advances in Data Management, 2009
ISBN 978-3-642-02189-3

Vol. 224. Amandeep S. Sidhu and Tharam S. Dillon (Eds.)
Biomedical Data and Applications, 2009
ISBN 978-3-642-02192-3

Vol. 225. Danuta Zakrzewska, Ernestina Menasalvas, and Liliana Byczkowska-Lipinska (Eds.)
Methods and Supporting Technologies for Data Analysis, 2009
ISBN 978-3-642-02195-4

Vol. 226. Ernesto Damiani, Jechang Jeong, Robert J. Howlett, and Lakhmi C. Jain (Eds.)
New Directions in Intelligent Interactive Multimedia Systems and Services - 2, 2009
ISBN 978-3-642-02936-3

Vol. 227. Jeng-Shyang Pan, Hsiang-Cheh Huang, and Lakhmi C. Jain (Eds.)
Information Hiding and Applications, 2009
ISBN 978-3-642-02334-7

Vol. 228. Lidia Ogiela and Marek R. Ogiela
Cognitive Techniques in Visual Data Interpretation, 2009
ISBN 978-3-642-02692-8

Vol. 229. Giovanna Castellano, Lakhmi C. Jain, and Anna Maria Fanelli (Eds.)
Web Personalization in Intelligent Environments, 2009
ISBN 978-3-642-02793-2

Vol. 230. Uday K. Chakraborty (Ed.)
Computational Intelligence in Flow Shop and Job Shop Scheduling, 2009
ISBN 978-3-642-02835-9

Vol. 231. Mislav Grgic, Kresimir Delac, and Mohammed Ghanbari (Eds.)
Recent Advances in Multimedia Signal Processing and Communications, 2009
ISBN 978-3-642-02899-1

Vol. 232. Feng-Hsing Wang, Jeng-Shyang Pan, and Lakhmi C. Jain
Innovations in Digital Watermarking Techniques, 2009
ISBN 978-3-642-03186-1

Vol. 233. Takayuki Ito, Minjie Zhang, Valentin Robu, Shaheen Fatima, and Tokuro Matsuo (Eds.)
Advances in Agent-Based Complex Automated Negotiations, 2009
ISBN 978-3-642-03189-2

Vol. 234. Aruna Chakraborty and Amit Konar
Emotional Intelligence, 2009
ISBN 978-3-540-68606-4

Vol. 235. Reiner Onken and Axel Schulte
System-Ergonomic Design of Cognitive Automation, 2009
ISBN 978-3-642-03134-2

Vol. 236. Natalio Krasnogor, Belén Melián-Batista, José A. Moreno-Pérez, J. Marcos Moreno-Vega, and David Pelta (Eds.)
Nature Inspired Cooperative Strategies for Optimization (NICSO 2008), 2009
ISBN 978-3-642-03210-3

Vol. 237. George A. Papadopoulos and Costin Badica (Eds.)
Intelligent Distributed Computing III, 2009
ISBN 978-3-642-03213-4

Vol. 238. Li Niu, Jie Lu, and Guangquan Zhang
Cognition-Driven Decision Support for Business Intelligence, 2009
ISBN 978-3-642-03207-3

Vol. 239. Zong Woo Geem (Ed.)
Harmony Search Algorithms for Structural Design Optimization, 2009
ISBN 978-3-642-03449-7

Vol. 240. Dimitri Plemenos and Georgios Miaoulis (Eds.)
Intelligent Computer Graphics 2009, 2009
ISBN 978-3-642-03451-0

Vol. 241. János Fodor and Janusz Kacprzyk (Eds.)
Aspects of Soft Computing, Intelligent Robotics and Control, 2009
ISBN 978-3-642-03632-3

Vol. 242. Carlos A. Coello Coello, Satchidananda Dehuri, and Susmita Ghosh (Eds.)
Swarm Intelligence for Multi-objective Problems in Data Mining, 2009
ISBN 978-3-642-03624-8

Vol. 243. Imre J. Rudas, János Fodor, and Janusz Kacprzyk (Eds.)
Towards Intelligent Engineering and Information Technology, 2009
ISBN 978-3-642-03736-8

Vol. 244. Ngoc Thanh Nguyen, Rados law Piotr Katarzyniak, and Adam Janiak (Eds.)
New Challenges in Computational Collective Intelligence, 2009
ISBN 978-3-642-03957-7

Vol. 245. Oleg Okun and Giorgio Valentini (Eds.)
Applications of Supervised and Unsupervised Ensemble Methods, 2009
ISBN 978-3-642-03998-0

Oleg Okun and Giorgio Valentini (Eds.)

Applications of Supervised and Unsupervised Ensemble Methods

Springer

Oleg Okun
Precise Biometrics AB
Scheelevagen 30, P.O. Box 798
220 07 Lund, Sweden
E-mail: oleg.okun@precisebiometrics.com,
olegokun@yahoo.com

Giorgio Valentini
Dipartimento di Scienze dell'Informazione
Università degli Studi di Milano
Via Comelico 39
20135 Milano
Italy
E-mail: valentini@dsi.unimi.it

ISBN 978-3-642-03998-0 e-ISBN 978-3-642-03999-7

DOI 10.1007/978-3-642-03999-7

Studies in Computational Intelligence ISSN 1860-949X

Library of Congress Control Number: 2009934300

© 2009 Springer-Verlag Berlin Heidelberg

This work is subject to copyright. All rights are reserved, whether the whole or part of the material is concerned, specifically the rights of translation, reprinting, reuse of illustrations, recitation, broadcasting, reproduction on microfilm or in any other way, and storage in data banks. Duplication of this publication or parts thereof is permitted only under the provisions of the German Copyright Law of September 9, 1965, in its current version, and permission for use must always be obtained from Springer. Violations are liable to prosecution under the German Copyright Law.

The use of general descriptive names, registered names, trademarks, etc. in this publication does not imply, even in the absence of a specific statement, that such names are exempt from the relevant protective laws and regulations and therefore free for general use.

Typeset & Cover Design: Scientific Publishing Services Pvt. Ltd., Chennai, India.

Printed in acid-free paper

9 8 7 6 5 4 3 2 1

springer.com

To Raisa, Gregory, and Antoshka
Oleg Okun
A Vito di Gesù, grande ricercatore dall'animo nobile, ma soprattutto un amico
– Giorgio Valentini

Preface

This book contains the extended papers presented at the 2nd Workshop on Supervised and Unsupervised Ensemble Methods and their Applications (SUEMA) held on 21–22 July, 2008 in Patras, Greece, in conjunction with the 18th European Conference on Artificial Intelligence (ECAI 2008). This workshop was a successor of the smaller event held in 2007 in conjunction with 3rd Iberian Conference on Pattern Recognition and Image Analysis, Girona, Spain. The success of that event as well as the publication of workshop papers in the edited book "Supervised and Unsupervised Ensemble Methods and their Applications", published by Springer-Verlag in Studies in Computational Intelligence Series in volume 126, encouraged us to continue a good tradition.

The scope of both SUEMA workshops (hence, the book as well) is the application of theoretical ideas in the field of ensembles of classification and clustering algorithms to real/life problems in science and industry. Ensembles, which represent a number of algorithms whose class or cluster membership predictions are combined together to produce a single outcome value, have already proved to be a viable alternative to a single best algorithm in various practical tasks under different scenarios, from bioinformatics to biometrics, from medicine to network security. The ensemble approach is caused to life by the famous "no free lunch" theorem, stating that there is no absolutely best algorithm to solve all problems. Although ensembles cannot be considered as absolute remedy of a single algorithm deficiency, it is widely believed that ensembles provide a better answer to "no free lunch" theorem than a single best algorithm. Statistical, algorithmical, representational, computational and practical reasons can explain the success of ensemble methods.

The purpose of this book is to encourage practitioners in various branches of science and techology to adopt the ensemble approach for their daily research work. We hope that fourteen chapters composing the book will be a good guide in the sea of numerous opportunities for ensemble methods.

The book has the following organization. Chapter 1 serves as a tutorial introduction into ensemble pruning (selection) methods. Chapters 2 and 3 concern classifier ensemble applications to email spam filtering. Chapter 4 describes facial expression recognition using ensembles of neural networks and error/correcting output coding. Chapter 5 deals with gene function prediction based on ensembles of support vector machines. Classification employing is the concept of partitioner trees introduced in Chapter 6. Chapter 7 proposes a new nearest neighbor-like technique to increase the diversity for ensembles of decision trees. Semi-supervised ensemble clustering of remote sensing data is considered in Chapter 8. Chapter 9 applies tenzor voting theory to image classification and inpainting. Chapter 10 focuses on clustering ensembles and semisupervised clustering under active constraints. Multi-class classification when misclassification costs vary from class to class is presented in Chapter 11. Ensemble selection for weather forecasting and air pollution dispersion prediction is explored in Chapter 12. Chapter 13 concentrates on embedding feature selection into classifier ensemble training. Chapter 14 reports a classifier ensemble application to decision support for intensive medical care.

The book is intended to be primarily a reference work. Hence, it could be a good complement to two excellent books on the general ensemble methodology – "*Combining pattern classifiers: methods and algorithms*" by Ludmila Kuncheva (John Wiley & Sons, 2004) and "*Decomposition methodology for knowledge discovery and data mining: theory and applications*" by Oded Maimon and Lior Rokach (World Scientific, 2005). Extra primal sources of information are proceedings of the biannual international workshop on Multiple Classifier Systems (MCS) published by Springer-Verlag, and proceedings of the International Conference on Information Fusion (FUSION) organized by the International Society of Information Fusion (http://www.isif.org/). Among other conferences of interest are International Conference on Machine Learning (ICML), European Conference on Machine Learning (ECML), International Joint Conference on Artificial Intelligence (IJCAI), European Conference on Artificial Intelligence (ECAI), and International Conference on Machine Learning and Data Mining (MLDM) (proceedings of ECML and MLDM are published by Springer-Verlag). Two international journals are largely devoted to the topic of our book are Information Fusion published by Elsevier and Journal of Advances in Information Fusion published by The International Society of Information Fusion, but most machine learning journals such as Machine Learning, the Journal of Machine Learning Research and the IEEE Transactions on Pattern Analysis and Machine Intelligence dedicate large room to papers on ensemble methods. This list, though comprehensive, but is certainly incomplete.

We would like to express our gratitude to several people and organizations who helped this book to appear. PASCAL 2 (Pattern Analysis, Statistical Modelling and Computational Learning) European Network of Excellence sponsorship of SUEMA'2008 is greatfully acknowledged.

We are thankful to Prof. Boi Faltings, Prof. Ioannis Vlahavas, and Prof. Pavlos Peppas for the opponutinity to hold SUEMA'2008 in the ancient Patras.

The contributions of SUEMA'2008 participants to this book made it to be born and we are grateful to all authors for their time, efforts, and warm support of our undertaking.

Prof. Janusz Kacprzyk and Dr. Thomas Ditzinger from Springer-Verlag deserved our special acknowledgment for warm welcome to our book and their support and a great deal of encouragement. Finally, we thank all other people in Springer who participated in the publication process.

Malmö (Sweden) and Genoa (Italy), June 2009

Oleg Okun
Giorgio Valentini

Contents

List of Contributors

Battista Biggio
Department of Electrical and Electronic Engineering,
University of Cagliari,
Piazza d'Armi, 09123
Cagliari, Italy
battista.biggio@diee.unica.it

Giorgio Fumera
Department of Electrical and Electronic Engineering,
University of Cagliari,
Piazza d'Armi,
09123 Cagliari, Italy
fumera@diee.unica.it

Fabio Roli
Department of Electrical and Electronic Engineering,
University of Cagliari,
Piazza d'Armi,
09123 Cagliari, Italy
roli@diee.unica.it

Angelo Ciaramella
Department of Applied Science,
University of Naples "Parthenope",
Isola C4, Centro
Direzionale I-80143,
Napoli, Italy
angelo.ciaramella@uniparthenope.it

Giulio Giunta
Department of Applied Science,
University of Naples "Parthenope",
Isola C4, Centro
Direzionale I-80143,
Napoli, Italy
giulio.giunta@uniparthenope.it

Angelo Riccio
Department of Applied Science,
University of Naples "Parthenope",
Isola C4, Centro
Direzionale I-80143,
Napoli, Italy
angelo.riccio@uniparthenope.it

Stefano Galmarini
European Commission - DG
Joint Research Centre,
Institute for Environment
and Sustainability, Ispra, Italy
stefano.galmarini@jrc.it

Pedro Gago
ESTG, Instituto Politécnico
de Leiria, Portugal
pgago@estg.ipleiria.pt

Manuel Filipe Santos
DSI, Universidade
do Minho, Portugal
mfs@dsi.uminho.pt

Francesco Gargiulo
Dipartimento di Informatica e
Sistemistica, University of
Naples Federico II, Italy
francesco.grg@unina.it

Antonio Penta
Dipartimento di Informatica e
Sistemistica, University of
Naples Federico II, Italy
a.penta@unina.it

Antonio Picariello
Dipartimento di Informatica e
Sistemistica, University of
Naples Federico II, Italy
picus@unina.it

Carlo Sansone
Dipartimento di Informatica e
Sistemistica, University of
Naples Federico II, Italy
carlosan@unina.it

Georg Krempl
Department of Statistics and
Operations Research,
University of Graz, Austria
georg.krempl@uni-graz.at

Vera Hofer
Department of Statistics and
Operations Research,
University of Graz, Austria
vera.hofer@uni-graz.at

Gabriele Lombardi
Dipartimento di Scienze
dell'Informazione,
Università degli
Studi di Milano,
Via Comelico 39-41,
Milano, Italy
lombardi@dsi.unimi.it

Elena Casiraghi
Dipartimento di Scienze
dell'Informazione,
Università degli
Studi di Milano,
Via Comelico 39-41,
Milano, Italy
casiraghi@dsi.unimi.it

Paola Campadelli
Dipartimento di Scienze
dell'Informazione, Università
degli Studi di Milano,
Via Comelico 39-41,
Milano, Italy
campadelli@dsi.unimi.it

Jesús Maudes
Department of Ingenieria
Civil, University of Burgos,
Escuela Politécnica
Superior C/Francisco
de Vitoria s/n 09006, Spain
jmaudes@ubu.es

Juan J. Rodríguez
Department of Ingenieria
Civil, University of Burgos,
Escuela Politécnica
Superior C/Francisco de
Vitoria s/n 09006, Spain
jjrodriguez@ubu.es

César García-Osorio
Department of Ingenieria
Civil, University of Burgos,
Escuela Politécnica
Superior C/Francisco
de Vitoria s/n 09006, Spain
cgosorio@ubu.es

Sebastian Nusser
School of Computer Science,
Otto-von-Guericke-University,
Universitätsplatz 2,
39106 Magdeburg, Germany
mail@seb-nusser.de

Clemens Otte
Siemens AG, Corporate Technology,
Information & Communications,
Learning Systems,
Otto-Hahn-Ring 6,
81730 Munich, Germany
clemens.otte@siemens.com

Werner Hauptmann
Siemens AG, Corporate Technology,
Information & Communications,
Learning Systems,
Otto-Hahn-Ring 6,
81730 Munich, Germany
werner.hauptmann@siemens.com

Erinija Pranckeviciene
Department of Human and
Medical Genetics,
Faculty of Medicine,
Vilnius University,
Santariskiu 2, LT08661
Vilnius, Lithuania,
erinija.pranckeviciene@mf.vu.lt

Muna Al-Razgan
George Mason University,
4400 University Drive,
MS 4A5, Fairfax VA 22030, USA
malrazga@gmu.edu

Carlotta Domeniconi
George Mason University,
4400 University Drive,
MS 4A5, Fairfax VA 22030, USA
carlotta@cs.gmu.edu

Matteo Re
Dipartimento di Scienze
dell'Informazione, Università
degli Studi di
Milano, Italy
re@dsi.unimi.it

Giorgio Valentini
Dipartimento di Scienze
dell'Informazione, Università
degli Studi di
Milano, Italy
valentini@dsi.unimi.it

Grigorios Tsoumakas
Department of Informatics,
Aristotle University of
Thessaloniki, 54124
Thessaloniki, Greece
greg@csd.auth.gr

Ioannis Partalas
Department of Informatics,
Aristotle University of
Thessaloniki, 54124
Thessaloniki, Greece
partalas@csd.auth.gr

Ioannis Vlahavas
Department of Informatics,
Aristotle University of
Thessaloniki, 54124
Thessaloniki, Greece
vlahavas@csd.auth.gr

Cédric Wemmert
LSIIT - University of Strasbourg,
Pôle API, Bd Sébastien
Brant - 67412 Illkirch, France
wemmert@lsiit.u-strasbg.fr

Germain Forestier
LSIIT - University of Strasbourg,
Pôle API, Bd Sébastien
Brant - 67412 Illkirch, France
forestier@lsiit.u-strasbg.fr

Sébastien Derivaux
LSIIT - University of Strasbourg, Pôle API, Bd Sébastien Brant - 67412 Illkirch, France
derivaux@lsiit.u-strasbg.fr

Terry Windeatt
CVSSP, University of Surrey, University, Guildford, Surrey, UK GU2 7XH
t.windeatt@surrey.ac.uk

An Ensemble Pruning Primer

Grigorios Tsoumakas, Ioannis Partalas, and Ioannis Vlahavas

Abstract. Ensemble pruning deals with the reduction of an ensemble of predictive models in order to improve its efficiency and predictive performance. The last 12 years a large number of ensemble pruning methods have been proposed. This work proposes a taxonomy for their organization and reviews important representative methods of each category. It abstracts their key components and discusses their main advantages and disadvantages. We hope that this work will serve as a good starting point and reference for researchers working on the development of new ensemble pruning methods.

1 Introduction

Ensemble methods [5, 13] have been a very popular research topic during the last decade. It has attracted the interest of scientists from several fields including Statistics, Machine Learning, Pattern Recognition, and Knowledge Discovery in Databases. The success of ensemble methods arises largely from the fact that they offer an appealing solution to several interesting learning problems of the past and the present, such as improving predictive performance, learning from multiple physically distributed data sources, scaling inductive algorithms to large databases and learning from concept-drifting data streams.

Typically, ensemble methods comprise two phases: the *production* of multiple predictive models and their *combination*. Recent work [2, 4, 7, 9, 15, 16, 17, 21, 23, 28, 35] considers an additional intermediate phase that deals with the reduction of the ensemble size prior to combination. This phase is commonly called *ensemble pruning* , while other names include *selective ensemble*, *ensemble thinning* and *ensemble selection*.

Ensemble pruning is important for two reasons: *efficiency* and *predictive performance*. Having a very large number of models in an ensemble adds a lot of

Grigorios Tsoumakas · Ioannis Partalas · Ioannis Vlahavas
Department of Informatics, Aristotle University of Thessaloniki, 54124 Thessaloniki, Greece
e-mail: {greg,partalas,vlahavas}@csd.auth.gr

O. Okun & G. Valentini (Eds.): Appli. of Supervised & Unsuper. Ensemble Meth., SCI 245, pp. 1–13.
springerlink.com © Springer-Verlag Berlin Heidelberg 2009

computational overhead. For example, decision tree models may have large memory requirements [16] and lazy learning methods have a considerable computational cost during execution. The minimization of run-time overhead is crucial in certain applications, such as stream mining. In addition, when models are distributed over a network, the reduction of the number of models leads to the reduction of communication costs.

Equally important is the second reason, predictive performance. An ensemble may consist of both high and low predictive performance models. The latter may negatively affect the overall performance of the ensemble. In addition, an ensemble may contain many models that are very similar to each other. This reduces its diversity and capability for error correction. Pruning low performance models while maintaining a high diversity among the remaining members of the ensemble is typically considered a proper recipe for an effective ensemble.

Note that ensemble pruning is different from *ensemble weighting* [34], where the decisions of all models in the ensemble are considered, but with a different weight. Ensemble weighting is concerned solely with increasing the predictive performance, as it needs to maintain all models of the ensemble.

One of the first ensemble pruning approaches is discussed in [24]. The 12 years that followed have witnessed the development of several diverse methods for ensemble pruning . This work proposes a taxonomy for their organization and reviews important representative methods of each category. It steers clear of a mere enumeration of particular approaches in the related literature and instead attempts to abstract their key components, to discuss their main advantages and disadvantages and to analyze their complexity whenever possible. We hope that this work will serve as a good starting point and reference for researchers working on the development of new ensemble pruning methods.

The remainder of this chapter is structured as follows. Section 2 contains background material on ensemble production and combination. Section 3 presents the proposed taxonomy, introduces notation and discusses issues that are common for all methods. Sections 4 to 7 review important representative methods of each category in the taxonomy. Finally, the conclusions of this work are presented in Sect. 8.

2 Background

This section provides background material on ensemble methods. More specifically, information about the different ways of producing models is presented as well as different methods for combining the decisions of the models.

2.1 Producing the Models

An ensemble can be composed of either *homogeneous* or *heterogeneous models*. Homogeneous models are derived from different executions of the same learning algorithm. Such models can be produced by using different values for the parameters of the learning algorithm, injecting randomness into the learning algorithm or through the manipulation of the training instances, the input attributes and the model outputs [6]. Popular methods for producing homogeneous models are *Bagging* [3] and *Boosting* [25] .

Heterogeneous models are derived from running different learning algorithms on the same data set. Such models have different views about the data, as they make different assumptions about it. For example, a properly trained neural network is robust to noise in contrast to a decision tree.

2.2 Combining the Models

Common methods for combining an ensemble of predictive models include *voting*, *stacked generalization* and *mixture of experts*.

In voting, each model outputs a class value (or ranking, or probability distribution) and the class with the most votes is the one proposed by the ensemble. When the class with the maximum number of votes is the winner, the rule is called *plurality voting* and when the class with more than half of the votes is the winner, the rule is called *majority voting*. A variant of voting is weighted voting where the models are not treated equally as each of them is associated with a coefficient (weight), usually proportional to its classification accuracy.

Let x be an instance and m_i, $i = 1,\ldots,k$ a set of models that output a probability distribution $m_i(x,c_j)$ for each class c_j, $j = 1,\ldots,n$. The output of the (weighted) voting method $y(x)$ for instance x is given by the following mathematical expression:

$$y(x) = \arg\max_{c_j} \sum_{i=1}^{k} w_i m_i(x,c_j) \ ,$$

where w_i is the weight of model i. In the simple case of voting (unweighted), the weights are all equal to one, that is, $w_i = 1, i = 1,\ldots,k$.

Stacked generalization [32], also known as *Stacking*, is a method that combines models by learning a meta-level (or level-1) model that predicts the correct class based on the decisions of the base level (or level-0) models. This model is induced on a set of meta-level training data that are typically produced by applying a procedure similar to k-fold cross validation on the training data. The outputs of the base-learners for each instance along with the true class of that instance form a meta-instance. A meta-classifier is then trained on the meta-instances. When a new instance appears for classification, the output of the all base-learners is first calculated and then propagated to the meta-classifier, which outputs the final result.

The mixture of experts architecture [12] is similar to the weighted voting method except that the weights are not constant over the input space. Instead there is a gating network which takes as input an instance and outputs the weights that will be used in the weighted voting method for that specific instance. Each expert makes a decision and the output is averaged as in the method of voting.

3 A Taxonomy of Ensemble Pruning Methods

We propose the organization of the various ensemble pruning methods into the following categories:

- *Ranking* based. Methods of this category are conceptually the simplest. They order the models of the ensemble once according to an evaluation function and select models in this *fixed* order.
- *Clustering* based. Methods of this category comprise two stages. Initially, they employ a clustering algorithm in order to discover groups of models that make similar predictions. Subsequently, each cluster is separately pruned in order to increase the overall diversity of the ensemble.
- *Optimization* based. Ensemble pruning can be posed as an optimization problem as follows: find the subset of the original ensemble that optimizes a measure indicative of its generalization performance (e.g., accuracy on a separate validation set). Exhaustive search of the space of ensemble subsets is infeasible for a moderate ensemble size.
- *Other*. This category includes methods that do not fall into one of the previous categories.

Before proceeding to the description of the main characteristics of each category, some common notation is introduced. The original ensemble is denoted as $H = \{h_t, t = 1,2,\dots,T\}$. All methods employ a function that evaluates the suitability of single models, model pairs or ensembles of more than two models for inclusion in the final ensemble. Evaluation is typically based on the predictions of the models on a set of data, which will be called the *pruning set*. The role of the pruning set can be performed by the training set, a separate validation set, or even a set of - naturally existing or artificially produced - instances with unknown value for the target variable. The pruning set will be denoted as $D = \{(\mathbf{x}_i, y_i), i = 1,2,\dots,N\}$, where $\mathbf{x}_i$ is a vector with feature values and y_i is the value of the target variable, which may be unknown.

4 Ranking-Based Methods

The main point of differentiation among the methods of this category is the evaluation measure used for model ranking. Using the predictive performance of individual models is too simplistic and does not achieve satisfying results [24, 33]. Information-theoretic measures were also used in [33] for the evaluation of Bayesian models, with equally disappointing results.

Kappa pruning [16] employs a diversity measure for evaluation. It ranks all *pairs* of classifiers in H based on the κ statistic of agreement calculated on the training set. Its time complexity is $O(T^2N)$. Kappa pruning could be generalized by accepting a parameter to specify any pairwise diversity measure for either classification or regression models, in place of the κ statistic. However, it would still beg for one fundamental theoretical question: Do two diverse pairs of models, lead to one diverse ensemble of four models? The intuitive answer is no. In fact, kappa pruning has been shown to be non-competitive for pruning classifier ensembles produced via Bagging [20] .

An efficient and effective ranking-based pruning method for ensembles of classifiers is orientation ordering [18]. A key concept in orientation ordering is the *signature vector* of a classifier h_t: an N-dimensional vector with elements taking the value +1 if $h_t(\mathbf{x}_i) = y_i$ and -1 if $h_t(\mathbf{x}_i) \neq y_i$. The average signature vector of all classifiers in an ensemble is called the *ensemble signature vector*. It is indicative of the ability of the ensemble to correctly classify each example in the pruning set (the training set in this method) using majority voting for classifier combination. The *reference vector* is a vector perpendicular to the ensemble signature vector that corresponds to the projection of the first quadrant diagonal onto the hyper-plane defined by the ensemble signature vector.

Orientation ordering ranks the classifiers by increasing value of the angle between their signature vector and the reference vector. Essentially this ordering gives preference to models that correctly classify those examples that are incorrectly classified by the full ensemble. Orientation ordering is among the fastest methods for ensemble pruning, with time complexity of $O(TN)$. In addition, its predictive performance is not significantly worse than state-of-the-art methods for pruning classifier ensembles produced via Bagging [20] .

Another interesting issue in ranking-based ensemble pruning methods concerns the choice of the final number of models from the obtained ranking. One approach is to use a fixed user-specified amount or percentage of models. In kappa pruning, for example, classifier pairs are selected in ascending order of agreement until a specified number of models has been reached. If the goal of pruning is to improve efficiency, then this approach can be used in order to obtain the desired amount of models, which may be dictated by constraints (memory and speed) in the application environment.

A second approach is to dynamically select the size based on the evaluation measure or the predictive performance of ensembles of different size. In orientation ordering, for example, only the classifiers whose angle is less than $\pi/2$ are included in the final ensemble, while in [33], the models whose evaluation measure is lower than the average of all models are pruned. This approach is more preferable when the goal of pruning is to improve predictive performance, as it is more flexible and can sacrifice efficiency for effectiveness.

5 Clustering-Based Methods

The first issue for the methods of this category is the choice of a clustering algorithm. Past approaches have used hierarchical agglomerative clustering [9], k means [8, 15], and deterministic annealing [1].

Clustering algorithms are based on the notion of distance. Therefore, the second issue for the clustering-based methods is the choice of an appropriate distance measure. The probability that classifiers do not make coincident errors on a separate validation set was used as a distance measure in [9]. This measure is actually equal to one minus the *double fault* diversity measure [14]. The Euclidean distance in the training set is used in [8, 15]. Actually, any distance measure suitable for nominal (classifiers) or numeric (regressors) output could be used. Note that there is no need for a labeled pruning set in this case [1]. Artificially generated data could be used instead.

Another important issue concerns the process of pruning each cluster. An elegant approach was used in [1], where a new model is trained for each cluster, using the cluster centroids as values of the target variable. Another interesting approach is to select from each cluster the single classifier that is most distant to the rest of the clusters [9]. The approach followed in [15] was to iteratively remove models from the least to the most accurate, until the accuracy of the entire ensemble starts to decrease. This, however, does not guarantee the selection of a single model from each cluster. The most accurate model of each cluster was selected in [8].

A final issue worth mentioning is the choice of the number of clusters. This could be determined based on the performance of the method on a validation set [8]. In [15], the number of clusters was gradually increased until the disagreement between the cluster centroids started to deteriorate.

6 Optimization-Based Methods

In the following subsections we look into ensemble pruning methods that are based on three different optimization approaches: *genetic algorithms*, *semi-definite programming* and *hill climbing*. The last approach is examined at a greater level of detail, as a large number of this kind of ensemble pruning methods have been recently proposed.

6.1 Genetic Algorithms

The Gasen-b method [36] performs stochastic search in the space of model subsets using a genetic algorithm. The ensemble is represented as a bit string, using one bit for each model. Models are included into or excluded from the ensemble, depending on the value of the corresponding bit. Gasen-b performs standard genetic operations such as mutations and crossovers and uses default values for the parameters of the genetic algorithm. The fitness function for an individual $S \subseteq H$ is the accuracy of S on a separate validation set using voting for model combination.

6.2 Semi-definite Programming

Zhang et al. [35] formulate the ensemble pruning problem as a mathematical problem and apply semi-definite programming (SDP) techniques. In particular, the authors initially formulated the ensemble pruning problem as a quadratic integer programming problem that looks for a fixed-size subset of k classifiers with minimum misclassification and maximum divergence.

They subsequently found that this quadratic integer programming problem is similar to the *max cut with size k* problem, which can be approximately solved using an algorithm based on SDP. Their algorithm requires the number of classifiers to retain as a parameter and runs in polynomial time.

6.3 Hill Climbing

Hill climbing search greedily selects the next state to visit from the neighborhood of the current state. States, in our case, are the different subsets of models and the neighborhood of a subset $S \subseteq H$ consists of those subsets that can be constructed by adding or removing one model from S. We focus on the directed version of hill-climbing that traverses the search space from one end (empty set) to the other (complete ensemble). An example of the search space for an ensemble of four models is presented in Fig. 1.

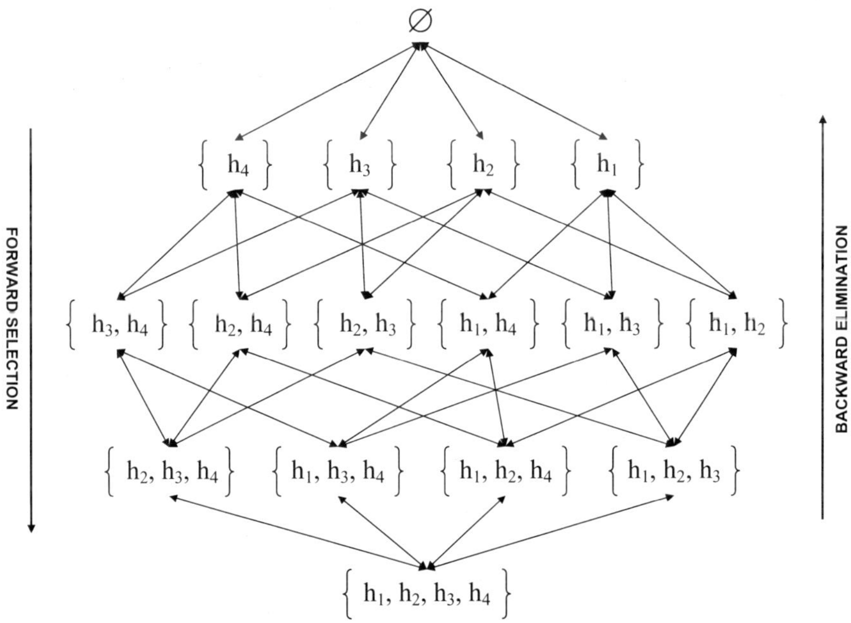

Fig. 1 An example of the search space of hill climbing ensemble pruning methods for an ensemble of 4 models

Depending on the direction of search, we have forward selection [4, 7, 16, 17, 33] and backward elimination [2, 21, 33] methods. In both cases, the traversal requires the evaluation of $\frac{T(T+1)}{2}$ subsets, leading to a time complexity of $O(T^2 g(T,N))$. The term $g(T,N)$ concerns the complexity of the evaluation process, which is linear with respect to N and ranges from constant to quadratic with respect to T, as we shall see in the rest of this section.

Similarly to ranking-based methods, the main component that differentiates hill climbing ensemble pruning methods is the evaluation measure. Evaluation measures can be grouped into two major categories: *performance* based and *diversity* based.

The goal of performance based measures is to find the model h_t that maximizes the performance of the ensemble produced by adding (removing) h_t to (from) the current ensemble. Their calculation depends on the method used for ensemble combination, which usually is voting. Accuracy was used as an evaluation measure in [7, 16, 33], while [4] experimented with several metrics, including accuracy, root-mean-squared-error, mean cross-entropy, lift, precision/recall break-even point, precision/recall F-score, average precision and ROC area. Another measure is *benefit* which is based on a cost model and has been used in [7].

The calculation of performance-based metrics requires the decision of the current ensemble S on all examples of the pruning set. Therefore, the complexity of these measures is $O(|S|N)$. However, this complexity can be optimized to $O(N)$ if the predictions of the current ensemble are updated incrementally each time a model is added to/removed from it.

It is generally accepted that an ensemble should contain diverse models in order to achieve high predictive performance. However, there is no clear definition of diversity, neither a single measure to calculate it. In their interesting study, Kuncheva and Whitaker [14] could not reach into a solid conclusion on how to utilize diversity for the production of effective classifier ensembles . In a more recent theoretical and experimental study on diversity measures [27], the authors reached to the conclusion that diversity cannot be explicitly used for guiding the process of hill climbing methods. Yet, certain approaches have reported promising results [2, 17, 21].

One issue worth mentioning here is how to calculate the diversity during the search in the space of ensemble subsets. For simplicity we consider the case of forward selection only. Let S be the current ensemble and $h_t \in H \setminus S$ a candidate classifier to add to the ensemble.

One could compare the diversities of ensembles $S' = S \cup h_t$ for all candidate $h_t \in H \setminus S$ and select the one with the highest diversity . Any pairwise and non-pairwise diversity measure can be used for this purpose. The time complexity of most non-pairwise diversity measures is $O(|S'|N)$, while that of pairwise diversity measures is $O(|S'|^2N)$. However, a straightforward optimization can be performed in the case of pairwise diversity measures. Instead of calculating the sum of the pairwise diversity for every pair of classifiers in each candidate ensemble S', one can simply calculate the sum of the pairwise diversities only for the pairs that include the candidate classifier h_t. The sum of the rest of the pairs is equal for all candidate ensembles. The same optimization can be achieved in backward elimination too. This reduces their time complexity to $O(|S|N)$.

Several methods [2, 17, 21, 27] use a different approach to calculate diversity during the search. They use pairwise measures to compare the candidate classifier h_t with the current ensemble S, which is viewed as a single classifier that combines the decisions of its members with voting. This way they calculate the diversity between the current ensemble as a whole and the candidate classifier. Such an approach has time complexity $O(|S|N)$, which can be optimized to $O(N)$ if the predictions of the current ensemble are updated incrementally each time a model is added to/removed from it.

In the past, the widely known diversity measures *disagreement*, *double fault*, *Kohavi-Wolpert variance*, *inter-rater agreement*, *generalized diversity* and *difficulty* were used for hill climbing ensemble pruning in [27]. Four diversity measures designed specifically for hill climbing ensemble pruning are introduced in [2, 17, 21]. We next present these measures using a common notation.

We can distinguish four events concerning the decision of the current ensemble and the candidate classifier:

$$\begin{aligned}
e_{tf}(\mathbf{x}_i) &: y = h_t(\mathbf{x}_i) \wedge y \neq S(\mathbf{x}_i) \\
e_{ft}(\mathbf{x}_i) &: y \neq h_t(\mathbf{x}_i) \wedge y = S(\mathbf{x}_i) \\
e_{tt}(\mathbf{x}_i) &: y = h_t(\mathbf{x}_i) \wedge y = S(\mathbf{x}_i) \\
e_{ff}(\mathbf{x}_i) &: y \neq h_t(\mathbf{x}_i) \wedge y \neq S(\mathbf{x}_i)
\end{aligned}$$

The *complementariness* [17] of a model h_k with respect to an ensemble S and a pruning set D is calculated as follows:

$$COM_D(h_k, S) = \sum_{i=1}^{N} I(e_{tf}(\mathbf{x}_i)) \; ,$$

where $I(true) = 1, I(false) = 0$ and $S(\mathbf{x}_i)$ is the classification of instance $\mathbf{x}_i$ from the ensemble S. This classification is derived from the application of an ensemble combination method to S, which usually is voting. The complementariness of a model with respect to an ensemble is actually the number of examples of D that are classified correctly by the model and incorrectly by the ensemble. A selection algorithm that uses the above measure, tries to add (remove) at each step the model that helps the ensemble to classify correctly the examples it gets wrong.

The *concurrency* [2] of a model h_k with respect to an ensemble S and a pruning set D is calculated as follows:

$$CON_D(h_k, S) = \sum_{i=1}^{N} \Big(2I(e_{tf}(\mathbf{x}_i)) + I(e_{tt}(\mathbf{x}_i)) - 2I(e_{ff}(\mathbf{x}_i)) \Big) \; .$$

This measure is very similar to complementariness with the difference that it takes into account two extra cases.

The *focused ensemble selection* method [21] proposes a measure that uses all the events and also takes into account the strength of the current ensemble's decision:

$$\begin{aligned}
FES(h_k, S) = \sum_{i=1}^{N} \Big(& NT_i I(e_{tf}(\mathbf{x}_i)) - NF_i I(e_{ft}(\mathbf{x}_i)) + \\
& + NF_i I(e_{tt}(\mathbf{x}_i)) - NT_i I(e_{ff}(\mathbf{x}_i)) \Big) \; ,
\end{aligned}$$

where NT_i denotes the proportion of models in the current ensemble S that correctly classify example $\mathbf{x}_i$, and $NF_i = 1 - NT_i$ denotes the number of models in S that classify it incorrectly.

The *margin distance minimization* method [17] is based on the same concepts as the orientation ordering ranking-based method (see Section 4). It searches for the ensemble S with the minimum distance between its signature vector $\mathbf{c}_S$ and a predefined vector $\mathbf{o}$ placed in the first quadrant of the N-dimensional hyperplane. Vector $\mathbf{o}$ corresponds to the ideal vector that correctly classifies all examples.

The method is based on a measure called *margin*. The margin, $MAR_D(h_k, S)$, of classifier h_k with respect to an ensemble S and a pruning set D is calculated as follows:

$$MAR_D(h_k, S) = d\left(\mathbf{o}, \frac{1}{|S|+1}\left(\mathbf{c}_{S\cup\{h_k\}}\right)\right),$$

where d is the Euclidean distance.

7 Other Methods

This category includes three approaches that do not belong to any of the previous categories. The first one is based on statistical procedures for directly selecting a subset of classifiers, the second is based on reinforcement learning and the third on Boosting.

7.1 Statistical Procedures

Tsoumakas et al. [28, 29] prune an ensemble of heterogeneous classifiers using statistical tests that determine whether the differences in predictive performance among the classifiers of the ensemble are significant. Only the classifiers with significantly better performance than the rest are retained and subsequently combined with the method of (weighted) voting.

Such statistical tests are called *multiple comparisons procedures* [10]. Three of those that were used in [29] are Tukey's test [30], Hsu's test [11] and Scott & Knott's procedure [26], with the last one offering the largest benefit.

The disadvantage of these methods is that they do not take the diversity of the ensemble into consideration. However, they could potentially play the role of a fast preliminary filtering of low performing models in a large ensemble, followed by a more advanced diversity-aware method.

7.2 Reinforcement Learning

Partalas et al. [22, 23] take a reinforcement learning approach to ensemble pruning. In particular, the problem of pruning an ensemble of T classifiers is modeled as an episodic task, where an agent takes T sequential actions, each corresponding to either the inclusion or exclusion of classifier $h_t, t = 1, \ldots, T$ to or from the final ensemble. The Q-learning algorithm [31] is then used to approximate the optimal policy for this task.

7.3 Boosting

An approach similar to Boosting was used for pruning an ensemble of classifiers produced via Bagging in [19]. The algorithm iteratively selects the classifier with the lowest weighted error on the training set. Instance weights are initialized and updated according to the AdaBoost algorithm. The only difference is that instead of terminating the process when the weighted error is larger than 0.5, the algorithm resets all instance weights and continues selecting models. The complexity of this approach is $O(T^2N)$.

This approach ranks individual classifiers, but it does so based on their weighted error on the training set. Since at each step of the algorithm the instance weights depend on the classifiers selected up to that step, we refrained from categorizing this approach to ranking-based methods, where each model can be independently evaluated and ranked independently of the currently selected models.

8 Conclusions

This work presented a taxonomy of ensemble pruning methods. We believe that such a taxonomy is necessary for researchers working on new methods. It will help them to identify the main categories of methods and their key points and to avoid duplication of work. Due to the large amount of existing methods and the different parameters of an ensemble selection framework (heterogeneous/homogeneous ensemble, algorithms used, size of ensemble, etc), it is possible to devise a new method, which may only differ in small, perhaps unimportant details from the existing methods. A generalized view of the methods, as offered from this work, will help to avoid work towards such small differences, and perhaps may lead to novel methods.

We do not argue that the proposed taxonomy is perfect. On the contrary, it is just the first step in abstracting and categorizing the different methods. We made an effort to include most of the important ensemble pruning methods, but no doubt, some high quality methods may have been left outside this study. For example, we have not considered instance-base ensemble pruning methods that dynamically prune the ensemble for each test instance.

This work refrained from performing experimental comparisons between the methods. However, we would like to stress the importance of the following guidelines for empirical ensemble pruning studies. Firstly, the ensemble should consist of a moderate size of models (e.g., 100 or more). For small ensemble sizes (e.g., 10 models), an exhaustive search for the best subset of models is computationally feasible, and perhaps even faster than some more complex methods of the literature. Secondly, the study should include a large number of datasets, and include appropriate statistical tests for the comparison of different methods in order to derive safe and useful conclusions.

References

1. Bakker, B., Heskes, T.: Clustering ensembles of neural network models. Neural Networks 16(2), 261–269 (2003)
2. Banfield, R.E., Hall, L.O., Bowyer, K.W., Kegelmeyer, W.P.: Ensemble diversity measures and their application to thinning. Information Fusion 6(1), 49–62 (2005)
3. Breiman, L.: Bagging predictors. Mach. Learn. 24(2), 123–140 (1996)
4. Caruana, R., Niculescu-Mizil, A., Crew, G., Ksikes, A.: Ensemble selection from libraries of models. In: Brodley, C.E. (ed.) Proc. 21st Int. Conf. Mach. Learn., Banff, AB, Canada. ACM, New York (2004)
5. Dietterich, T.G.: Machine-learning research: four current directions. AI Magazine 18(4), 97–136 (1997)
6. Dietterich, T.G.: Ensemble methods in machine learning. In: Kittler, J., Roli, F. (eds.) MCS 2000. LNCS, vol. 1857, pp. 1–15. Springer, Heidelberg (2000)
7. Fan, W., Chu, F., Wang, H., Yu, P.S.: Pruning and dynamic scheduling of cost-sensitive ensembles. In: Proc. 18th Natl. Conf. Artif. Intell., Edmonton, AB, Canada, pp. 146–151. AAAI, Menlo Park (2002)
8. Fu, Q., Hu, S.-X., Zhao, S.Y.: Clustering-based selective neural network ensemble. J. Zhejiang University Science 6A(5), 387–392 (2005)
9. Giacinto, G., Roli, F., Fumera, G.: Design of effective multiple classifier systems by clustering of classifiers. In: Proc. 15th Int. Conf. Pattern Recogn., Barcelona, Spain, pp. 160–163 (2000)
10. Hochberg, Y., Tamhane, A.C.: Multiple Comparison Procedures. John Wiley & Sons, Hoboken (1987)
11. Hsu, J.C.: Constrained simultaneous confidence intervals for multiple comparisons with the best. Annals of Statistics 12(3), 1136–1144 (1984)
12. Jacobs, R.A., Jordan, M.I., Nowlan, S.J., Hinton, G.E.: Adaptive mixtures of local experts. Neural Computation 3(1), 79–87 (1991)
13. Kuncheva, L.I.: Combining Pattern Classifiers: Methods and Algorithms. John Wiley & Sons, Hoboken (2004)
14. Kuncheva, L.I., Whitaker, C.J.: Measures of diversity in classifier ensembles and their relationship with the ensemble accuracy. Mach. Learn. 51(2), 181–207 (2003)
15. Lazarevic, A., Obradovic, Z.: The effective pruning of neural network classifiers. In: Proc. 2001 IEEE/INNS Int. Joint Conf. Neural Networks, Washington, DC, pp. 796–801. IEEE Comp. Soc., Los Alamitos (2001)
16. Margineantu, D., Dietterich, T.: Pruning adaptive boosting. In: Fisher, D.H. (ed.) Proc. 14th Int. Conf. Mach. Learn., Nashville, TN, pp. 211–218. Morgan Kaufmann, San Francisco (1997)
17. Martínez-Muñoz, G., Suárez, A.: Aggregation ordering in bagging. In: Proc. IASTED Int. Conf. Artif. Intell. and Appl., pp. 258–263. Acta Press, Calgari (2004)
18. Martínez-Muñoz, G., Suárez, A.: Pruning in ordered bagging ensembles. In: Cohen, W.W., Moore, A. (eds.) Proc. 23rd Int. Conf. Mach. Learn., Pittsburgh, PA, pp. 609–616. ACM, New York (2006)
19. Martínez-Muñoz, G., Suárez, A.: Using boosting to prune bagging ensembles. Pattern Recogn. Lett. 28(1), 156–165 (2007)

20. Martínez-Muñoz, G., Hernández-Lobato, D., Suárez, A.: An analysis of ensemble pruning techniques based on ordered aggregation. IEEE Trans. Pattern Analysis and Mach. Intell. 31(2), 245–259 (2009)
21. Partalas, I., Tsoumakas, G., Vlahavas, I.: Focused ensemble selection: a diversity method for greedy ensemble selection. In: Ghallab, M., Spyropoulos, C.D., Fakotakis, N., Avouris, N.M. (eds.) Proc. 18th European Conf. Artif. Intell., Patras, Greece, pp. 117–121. IOS Press, Amsterdam (2008)
22. Partalas, I., Tsoumakas, G., Vlahavas, I.: Pruning an ensemble of classifiers via reinforcement learning. Neurocomputing (in press, 2009)
23. Partalas, I., Tsoumakas, G., Katakis, I., Vlahavas, I.: Ensemble pruning using reinforcement learning. In: Proc. 4th Hellenic Conf. Artif. Intell., Heraclion, Greece, pp. 301–310 (2006)
24. Partridge, D., Yates, W.B.: Engineering multiversion neural-net systems. Neural Computation 8(4), 869–893 (1996)
25. Schapire, R.E.: The strength of weak learnability. Mach. Learn. 5(2), 197–227 (1990)
26. Scott, A.J., Knott, M.: A cluster analysis method for grouping means in the analysis of variance. Biometrics 30, 507–512 (1974)
27. Tang, E.K., Suganthan, P.N., Yao, X.: An analysis of diversity measures. Mach. Learn. 65(1), 247–271 (2006)
28. Tsoumakas, G., Katakis, I., Vlahavas, I.: Effective voting of heterogeneous classifiers. In: Boulicaut, J.-F., Esposito, F., Giannotti, F., Pedreschi, D. (eds.) ECML 2004. LNCS (LNAI), vol. 3201, pp. 465–476. Springer, Heidelberg (2004)
29. Tsoumakas, G., Angelis, I., Vlahavas, I.: Selective fusion of heterogeneous classifiers. Intell. Data Analysis 9(6), 511–525 (2005)
30. Tukey, J.W.: The problem of multiple comparisons. Unpublished manuscript (1953)
31. Watkins, C.J., Dayan, P.: Q-learning. Mach. Learn. 8(3-4), 279–292 (1992)
32. Wolpert, D.H.: Stacked generalization. Neural Networks 5(2), 241–259 (1992)
33. Yang, Y., Korb, K., Ting, K., Webb, G.: Ensemble selection for superparent-one-dependence estimators. In: Zhang, S., Jarvis, R.A. (eds.) AI 2005. LNCS (LNAI), vol. 3809, pp. 102–112. Springer, Heidelberg (2005)
34. Yang, Y., Webb, G.I., Cerquides, J., Korb, K.B., Boughton, J., Ting, K.M.: To select or to weigh: a comparative study of linear combination schemes for superparent-one-dependence estimators. IEEE Trans. Knowl. Data Engin. 19(12), 1652–1665 (2007)
35. Zhang, Y., Burer, S., Street, W.N.: Ensemble pruning via semi-definite programming. J. Mach. Learn. Res. 7, 1315–1338 (2006)
36. Zhou, Z.H., Tang, W.: Selective ensemble of decision trees. In: Wang, G., Liu, Q., Yao, Y., Skowron, A. (eds.) RSFDGrC 2003. LNCS (LNAI), vol. 2639, pp. 476–483. Springer, Heidelberg (2003)

Evade Hard Multiple Classifier Systems

Battista Biggio, Giorgio Fumera, and Fabio Roli

Abstract. Experimental and theoretical evidences showed that multiple classifier systems (MCSs) can outperform single classifiers in terms of classification accuracy. MCSs are currently used in several kinds of applications, among which security applications like biometric identity recognition, intrusion detection in computer networks and spam filtering. However security systems operate in *adversarial environments* against intelligent adversaries who try to evade them, and are therefore characterised by the requirement of a high robustness to evasion besides a high classification accuracy. The effectiveness of MCSs in improving the hardness of evasion has not been investigated yet, and their use in security systems is mainly based on intuitive and qualitative motivations, besides some experimental evidence. In this chapter we address the issue of investigating why and how MCSs can improve the hardness of evasion of security systems in adversarial environments. To this aim we develop analytical models of adversarial classification problems (also exploiting a theoretical framework recently proposed by other authors), and apply them to analyse two strategies currently used to implement MCSs in several applications. We then give an experimental investigation of the considered strategies on a case study in spam filtering, using a large corpus of publicly available spam and legitimate e-mails, and the SpamAssassin, widely used open source spam filter.

Keywords: multiple classifier systems, adversarial classification, hardness of evasion, spam filtering.

1 Introduction

During the past ten years multiple classifier systems (MCS) have become an established approach to design pattern classification systems. A large body of both

Battista Biggio · Giorgio Fumera · Fabio Roli
Department of Electrical and Electronic Engineering, University of Cagliari,
Piazza d'Armi, 09123 Cagliari, Italy
e-mail: {battista.biggio, fumera, roli}@diee.unica.it

O. Okun & G. Valentini (Eds.): Appli. of Supervised & Unsuper. Ensemble Meth., SCI 245, pp. 15–38.
springerlink.com © Springer-Verlag Berlin Heidelberg 2009

experimental and theoretical evidence shows that MCSs can outperform a single classifier in several real applications, in terms of classification accuracy (see, for instance, [6, 8]). In particular, several authors showed that MCSs can allow to improve the detection capability also in security applications like biometric authentication and intrusion detection in computer networks [4, 11]. It is also worth noting that the MCS classifier architecture is also used in commercial and open source spam filters. However, attaining a high classification accuracy or detection capability is not sufficient in security applications, and, in particular, in so-called *adversarial environments*, in which a security system faces an intelligent, adaptive adversary who exploits the available or acquired knowledge about the system just to evade it. Typical examples of this kind of applications are intrusion detection in computer networks and spam filtering. A crucial issue in this kind of applications, besides detection capability, is the *hardness of evasion*, which can be qualitatively defined as the effort required to the attacker to evade the system. While the effectiveness of MCSs in improving the detection capability has been deeply investigated so far, no work has formally analysed yet their effectiveness in improving the hardness of evasion of a classification system, although MCSs are widely used in adversarial classification tasks as mentioned above. Actually, the issue of the hardness of evasion has been addressed only recently in the machine learning and pattern recognition literature, often with respect to specific tasks, classification systems and type of attacks [5, 7] (for instance, the so-called *good words* attack against the text classifier used in most spam filters), and only in a few works under a more general perspective [1, 3, 9] aimed at developing analytical models of adversarial classification problems. With regard to the use of MCSs explicitly for the specific goal of improving the hardness of evasion, to our knowledge it was proposed by some authors for biometric tasks (see, for instance, [11]), but only by Perdisci et al. [10] for intrusion detection tasks, while no works considered MCSs for this specific goal in spam filtering tasks.

The aim of this chapter is to make a first step towards a better understanding of why and how MCSs can improve the hardness of evasion of a security system in adversarial classification problems, with respect to the use of a single classifier. We focus, in particular, on two strategies commonly used to design MCSs, and applied also in security applications. The first strategy is often applied when a set of heterogeneous features is available (as happens for instance in multimodal biometric identity verification tasks). In this case, it could be better to combine different classifiers trained on disjoint and heterogeneous subsets of features (for instance, a face classifier and a fingerprint classifier) instead of designing a single, "monolithic" classifier based on all the available features [8]. This strategy is also used when the feature set is very large (and not necessarily heterogeneous), since it is known that it can help avoiding over-fitting. In this chapter we investigate whether such strategy can be more effective than the use of a single classifier also in terms of hardness of evasion, in adversarial classification tasks.

The second strategy we consider is commonly used to update real spam filters as new kinds of attacks are detected, and was proposed in [10] explicitly for improving the hardness of evasion in intrusion detection tasks. This strategy consists in adding detectors (often, made up as classifiers) based on new features to a classification system. In this chapter we investigate whether this approach can be really effective to improve the hardness of evasion in adversarial classification tasks, trying to provide some argument more formal than the intuitive one proposed in [10]. To this aim, we will exploit a theoretical framework proposed by Dalvi et al. [3] for adversarial classification problems.

Then, to experimentally investigate the above issues we consider a spam filtering task as a case study, using a large corpus of publicly available spam and legitimate e-mails, and a real and widely used open source spam filter, SpamAssassin.

The chapter is organised as follows. In Sect. 2.1 we give an overview of the use of MCSs in security applications, and of the theoretical framework proposed in [3] for adversarial classification problems. In Sect. 3 we describe how we model an adversarial learning task in which the classifier is a MCS using the framework in [3], and how we model such kinds of tasks to investigate the effectiveness of a MCS against a single classifier, when both classifiers use the same feature set. The experimental results are reported in Sect. 4.

2 Related Work

In this section we first give an overview of past work on MCSs for security applications, and then summarise a formal framework proposed in [3] to model adversarial classification problems. Such framework will be exploited in Sect. 3.1 to analyse one of the two strategies mentioned in the introduction for improving the hardness of evasion of a security system.

2.1 Previous Works on Multiple Classifiers for Security Applications

The use of MCSs to improve the detection accuracy has been recently proposed by several authors for security applications, and in particular biometric authentication and verification [6, 8, 11] and intrusion detection in computer networks [4]. A common scenario in these applications is the availability of heterogeneous features coming from distinct pattern representations (for instance, features extracted from face and fingerprint images in biometric tasks). In this case it is natural to design a detection system based on the combination of different classifiers trained on disjoint feature subsets corresponding to the different pattern representations. This is a well known approach in the MCS field: if different sets of heterogeneous (and possibly loosely correlated) features are available, designing a MCS as described above can be simpler and more effective than designing a single classifier using all the available features (see for instance [8]). Moreover, if the overall number of features is large, a single classifier could be more prone to over-fitting than a MCS. Another motivation in the context of intrusion detection systems was pointed out in [4]: the ensemble approach "reflects the behaviour of network security experts, who usually look at different traffic statistics in order to produce reliable attack signatures."

The above arguments support the use of classifier ensembles to improve the effectiveness of security systems, in terms of attaining high detection rates. MCSs have also been explicitly proposed to improve the hardness of evasion in biometric tasks (see for instance [11]). To our knowledge, the only work which explicitly proposes MCSs to this aim for intrusion detection tasks is [10], while we are unaware of any work for spam filtering tasks. The approach followed in the mentioned works to improve the hardness of evasion is to add to a MCS one or more classifiers trained on new and different features. The motivations for biometric applications are very intuitive: for instance, in [11] it is claimed that using different biometrics like face, fingerprints, and speech allows to discourage attempts to evade a verification system, since this would require the construction of different kinds of fake biometric traits instead of only one. Similar qualitative arguments are given in [10] for intrusion detection systems: combining classifiers trained on different feature spaces "forces the attacker to devise a mimicry attack that evades multiple models of normal traffic at the same time, which is intuitively harder than evading just one model". We point out that, besides experimental evidences, all the above motivations in favour or the use of MCSs for the specific goal of improving hardness of evasion are only intuitive and qualitative, and are not based on formal and more compelling arguments.

Looking to real security systems, it turns out that the design of many spam filters and intrusion detection systems follows the approach based on combining an ensemble of detectors. Consider for instance two well-known open source systems: the SpamAssassin spam filter (http://spamassassin.apache.org) and the Snort intrusion detection system (http://www.snort.org/). Both SpamAssassin and Snort consist of a set of "tests" which check for different characteristics of input patterns (respectively e-mails and network packets) to detect the presence of "signatures" denoting a malicious origin of the pattern. Tests are often focused on specific signatures of known attacks. They can be of very different kinds, ranging from simple and fixed feature detectors (like a keyword detector in a spam filter) to arbitrarily complex classifiers (like the text classifiers used in spam filters, also know as "bayesian classifiers" in the spam filtering jargon). The outputs of all tests are then properly combined to obtain a decision on the input pattern (either "legitimate" or not). In the case of SpamAssassin, a score (a real number) is associated to each test. The scores of the tests which are satisfied by an e-mail are first summed up, and then the e-mail is labelled as spam, if the overall score exceeds a predefined threshold; otherwise the e-mail is labelled as legitimate. In the case of Snort, a logic OR is computed on the boolean outcomes of all tests (in other words, a network packet is considered as an attack, if it satisfies at least one of the tests). The SpamAssassin and Snort architectures make it easy to add new tests based on different features as new kinds of attacks come up, and to delete existing tests related to attacks that are no more used. These architectures are supported by experience and intuition that suggest the designer of these kinds of security systems that the characteristics which allow detecting malicious patterns can be very different and heterogeneous, and can change over time due to new tricks used by spammers and hackers to defeat spam filters and intrusion detection systems.

We conclude by pointing out again that so far the hardness of evasion of security systems based on MCSs for adversarial classification problems, in particular for intrusion detection and spam filtering tasks, has been motivated only with intuitive and qualitative arguments.

2.2 A Theoretical Framework for Adversarial Classification Problems

As explained above, the few works that proposed so far the use of MCSs for improving the hardness evasion in adversarial classification tasks were based only on informal and empirical motivations. This is actually true also for most works that proposed classification systems based on single classifiers. As mentioned in the introduction, just a few works addressed so far the hardness of evasion of machine learning systems under a more general perspective [1, 3, 9], in particular, to formally analyse it. In this section we focus on the work by Dalvi et al. [3], who developed a formal framework (the only one so far, to our knowledge) for adversarial classification problems. In the following we summarise this framework, which will be applied in Sect. 3 to model and analyse one of the MCS-based strategies considered in this chapter for improving the hardness of evasion (namely, adding classifiers trained on new features).

When machine learning or pattern recognition techniques are used in applications like spam filtering, intrusion detection, biometric authentication, etc., their task can be formalised as a two-class classification problem. Denoting with y the class label, instances belong either to a positive class made up of malicious instances ($y = +$), or to a negative class made up of innocent or legitimate instances ($y = -$). Instances are represented as vectors of N feature values, and are considered as random variables $X = (X_1, \dots, X_i, \dots, X_N)$. A realisation of such a random variable is denoted as $x = (x_1, \dots, x_i, \dots, x_N)$, where x_i is a possible value of the feature X_i. It is assumed that instances are generated i.i.d. according to a given distribution $P(X)$, which can be rewritten as $P(X) = P(X|+)P(+) + P(X|-)P(-)$. The feature space $\mathscr{X}$ is defined as the set of all possible realisations of X.

The framework in [3] considers tasks in which the adversary can modify positive instances (the ones generated by him) at the operation phase to make them being misclassified as legitimate by the classifier, but it can not modify any negative instance nor positive instances belonging to the training set. This happens in several real applications. For instance, if a spam filter is trained *off-line* on a hand-labelled corpora of e-mails, spammers can only modify their own spam e-mails to evade the filter, but can not modify legitimate e-mails nor any spam e-mail in the training set. In other cases, like intrusion detection systems trained *online*, the adversary can also modify training instances: the model in [3] can not be directly applied to these kinds of tasks.

In [3] it is further assumed that the classifier and the adversary act according to given utility and cost functions. Denoting with $y_{\mathscr{C}}(x)$ the decision function of the classifier, whose output is intended to be the label assigned to the instance x, the classifier's utility function is denoted as $U_C(y_{\mathscr{C}}, y)$, and represents the utility accrued by assigning to class $y_{\mathscr{C}}(x)$ an instance x belonging to class y. It is reasonable to assume that classifier's utility is positive for correctly classified instances and negative for misclassified ones, that is, $U_C(+,+) > 0$, $U_C(-,-) > 0$, $U_C(+,-) \leq 0$ and $U_C(-,+) \leq 0$. The cost for the classifier is assumed to be the one incurred for measuring feature values. In the following the cost for measuring the i-th feature of an input instance is denoted as V_i. The expression of the expected utility over the distribution $P(x,y)$ (the joint probability of pattern x being generated with true label y) can be obtained taking into account that the adversary acts by first generating a (positive) instance with a corresponding feature representation x, and then possibly modifying it to some different instance x' (if deemed necessary), with the aim of evading the classifier. Such modification is denoted as a function $\mathscr{A}(x)$. For example, spammers could add words which look legitimate to the body text of a spam e-mail, and this could result in a different e-mail feature representation (depending on the features used by the classifier). It follows that the expected utility for the classifier is given by:

$$U_{\mathscr{C}} = \sum_{(x,y)\in\mathscr{X}\times\mathscr{Y}} P(x,y)[U_C(y_{\mathscr{C}}(\mathscr{A}(x)),y) - \sum_{i=1}^{N} V_i], \tag{1}$$

where $\mathscr{Y} = \{+,-\}$ (note that, by the above assumptions, $\mathscr{A}(x) = x$ if $y = -$, namely for each negative instance).

The adversary's utility function is denoted similarly with $U_A(y_c, y)$. In this case a reasonable assumption is that $U_A(y_c, y)$ takes on a positive value for positive instances misclassified by the classifier as legitimate ($U_A(-,+) > 0$), a negative value for correctly classified positive instances ($U_A(+,+) \leq 0$), and a zero value for negative instances ($U_A(-,-) = U_A(+,-) = 0$), whatever the label assigned by the classifier (in other words, the adversary's utility is not affected by the correct or incorrect classification of negative instances). The cost for the adversary is the one incurred for modifying an instance x, according to the function $\mathscr{A}(x)$. It is assumed that the cost is given by $W(x,\mathscr{A}(x)) = \sum_{i=1}^{N} W_i(x,\mathscr{A}(x))$, being W_i the cost for modifying the i-th feature. Of course, $W_i = 0$ if the i-th feature is not changed, $W_i > 0$ otherwise. The expected utility for the adversary is thus:

$$U_{\mathscr{A}} = \sum_{(x,y)\in\mathscr{X}\times\mathscr{Y}} P(x,y)[U_A(y_{\mathscr{C}}(\mathscr{A}(x)),y) - W(x,\mathscr{A}(x))]. \tag{2}$$

Using the above model, the adversarial classification problem is formulated as a game between classifier and adversary, in which the two players make one move at a time. A move by the classifier consists in choosing a decision function $y_{\mathscr{C}}(\cdot)$ to maximise his expected utility, taking into account both the training set and any knowledge it may have about the strategy $\mathscr{A}(\cdot)$ chosen in the previous move by the adversary. Analogously, a move by the adversary consists in choosing a strategy $\mathscr{A}(\cdot)$ to maximise his own expected utility (Eq. (2)), taking into account the available knowledge about the decision function chosen by the classifier in the previous move. Although game theory could, in principle, be applied to find the optimal sequence of moves by both players according to their utility and cost functions, it was shown in [3] that this is computational intractable, and anyway it requires the knowledge of the distribution $P(x,y)$, which is unrealistic. Therefore, a simplified single-shot version of the game was considered in [3]. Initially, the classifier constructs a decision function using a given training set, assuming it is untainted. Then the adversary chooses his strategy $\mathscr{A}(\cdot)$, assuming he has perfect knowledge of the utility and cost functions of the classifier, and also of his classification algorithm and the training set used. Finally, classifier moves again by choosing a new decision function, assuming he has perfect knowledge of the adversary's utility and cost functions, and that he also knows that the adversary has just made his move based on perfect knowledge on the classifier. Under these assumptions, the optimal adversary's strategy for choosing $\mathscr{A}(\cdot)$ turns out to be the following: for each positive instance x, the adversary has to find a modification x' which maximises the corresponding summand in the expression of the adversary's expected utility (2):

$$\mathscr{A}(x) = \arg\max_{x' \in \mathscr{X}} [U_A(y_{\mathscr{C}}(\mathscr{C}(x'), +) - W(x,x')]. \tag{3}$$

Given the above definition of the adversary's utility and cost functions, it is easy to see that the adversary will change any given instance x, only if it is correctly classified by the classifier as positive, and if there is any instance $\mathscr{A}(x) \neq x$ which is misclassified by the classifier as negative, and the modification cost $W(x,\mathscr{A}(x))$ is lower than the utility gain $U_A(-,+) - U_A(+,+)$. Otherwise, the best strategy is to leave the instance x unchanged, namely $\mathscr{A}(x) = x$.

In [3] the above framework was applied to find the optimal strategies of the adversary and the classifier, when the classifier is a Naive Bayes. Experiments on a spam filtering task quantitatively showed that the classifier performance can significantly degrade, if the adversarial nature of this task is not taken into account, while an adversary-aware classifier can perform much better.

The assumption that the adversary and the classifier have perfect knowledge of each other is rather unrealistic in practical applications, as well as the assumption that their behaviour can be modelled in terms of utility functions whose expected value they aim to maximise. Despite this, we point out that this framework is the first one proposed to model adversarial learning problems in the machine learning field, and it is thus worth taking it into account to formally analyse strategies to improve the hardness of evasion as the ones considered in this chapter.

3 Are Multiple Classifier Systems Harder to Evade?

In this section we develop formal models of adversarial classification problems in which the classification system is made up of an MCS, with the aim of investigating whether MCSs could improve the hardness of evasion. In particular, the model in Sect. 3.1 is based on the framework developed in [3].

3.1 Adding Features to a Classification System

In this section we consider a strategy proposed by several authors to design classifiers for different security applications, and commonly used in real security systems, as explained in Sect. 2.1. The strategy consists in adding to a given classifier ensemble new classifiers trained on different features. We focus in particular on a simple kind of combining function, which consists in thresholding the weighted sum of the outputs provided by each classifier. The reason is that this combining function is equivalent to the functions commonly used in spam filters and intrusion detection systems, as the SpamAssassin and Snort software described in Sect. 2.1. In the following, we apply the framework of [3] to model a classifier based on this strategy, and analyse whether and how such strategy can allow to improve the hardness of evasion of the classifier, according to this framework.

We consider a classifier ensemble made up of N classifiers trained on different feature subsets. We denote with $s_i(x)$ the output provided by the i-th classifier for the instance x (x can be considered the whole feature vector, while each classifier uses only a subset of these features), and with t as the decision threshold. The decision function we consider can be defined as:

$$y_{\mathscr{C}}(x) = \begin{cases} +, \text{ if} \sum_{i=1}^{N} s_i \geq t, \\ -, \text{ if} \sum_{i=1}^{N} s_i < t. \end{cases} \quad (4)$$

We also consider V_i and $W_i(x, \mathscr{A}(x))$ as the cost incurred respectively by the classifier for measuring the values of the i-th feature set of the instance x, and by the adversary for modifying the same features of x.

In the framework of [3] the classifier strategy against the adversary leads to choose a new decision function at each move. To apply the framework of [3] to our case, we add the constraint that the new decision function is obtained by adding one or more new classifiers to the previous ensemble. As in [3], we assume that the initial classifier ensemble is trained on a given training set of untainted instances, namely $\mathscr{A}(x) = x$. Then the adversary reacts by devising a strategy $\mathscr{A}(x)$, which is likely to decrease the classifier effectiveness. Next, some new classifiers are added to the previous classifier ensemble. The adversary can in turn react again by devising a new strategy to defeat the new version of the MCS, and so on. The question we will try to answer is: does adding new classifiers to a previous ensemble makes it harder to evade?

Let us now define in details the adversary strategy, namely the optimal way in which the adversary should choose the function $\mathscr{A}(x)$ against a given ensemble of N classifiers. To this aim, we assume that the adversary knows the feature set used by each of the individual classifiers, the score $s_i(x), i = 1, \ldots, N$ provided by each individual classifier for any positive instance x, and the threshold t. For the sake of simplicity, we also assume that the cost $W_i(x, \mathscr{A}(x))$ for modifying the i-th feature set is equal to the absolute difference between the corresponding scores $s_i(x)$ and $s_i(\mathscr{A}(x))$. This means that the total cost $W(x, \mathscr{A}(x))$ equals the Manhattan distance in the N-dimensional score space between the corresponding score vectors. In other words, the higher the score reduction the adversary would like to attain, the more the changes he has to make to his positive instance. We point out that this is only a simplifying assumption, since in practice a given reduction of the overall score could be attained by different modifications to the same instance at the expense of a different cost incurred by the adversary. However, modelling this fact is much more complex, without any specific assumption about the nature of instances, of the features used by the classifier and of the kinds of attacks used by the adversary. It should be noted that an assumption similar to ours about the cost of modifying an instance in a given feature space was made in [9]: in that work, the cost was assumed to be a function of the distance between x and $\mathscr{A}(x)$ in the feature space.

The optimal strategy of the adversary in [3] is defined by Eq. (3). In our case, denoting with ΔU_A the difference $U_A(-,+) - U_A(+,+)$, with the above definition of the adversary's cost function the optimal strategy against an ensemble of N classifiers can be rewritten as follows:

$$\mathscr{A}(x) = \begin{cases} x' \neq x, \text{ if } \exists x' : y_{\mathscr{C}}(x') = -, \Delta U_A > W(x,x'), \\ \qquad x' = \arg\max_{x'' \in \mathscr{X}} \Delta U_A - W(x,x''), \\ x, \qquad \text{otherwise.} \end{cases} \tag{5}$$

The above optimal strategy can be rephrased as finding, for any given instance x which is correctly classified as positive by the classifier, namely $\sum_{i=1}^{N} s_i(x) \geq t$, an instance $\mathscr{A}(x)$ which is misclassified as negative by the classifier, namely $\sum_{i=1}^{N} s_i(\mathscr{A}(x)) < t$, and for which the utility gain ΔU_A exceeds the cost for making the modification, which by the above assumptions is given by:

$$W(x, \mathscr{A}(x)) = \sum_{i=1}^{N} |s_i(x) - s_i(\mathscr{A}(x))|. \tag{6}$$

If no such instance can be found, then x is left unchanged. It is not difficult to see that the minimum cost the adversary has to incur, so that the modified instance is misclassified as negative, equals the difference between the total score given to x by the classifier and the decision threshold t: $\sum_{i=1}^{N} s_i(x) - t$.

It is now possible to give a formal explanation, according to the framework in [3], of the reasons why adding new classifiers to a given ensemble could improve the hardness of evasion, as well as the detection capability. We consider the simplest case in which the previous classifiers and the decision threshold t are left unchanged at each move. A consequence of adding M new classifiers ($M \geq 1$) to an existing ensemble of N classifiers is that the score of any positive instance could increase, under the reasonable assumption that the individual classifiers are well trained. In particular, consider the optimal strategy $\mathscr{A}(x)$ against N classifiers. As seen above, when this strategy leads the adversary to modify an instance x to a different instance $\mathscr{A}(x) \neq x$, this allows to evade the classifier. Denoting with $s^k(x)$ the score provided by an ensemble of k classifiers, this means that $s^N(x) = \sum_{i=1}^{N} s_i(\mathscr{A}(x)) < t$. However, the modified instance obtained with the strategy devised against N classifiers is not guaranteed to evade a new ensemble comprising M new classifiers. Indeed the new score $s^{N+M}(x)$ will be given by the sum of the previous score and of the scores of the new classifiers, $s^N(x) + \sum_{i=N+1}^{N+M} s_i(x)$, which could exceed t. This means that the optimal strategy of the adversary against N classifiers is not guaranteed to be optimal against $M+N$ classifiers. Accordingly, the detection capability can improve by adding new classifiers. Moreover, the evasion cost could increase by adding new classifiers. Indeed, if the score for some positive instance x correctly classified by the new classifier ensemble increases from $s^N(x)$ to $s^{N+M}(x) > s^N(x)$, such increase could make the difference $s^{N+M}(x) - t$ larger than the utility, $U_A(-,+)$, that the adversary would gain by modifying x so that it is misclassified as negative. This implies that there could be some positive instances that the adversary can afford to modify to evade N classifiers, but not to evade $N+M$ classifiers. This means that the classifier has become harder to evade.

In the above model it is assumed that the adversary can modify only positive instances, and the analysis was focused only on the classifier's detection capability on positive instances (namely, on the false negative error rate). Before concluding this section it is worth discussing also the possible effects of adding classifiers, to the classification accuracy on negative instances (the false positive error rate). Under the same reasonable assumption above that the individual classifiers are well trained, a negative instance x which is correctly classified as negative by an ensemble of N classifiers (namely, $s^N(x) < t$) is likely to be classified as negative also by an ensemble of $N+M$ classifiers, if the sum of the new scores $\sum_{i=N+1}^{N+M} s^i(x)$, is lower than $t - s^N(x)$. Moreover, instances erroneously classified as positive by N classifiers (namely, $s^N(x) \geq t$) could be correctly recognised as negative by a larger ensemble, if $\sum_{i=N+1}^{N+M} s^i(x)$ is negative and lower than $t - s^N(x)$.

To sum up, the above analysis provides the first, formal support to the strategy of adding new classifiers trained on different features to a given classifier ensemble, to improve both the detection capability and the hardness of evasion. In Sect. 4.1 we will give an experimental evaluation on a case study related to spam filtering.

3.2 Splitting Features across an Ensemble of Classifiers

The second MCS-based strategy we consider is a design approach applied in several applications to simplify the classifier design and to improve classification accuracy, when the feature set is very large or is made up of heterogeneous feature subsets. The approach consists in combining different classifiers trained on disjoint feature subsets, instead of designing a single, "monolithic" classifier based on all the available features. It can be implemented naturally on heterogeneous features: a typical example is the combination of a face classifier and a fingerprint classifier in biometric tasks. The issue we address is the following: could this design approach be exploited to improve the hardness of evasion of a security system in adversarial classification tasks?

In this chapter we try to give a first answer to this question, without focusing to any specific application. We will not apply the model in [3] as made for the MCS-based strategy analysed in Sect. 3.1, since in this case we are not analysing a defence strategy, but we are comparing two different classifier design approaches. We develop instead a simple, general model of a classification system based on these two classifier architectures (either a single classifier trained on a given feature set, or an ensemble of classifiers trained on disjoint subsets of the same feature set), and a method for evaluating the corresponding hardness of evasion.

We first assume that a fixed feature set $x_1, \ldots, x_n$ is available, and that all the features are binary and take on the values 0 and 1. Without loosing generality we assume that the value 1 of any feature denotes the presence of some "malicious" characteristic in the input instance, while a 0 value denotes its absence. In the case of a single classifier, denoting with x a feature vector and with $y_{\mathscr{C}}(x)$ the decision function, we assume that $y_{\mathscr{C}}(x)$ is a thresholded weighted sum of the input features $x_1, \ldots, x_n$, with weights $w_1, \ldots, w_n$:

$$y_{\mathscr{C}}(x) = \begin{cases} +, \text{ if} \sum_{i=1}^{n} w_i x_i \geq t, \\ -, \text{ if} \sum_{i=1}^{n} w_i x_i < t. \end{cases} \tag{7}$$

Note that this kind of decision function, as well as the above assumption on the features, fits several real classification systems for security tasks, like SpamAssassin and Snort.

In the case of an MCS made up of N classifiers trained on N disjoint subsets of the same n features, we assume that the individual classifiers have the same kind of decision function (7). As a combining function, we consider the logical OR between the N boolean outputs of the individual classifiers, where the logical value *true* is assumed to denote the positive class $y = +$ (in other words, for an input pattern being labelled as positive by the MCS, it is sufficient that at least one of the individual classifiers labels it as positive). We consider a non-linear combining function because a linear one (a linear combination of the soft outputs of the individual classifiers) would lead to the same kind of decision function as the one of the monolithic classifier (since also the decision functions of individual classifiers is linear). We consider in particular the logical OR because of its simplicity, and because it is particularly suited to keep the false negative error rate low. We remind the reader that this combining function is used in Snort. In principle, it could also be used to combine different spam filters or different intrusion detection systems, whose outputs can be viewed as the features. Given that the value 1 of any feature denotes the presence of some "malicious" characteristic in the input instance, it follows that all the weights of both the monolithic classifier and the N individual classifiers of the ensemble are non-negative (because it is reasonable that the presence of a "malicious" characteristic must not decrease the overall score of a classifier). A scheme of the two classifier architectures is shown in Fig. 1.

In the two classifier architectures above, the parameters to be set during the training phase are the number of individual classifiers in the MCS, the feature subset associated to each individual classifier, and the values of the weights and of the decision thresholds in the decision functions of the linear classifiers. These choices will affect the effectiveness of the classifiers. The effectiveness has to be measured in terms of both the classification accuracy and the hardness of evasion. Note that in adversarial classification problems the classification accuracy should be intended as a "static" characteristic of a classifier, in the sense that it is related to a *fixed* strategy used by the adversary. Such strategy can be considered as represented by

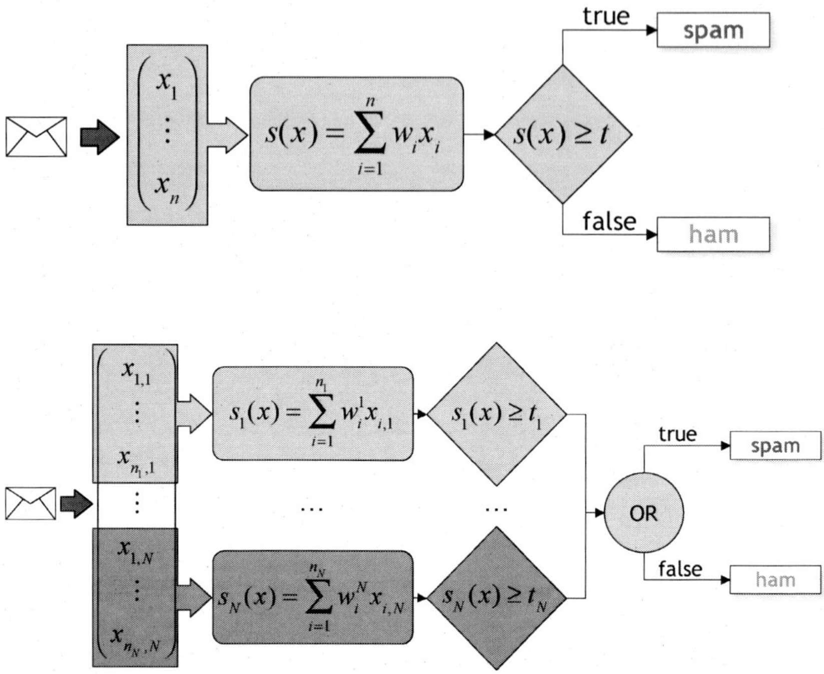

Fig. 1 The two classifier architectures considered in this section. A single, linear classifier working on n features (top). N linear classifiers working on disjoint subsets of the same n features, whose decisions are combined using the OR logical function (bottom)

the training instances. The hardness of evasion measures instead how easy is for the adversary to evade the classifier using one or more different specific strategies. In other words, it measures how vulnerable the classifier is to specific kind of attacks, different than the ones represented in the training set. Therefore, when comparing different security systems both measures have to be taken into account, as in the scheme of Fig. 2. Ideally, a security system should be characterised both by a high accuracy and a high hardness of evasion. In practice, a trade-off between the two goals could be needed.

The classification accuracy can be evaluated in terms of the false positive and false negative classification rates. Usually, the suitable trade-off between these misclassification rates is application-dependent. How to measure the hardness of evasion is clearly application-dependent as well. In particular, it could depend on the kind of classification system, on the knowledge the adversary has about it, and on the kinds of attacks he could make. However in this section we consider a measure of the hardness of evasion focused on comparing the two classifier architectures we are interested in, without making any specific assumption on the application. Concerning the adversary, we consider the worst case scenario for the classification system, as in the framework in [3]: we assume that the adversary has full knowledge of the classifier architecture, of the features and of the parameter values, and

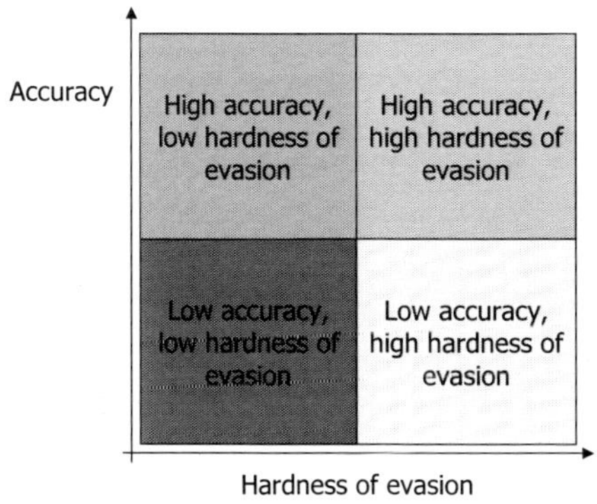

Fig. 2 An example of the two measures which should be used to evaluate the performance of a classifier in a security system: the classification accuracy against a given strategy used by the adversary (represented by training instances), and the hardness of evasion against new kinds of attacks. Ideally, the classifier should exhibit both a higher accuracy and a high hardness of evasion (upper right region in the accuracy-hardness of evasion plane)

is capable to evade any feature (namely, to turn the value of any feature from 1 to 0). We further assume that the adversary has to make the same effort to evade any feature. We point out that this last assumption could not be true in practice. However, taking into account different costs for evading different features would make our model much more complex, which is out of the scope of this chapter. Under the above assumptions, the hardness of evasion can be defined as follows:

For a given feature set, the hardness of evasion is defined as the expected value of the minimum number of features which have to be modified to evade the classifier.

Accordingly, a classifier A will be harder to evade than a classifier B, if the average minimum number of features the adversary has to evade for evading A is higher than for evading B. The whole classifier performance could thus be measured in the accuracy-hardness of evasion plane of Fig. 2 by using a proper combination of false positive and false negative classification rates in the Y axis (note that, in this case, the accuracy increases for *decreasing* values in the corresponding axis) and the average minimum number of features to evade for evading the whole classifier in the X axis.

It is now possible to discuss, at least informally, whether and how the MCS classifier architecture discussed above could be harder to evade than the monolithic classifier. Consider a given positive instance whose feature vector x is correctly classified by the monolithic classifier, namely, $s(x) \geq t$ (see Fig. 1). Under the above assumptions, to evade such classifier the adversary will have to modify such instance to some instance with feature vector x' with the aim of turning to 0 those features which exhibit in x a value of 1 and are associated to the largest weights, until the overall output of the classifier becomes lower than the threshold t: $s(x') < t$. Instead, to evade the MCS the adversary has to evade all individual classifiers which correctly classify an instance as positive, since they are combined with the logical

OR function. Since the individual classifiers are assumed to implement the same kind of decision function (7) as the monolithic classifier, the adversary will have to apply the same strategy above against *all* the individual classifiers. More precisely, let us denote with x_m the feature vector of the m-th individual classifier. Assuming it correctly classifies x_m as positive (namely, $s_m(x_m) \geq t_m$), the adversary will have to modify his original instance to some other instance with feature vector x'_m, in which the features exhibiting in x_m a value of 1 and are associated to the largest weights are turned to 0, until the overall output of the m-th classifier becomes lower than the threshold t_m: $s_m(x'_m) < t_m$. We point out again that this has to be done for each individual classifier, which correctly classifies an instance as positive. It follows that a proper choice of the feature subsets could force the adversary to evade on average a higher number of features to evade an MCS with the above architecture, than to evade the monolithic classifier. It should however be noted that the kind of MCS considered in this section could exhibit a higher false positive error rate than the monolithic classifier, since each individual classifier of the MCS is trained with a smaller feature set, and an input instance is labelled as positive if at least one individual classifier labels it as positive. Accordingly, the attainable advantage of the MCSs in terms of hardness of evasion could need to be traded-off for an increase in false positives. In the following section, we experimentally investigate the hardness of evasion of these two classifier architectures on the same case study as the one considered for the MCS-based strategy analysed in Sect. 3.1.

4 A Case Study in Spam Filtering

In this section we apply the two formal models of Sect. 3 to a case study of a spam filtering task, to experimentally analyse the hardness of evasion of the two corresponding MCS-based strategies considered in this chapter. For our experiments we use the well known open source SpamAssassin spam filter, whose architecture has been described in Sect. 2.1, and a large and publicly available corpus of real spam and legitimate e-mails.

We used the latest versions of SpamAssassin available at the time of carrying out our experiments: version 3.2.4 for the experiments in Sect. 4.1, and 3.2.5 for the ones in Sect. 4.2. We used the filter configuration named "bayes+net", which includes all the available tests (several hundreds). The outputs of all tests are binary (either 0 or 1). Nine of the tests are associated to a text classifier with features corresponding to terms in the e-mails' header and body. The continuous-valued output of the text classifier is discretized by default into nine disjoint intervals, each of which is associated with a binary test. All the remaining tests consist in fixed feature detectors. SpamAssassin is deployed with a default value for the weight of each test, and a default value of 5 for the detection threshold. All these values can be modified by the user. All the details about SpamAssassin, including the description of its tests, can be found in http://spamassassin.apache.org/

The e-mail corpus we used is the TREC 2007 e-mail data set, available at http://plg.uwaterloo.ca/~gvcormac/treccorpus07/. It is made up of 75,419 real e-mail messages, received by a mail server between April 2007 and July 2007, and contains 25,220 legitimate and 50,199 spam messages.

In Sects. 4.1 and 4.2 we describe the experiments aimed at evaluating the hardness of evasion attainable respectively by adding new classifiers to a classifier ensemble, and by using an ensemble of classifiers trained on disjoint subsets of given feature set, instead of a single classifier trained on the whole feature set.

4.1 Adding Features to a Spam Filter

In this section we evaluate the hardness of evasion of the SpamAssassin filter, attainable by adding new tests (which can be thought as classifiers, as explained above) to a previous set of tests (equivalently, to a previous classifier ensemble). We point out that in our experiments the SpamAssassin tests can be considered as classifiers, although most of them are fixed feature detectors, because we consider the case in which the previous classifiers of the ensemble are *not* retrained, and neither their weight nor the decision threshold is changed, after a new classifier are added. For these experiments we used the first 10,000 messages of the TREC 2007 corpus, in chronological order (1,969 legitimate e-mails and 8,031 spam e-mails), to train the SpamAssassin text classifier. The remaining 65,419 e-mails were used as a test set.

In the model of Sect. 3.1 the adversary was assumed to be capable to modify his instances to attain any modification he would like on the classifier's outputs. In practice this could be not always possible. However, it was very difficult to check whether this is possible or not for all SpamAssassin's tests. For the sake of simplicity, we kept the above assumption and avoided to devise real modifications to e-mails to attain the desired output changes. We point out that this assumption is totally in favour of the adversary, since we are not setting any constraint on the actual modifications which can be made on spam e-mails by him.

In the experiments we considered the following utility function of the classifier: $U_C(+,+) = 1, U_C(-,+) = -10, U_C(+,-) = -1, U_C(-,-) = 1$, namely, it gains 1 for correct classifications, looses 1 for misclassifying a spam e-mail as legitimate, and looses 10 for misclassifying a legitimate e-mail as spam. This is coherent with the considered application, in which it is generally agreed that false positive errors are much more harmful than false negative ones. The utility function of the adversary was set to 0, except for the gain accrued for evading the classifier, namely for spam e-mails misclassified as legitimate, for which two different values (1 and 5) were considered: $U_A(+,+) = U_A(+,-) = U_A(-,-) = 0, U_A(-,+) = 1,\ 5$. We considered two different values of $U_A(-,+)$ to evaluate scenarios characterised by a different value of the maximum cost the adversary can (or wants) to pay to evade a spam filter. We point out that the above choices of the relative values of the utility functions is somewhat arbitrary, besides the obvious constraints mentioned above, due to the fact that such costs can not be precisely evaluated in practice (and it could also be questionable that the *real* behaviour of a classification system and of an adversary can be modelled in terms of such utility functions, as pointed out in Sect. 2.2). However, we are interested here to the qualitative behaviour of the classifier's and adversary's performance (in terms of its expected utility), and different choices of the utility values would affect only the absolute values of their expected utilities, not their qualitative behaviour.

Finally, we assumed that the cost V_i faced by the classifier for measuring the features associated to the i-th text (or classifier) is zero, since such cost is just a negative constant added to the expected value of the utility function in the framework of Sect. 3.1. The cost for the adversary was defined as explained in Sect. 3.1, as the manhattan distance in classifiers' outputs space between the outputs given to a positive instance after and before the modification made by the adversary.

The addition of new classifiers at each move of the game was modelled as follows. Since the number of SpamAssassin tests is rather high (several hundreds), we did not add just one test at each move. Instead we subdivided them into n disjoint subsets $S_1, \ldots, S_n$, and added at each step all the tests of a given subset. For the purposes of these experiments we chose $n = 6$. The number of test was set to 119 for S_1 and to 100 for all the other subsets, for a total of 619 tests. This choice was made since only 619 out of all tests gave an output value of 1 for at least one of the e-mails in our data set: we considered therefore only these 619 tests. In the real SpamAssassin filter new tests are usually added in response to new spammers' tricks. Accordingly, it would have been reasonable to subdivide the tests taking into account their chronological order. Unfortunately the time in which each test was introduced is not reported in the SpamAssassin documentation. So we had to resort to a random subdivision. To make experiments easily reproducible, we sorted all tests alphabetically according to their names as listed in http://spamassassin.apache.org/tests_3_2_x.html. The only exception were the nine tests related to the text classifier, which were included in the first subset since it is known that text classifiers are used in spam filters since several years.

The moves of the classifier and the adversary at each step of the game were implemented according to the following procedure. At each step, we first evaluate the performance of the classifier and the adversary after a new set of tests is added to the classifier, and the adversary uses the strategy which was optimal against the previous set of tests (in the first step, this means that the adversary does not modify his instances). This simulates what happens in real cases, as soon as a spam filer is updated. Then the optimal strategy of the adversary against the new set of tests is computed, and the performances of the classifier and the adversary are evaluated again. This simulates what happens when spammers devise new tricks to evade the last version of a spam filter. This procedure can be formalised as follows:

1. $R \leftarrow \emptyset$, $\mathscr{A}^0(x) = x$ for all x
2. For $k = 1, \ldots, n$:

   2.1 $R \leftarrow R \bigcup S_k$

   2.2 Evaluate the expected utility of the classifier and of the adversary, when the classifier uses the tests in R and the adversary uses the strategy $\mathscr{A}^{k-1}(x)$ which was optimal for the previous set of tests

   2.3 Compute the optimal adversarial strategy $\mathscr{A}^k(x)$ against tests in R

2.4 Evaluate the expected utility of the classifier and of the adversary, when the classifier uses the tests in R and the adversary uses the corresponding optimal strategy $\mathscr{A}^k(x)$

The adversary's optimal strategy $\mathscr{A}^k(x)$ at step k was computed as follows, according to Sect. 3.1. Denoting the set of tests $S_1 \bigcup \ldots \bigcup S_k$ used by the classifier as R, for any positive instance x correctly classified by the filter (namely $y_{\mathscr{C}}(x) = +$, or equivalently $\sum_{i \in R} s_i(x) \geq t$), we compute the set of feasible values s'_i for the scores of tests in R which would correspond to an instance x' classified as negative (namely $\sum_{i \in R} s'_i < t$), such that the corresponding cost $W(x, x') = \sum_{i \in R} |s'_i - s_i(x)|$ is minimum and is lower than the utility gain. If such scores exist, then we assume that the adversary evades the filters by modifying x, otherwise it is assumed that the adversary can not afford to modify x to evade the filter.

The results are shown in Figs. 3 and 4, for both utility functions considered for the adversary, in terms of the expected utility of the adversary and of the classifier, as a function of the number of tests used in SpamAssassin. The results in the top-left graph refer to the case in which the adversary does not modify his instances. As one could expect, the expected utility of the classifier increases as the number of test increases while the opposite happens for the adversary. This means that adding new classifiers (tests) based on different features (without modifying the previous ones nor the detection threshold) allowed to improve the detection capability. The only exception is when going from 419 to 519 tests. The bottom-left graph shows what happens when the adversary uses the optimal strategy against each set of tests. The expected utility of the adversary significantly improves with respect to the previous case. The expected utility of the classifier still increases as the number of tests increases, but obviously attains lower values than in the previous case. However, it is worth noting that the improvement attained by the adversary, reported in the top-right graph, tends to decrease as the number of tests increases. Similarly, the decrease in the classifier's expected utility tends to be higher for smaller number of tests. The reason is that the modification cost the adversary has to face to evade the classifier increases as the number of tests increases, until it exceeds the utility gain for some positive instances, making it no more convenient to modify them. This is a clear evidence that adding new classifiers based on different features can allow to improve not only the classifier's discriminant capability, but also its hardness of evasion. Consider finally the bottom-right graph, corresponding to the case when the classifier adds new tests, and the adversary uses the strategy which was optimal against the *previous* set of tests. For smaller number of tests (up to 319), the expected utility of the adversary is between the ones of the first two graphs: this is reasonable, because it is now trying to evade only *some* of the tests used by the classifier. However, for larger number of tests its expected utility is even worse than the one it attained *without* trying to evade any test. The expected utility of the classifier is instead very close to the one it attained when the adversary did not try to evade any test. This means that the addition of new tests allowed to compensate the actions made by the adversary to evade the previous tests. In other words, most spam e-mails which evaded the previous version of the filter were detected by the new tests.

The behaviour of the expected utility for the two different values of the gain accrued by the adversary for evading the classifier (Fig. 3 vs. Fig. 4) is similar, with the obvious difference that the expected utility of the adversary is higher in the graphs of Fig. 4 than in Fig. 3, since it can afford a higher cost to modify instances. The opposite happens for the classifier.

These experimental results on a real case study give thus a quantitative confirmation to the theoretical explanation given in Sect. 3.1 on the effectiveness of adding new classifiers based on different features in improving both the detection capability and the hardness of evasion of a security system like a spam filter.

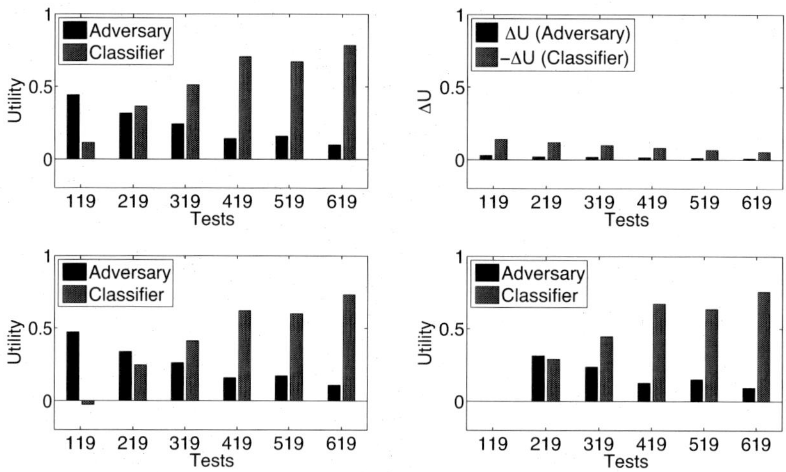

Fig. 3 Expected utility for the adversary and the classifier, as a function of the size of the classifier ensemble, when $U_A(-,+) = 1$. Top-left: the adversary does not modify his instances. Bottom-left: the adversary uses the optimal strategy against the classifier. Top-right: the gain and the loss in expected utility attained respectively by the adversary and the classifier, when passing from the situation in the top-left graph to that in the bottom-left one, as a function of the ensemble size. Bottom-right: for each ensemble size, the adversary uses the optimal strategy against the *previous* set of rules

4.2 *Splitting the Features of a Spam Filter across an Ensemble of Classifiers*

In this section we give an experimental evaluation of the classification accuracy and the hardness of evasion of the two classifier design architectures modelled in Sect. 3.2: a single linear classifier trained on a given set of features, and an ensemble of linear classifiers trained on disjoint subsets of the same features and combined with the OR logical function. The experiments were carried out on the TREC 2007

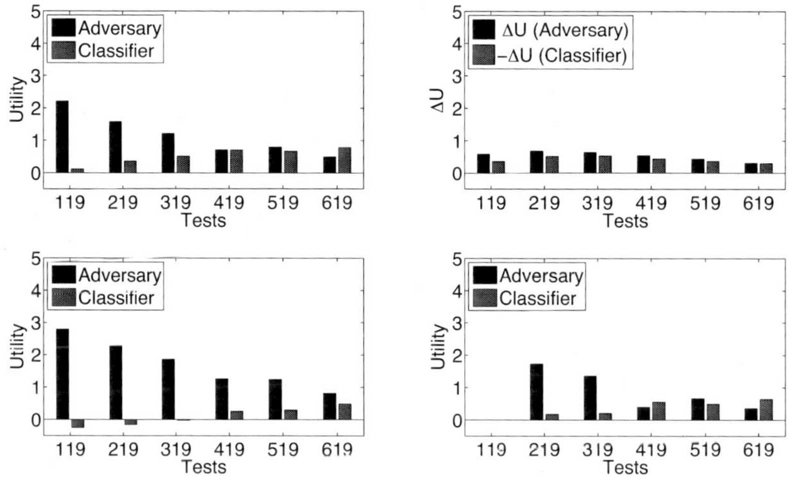

Fig. 4 Expected utility for the adversary and the classifier, as a function of classifier ensemble size, when $U_A(-,+) = 5$. See caption of Fig. 3 for the other details

e-mail corpus described in Sect. 4.1. We used as feature set the tests of the SpamAssassin filter (version 3.2.5) which gave an output value of 1 for at least one of the e-mails of this data set. Their number was 549. To implement the monolithic linear classifier and the individual classifiers of the ensemble we used a support vector machine (SVM) with the linear kernel. Since nine of the SpamAssassin tests are associated with a text classifier, they were not used as features of the MCS. The text classifier itself was instead used as one of the individual classifiers of the ensemble. In this case, given that its output is a real number in the interval $[0,1]$ (with small values denoting legitimate e-mails), we set a decision threshold of 0.5.

The first 10,000 e-mails of the data set, in chronological order (1,969 legitimate e-mails and 8,031 spam e-mails), were used to train the SpamAssassin text classifier and the individual classifiers of the MCS. The next 10,000 e-mails were used to train the monolithic classifier (we avoided using to this aim the same first 10,000 e-mails used to train the text classifier, since its outputs were used as features of the monolithic classifier). The remaining 55,419 e-mails were used as a test set. The SVMs were trained using the publicly available *libsvm* software [2]. To carry out multiple runs of the experiments, all the classifiers were trained on 2,000 e-mails randomly extracted from the corresponding training sets described above.

The SVM parameters of the monolithic classifier were set through a 5-fold cross validation on the training set, by minimising an objective function given by $100 \times FP + FN$, being FP and FN the two kinds of error rates. In other words, the cost of false positive errors was fixed to be one hundred times higher than the cost of false negative errors. The decision threshold of the SVM was fixed by minimising the same objective function. The same procedure was used to set the parameters and the decision threshold of the individual classifiers of the MCS. However in this case we fixed the cost of false positive errors to be *one thousand* times higher than

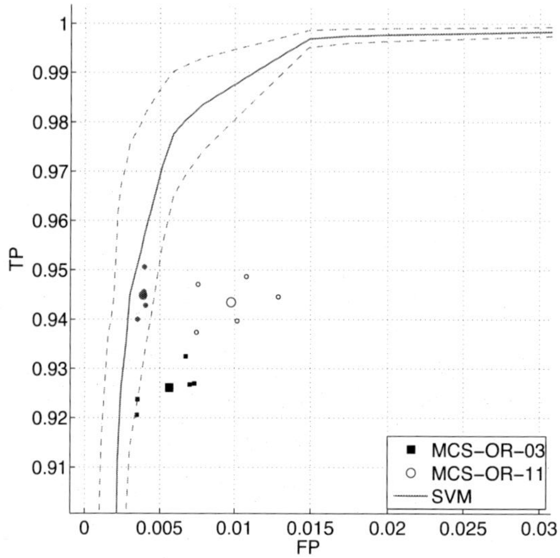

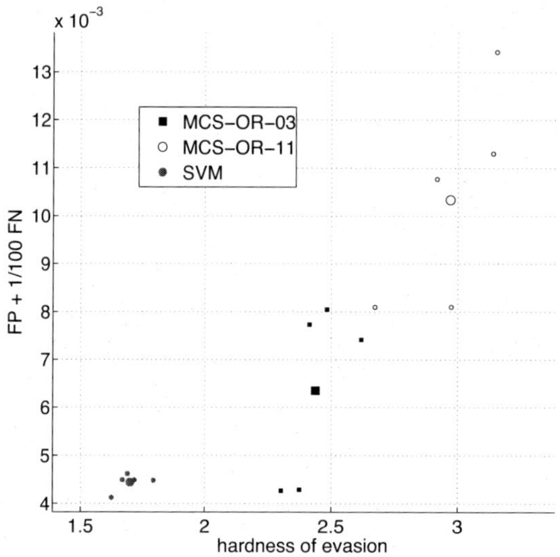

Fig. 5 Top: classification accuracy of the monolithic classifier (solid circles) and of the two MCSs (squares: three classifiers; white circles: eleven classifiers) in the TP, FP (ROC) plane. Small circles and small squares represent the values attained in the five runs of the experiments, while large ones represent the corresponding average values. The average ROC curve of the monolithic classifier is also shown (solid line), together with its standard deviation (dashed lines). Bottom: average FN rates with standard deviation as a function of the maximum number of features the adversary can evade. The FN rates when no feature is evaded correspond to the ones in the left plot

the cost of false negative errors. The reason is that, differently from the features of the monolithic classifier, the individual classifiers of the MCS have been combined with the OR function, implying that the MCS labels an e-mail as spam, if at

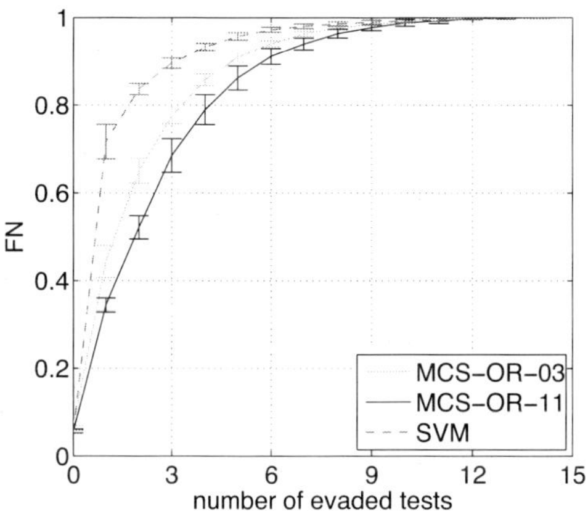

Fig. 6 Trade-off between the average classification accuracy (Y axis) and the average hardness of evasion (X axis) of the monolithic classifier (dashed line) and of the two MCSs (solid lines), over five runs of the experiments. Classification accuracy is measured as the false negative error rate. The hardness of evasion is measured in terms of the average minimum number of features the adversary has to evade, for evading the whole classifier. Horizontal bars represent the standard deviation of classification accuracy over the five runs of the experiments. Note that the area of the plot corresponding to the best accuracy-hardness of evasion trade-off is the bottom-right one

least one of the individual classifiers labels it as spam; it is therefore necessary to keep the false positive error rate of the individual classifiers of the MCS as low as possible. The parameters of the SVMs were the regularisation parameter C and the relative cost of false positive errors with respect to false negative ones, used in the *libsvm* objective function of the SVM learning algorithm (denoted in the following as w_{FP}).[1] For the monolithic classifier, the parameter values were chosen among all the possible combinations of the C values $\{0.001, 0.01, 0.1, 1, 10, 100\}$, and w_{FP} values $\{2, 5, 7, 10, 50, 100\}$. For the individual classifiers of the MCS we considered the same C values above and the w_{FP} values $\{10, 50, 100, 500, 1000\}$.

As mentioned above, the MCS based on SpamAssassin tests was made up of the SpamAssassin text classifier and of $N-1$ linear classifiers trained on disjoint subsets of the 541 tests not associated to the text classifier. We considered two different ensemble sizes: $N = 3$ (namely, the text classifier and two linear classifiers) and $N = 11$. The 541 available tests (features) were distributed uniformly among the linear classifiers. The choice of which features associate to each classifier should be made by taking into account the kinds of the features. For instance, heterogeneous features could be fed to different classifiers. For the sake of simplicity, in our experiments we randomly split the features into N disjoint subsets (we just checked whether the false positive error rate of the MCS, estimated on the 2,000 e-mails of the training

[1] We point out that the cost parameter w_{FP} of the SVM learning algorithm could not reflect the *real* cost considered in the task at hand: therefore its optimal value could be different from the real cost.

set, was higher than 0.01: in that case we disregarded the corresponding feature splitting).

The experiments described above were repeated five times on different randomly extracted training sets of the SVMs. Let us consider first the accuracy of the two kinds of classifiers (the monolithic classifier and the MCS). We report it in the Receiver Operating Characteristic (ROC) plane, in which the Y axis corresponds to the true positive classification rate (TP) and the X axis to the false positive rate (FP). In Fig. 5 (top) we show the TP and FP values of the monolithic classifier and of the two MCSs on the e-mails in the test set, obtained in the five runs of the experiments, as well as their average values across the five runs. For the sake of completeness we also report the whole ROC curve of the monolithic classifier (obtained by varying the decision threshold of the SVM). It can be seen that the accuracy of the monolithic classifier is slightly higher than the one of the two MCSs (it exhibits both higher TP and lower FP values). We remind the reader that this accuracy refers to the case when the adversary does not attack the classifiers. The hardness of evasion is instead shown in Fig. 5 (bottom). Since we are assuming that the adversary can only modify positive instances, only the FN rate can change under attack. Accordingly, in Fig. 5 (bottom) we report the FN rate as a function of the maximum number of features the adversary can evade. The FN rates for zero evaded tests correspond to the values reported in Fig. 5 (top). Figure 5 (bottom) clearly shows that, although the MCSs have a worse FN rate than the monolithic classifier when they are not under attack, they are harder to evade. For instance, if the adversary evades at most one feature, the FN rate of the two MCSs is between about 0.35 and 0.45, while the FN rate of the monolithic classifier is about 0.70, and so on. In other words, evading an MCS in which features are split across different classifiers required to evade a higher number of features than in the case they were processed by a monolithic classifier, as argued in Sect. 3.2, although this comes at the expense of a slight increase in the false positive rate.

Consider finally a comprehensive plot showing the trade-off between the accuracy (when the adversary does not attack) and the hardness of evasion, as in the scheme of Fig. 2. The accuracy (Y axis) is measured using the same trade-off between FP and FN rates as in the objective function of the monolithic classifier: $100 \times FP + FN$. The hardness of evasion is measured as explained in Sect. 3.2, as the average minimum number of features that have to be evaded to evade the whole classifier. The accuracy-hardness of evasion trade-off attained by the monolithic classifier and by the MCSs is shown in Fig. 6. From this plot it is easy to see that the monolithic classifier attains a slightly higher accuracy (two to three times better than the MCSs), at the expense of a lower hardness of evasion (up to two times lower than that of the MCSs).

To sum up, the results presented in this section can be considered as the first experimental evidence, based on a formal setting, that the MCS architecture based on splitting features across different classifiers can be exploited in security tasks to improve the hardness of evasion in security systems.

5 Conclusions

Taking into account explicitly the presence of an intelligent, adaptive adversary in the design of classification systems for security applications, with the aim of making a classifier harder to evade, is a topic which has been addressed only recently in the machine learning and pattern recognition literature. So far no general frameworks exist yet to deal with this problem. In this chapter we addressed this issue focusing on tasks like spam filtering and intrusion detection in computer networks, and on a classifier architecture based on an ensemble of classifiers. This architecture has been recently proposed by several authors and is used in commercial and open source products, but is supported so far only on by empirical and intuitive motivations. We tried to give a first answer, based on more formal motivations, to the questions of whether and how multiple classifier systems could allow to improve the hardness of evasion of a classifier. We considered in particular a defence strategy consisting in adding classifiers based on new features to a previous ensemble (as usually done in spam filters and intrusion detection systems to deal with new kinds of attacks), and to a design approach based on combining classifiers trained on disjoint subsets of features, instead of designing a monolithic classifier trained on the same features. We developed formal models of the corresponding classification systems and of possible adversary's strategies used to attack them (exploiting the framework developed in [3] to analyse the former strategy). We then gave an experimental evaluation on a case study related to the spam filtering task, using a real spam filter and a large and publicly available corpus of real spam e-mails.

Our results can be exploited as a starting point of future works aimed at formulating practical guidelines for the design of more robust classification systems in security applications.

Acknowledgements. We would like to thank Nilesh Dalvi and Mausam for providing us the code used in [3].

References

1. Barreno, M., Nelson, B., Sears, R., Joseph, A.D., Tygar, J.D.: Can machine learning be secure? In: Proc. 2006 ACM Symp. Inf., Computer and Communications Security, Taipei, Taiwan, pp. 16–25. ACM, New York (2006)
2. Chang, C.C., Lin, C.J.: LIBSVM: a library for support vector machines (2001), http://www.csie.ntu.edu.tw/~cjlin/libsvm
3. Dalvi, N., Domingos, P., Mausam, S.S., Verma, D.: Adversarial classification. In: Proc. 10th ACM SIGKDD Int. Conf. Knowledge Discovery and Data Mining, Seattle, WA, pp. 99–108. ACM, New York (2004)
4. Giacinto, G., Roli, F., Didaci, L.: Fusion of multiple classifiers for intrusion detection in computer networks. Pattern Recognition Letters 24(12), 1795–1803 (2003)
5. Globerson, A., Roweis, S.T.: Nightmare at test time: robust learning by feature deletion. In: Cohen, W.W., Moore, A. (eds.) Proc. 23rd Int. Conf. Mach. Learn., Pittsburgh, PA, pp. 353–360. ACM, New York (2006)

6. Haindl, M., Kittler, J., Roli, F. (eds.): MCS 2007. LNCS, vol. 4472. Springer, Heidelberg (2007)
7. Jorgensen, Z., Zhou, Y., Inge, M.: A multiple instance learning strategy for combating good word attacks on spam filters. J. Mach. Learn. Research 9, 1115–1146 (2008)
8. Kittler, J., Hatef, M., Duin, R.P., Matas, J.: On combining classifiers. IEEE Trans. Pattern Analysis and Mach. Intell. 20(3), 226–239 (1998)
9. Lowd, D., Meek, C.: Adversarial learning. In: Press, A. (ed.) Proc. 11th ACM SIGKDD Int. Conf. Knowledge Discovery and Data Mining, Chicago, IL, pp. 641–647. ACM, New York (2005)
10. Perdisci, R., Gu, G., Lee, W.: Using an ensemble of one-class svm classifiers to harden payload-based anomaly detection systems. In: Proc. IEEE Int. Conf. Data Mining, Hong Kong, pp. 488–498. IEEE Comp. Soc., Los Alamitos (2006)
11. Ross, A.A., Nandakumar, K., Jain, A.K.: Handbook of Multibiometrics. Springer, Heidelberg (2006)

A Personal Antispam System Based on a Behaviour-Knowledge Space Approach

Francesco Gargiulo, Antonio Penta, Antonio Picariello, and Carlo Sansone

Abstract. In their daily work and common life, people suffer serious problems with Unsolicited Commercial E-mails (UCE), commonly known as *spam*: common people, small companies and large public or private institutions feel that spam has weakened the reliability and effectiveness of email as an efficient tool for communicating. To establish simple, fast and effective countermeasures against spam attacks is a necessary strategy of a modern mailing management system. In this chapter we describe a novel method for detecting spam messages, analyzing both text and image attached components. In particular, we describe an architecture for deploying a personal antispam system able to overcome some problems that are still besetting the state-of-the-art spam filters. Text analysis is accomplished by considering recent advances in both semantic and syntactic analysis; in addition, spammers tricks based on images are also taken into account. A Behaviour Knowledge Space approach for fusing the different results coming from the analysis of the different parts of the e-mails enhances the performance of the proposed system, as described by the experiments we have carried out.

Keywords: text-based and image-based spam, Singular Value Decomposition, Latent Semantic Analysis, Pattern Recognition, Image Analysis, combining classifiers, Behaviour Knowledge Space.

1 Introduction

It is a well-known story that e-mail has grown from a tool used by few universities and scientists to a ubiquitous communication tool, evolving from simple plain text into a powerful multimedia message. At the same time, following the growth of e-mail production and diffusion, spam has changed from a minor and sometimes

Francesco Gargiulo · Antonio Penta · Antonio Picariello · Carlo Sansone
Dipartimento di Informatica e Sistemistica, University of Naples Federico II, Italy
e-mail: {francesco.grg, a.penta, picus, carlosan}@unina.it

O. Okun & G. Valentini (Eds.): Appli. of Supervised & Unsuper. Ensemble Meth., SCI 245, pp. 39–57.
springerlink.com © Springer-Verlag Berlin Heidelberg 2009

bothering problem into a multi-billion dollar problem. The presence of spam, in fact, can seriously compromise normal user activities, forcing to navigate through mailboxes to find the - relatively few - interesting e-mails, thus wasting time and bandwidth and occupying huge storage space.

The types of those messages vary: some of them contains advertisements, other e-mails provide winning notifications, and sometimes we get messages with executable files, which finally emerge as malicious codes, such as viruses and Trojan horses. In addition, spam e-mails may often have unsuitable content (as a pornographic material advertising) that is illegal and sometimes dangerous for non adult users.

The recognition of spam content is not a trivial task: there are some factors that are related to human perception, economic behavior, legal context, and that are hardly measurable or summarized in adequate features. The very definition of *spam e-mails* requires a common agreement that is not easy to find.

In our opinion, *all* kind of spam e-mails have several common characteristics, such as: *i*) they are unsolicited, *ii*) they have a commercial content, even though the content itself is continuously evolving, trying to outsmart the classical countermeasures adopted by anti-spam filters. Consequently, a great variety of technical methodology have been implemented in current anti-spam systems [6]. The common technical solutions propose filtering strategies based on sender address and/or body content. We focused our attention on that measures related to e-mail contents, in particular both *texts and images*, rather then on networking and identity strategies [27], since our goal is to develop a personal antispam system.

In this chapter we combine the visual clues with the semantic information related to the e-mail body, to determine whether a message is spam . In order to address the problem of combining a non-constant number of modules, since it is not possible to *a priori* know if there is one or more images attached to the e-mail and/or there is textual information to be processed, we propose the use of a *Behaviour Knowledge Space* [19] approach. This also allows us to easily include new modules into our architecture that could be required for addressing new spammers' tricks.

The chapter is organized as follows. After an overview of the related literature in Sect. 2, Sect. 3 describes at a glance the main component of the proposed system. In Sect. 4 and 5 we describe text and image features respectively, while in Sect. 6 we show how to combine them. In Sect. 7 several experiments are discussed, and finally in Sect. 8 we report the conclusions of our work.

2 Related Work

In this work we approached the problem of detecting spam diffused through texts and/or images. So, this section is divided into two parts, the first one related to text-based spam detection and the second one to image-based spam detection.

Text-based: Textual filtering methods are widely deployed; they vary in the inspected content and the proposed methodology. Some filters consider only the header or the body of an e-mail, while other ones take both. These approaches use different models, considering word-tokens, their frequencies and their combinations. In *rule based-filters* [8] the users define some rules related to the headers or the bodies, considering particular words as *sign* of spam content; anyway, this simple solution is strongly dependent on how the words used by spammers can change.

Differently, *Signature-based* methods do not really deal with whole messages or specific tokens, transforming the message into a *signature*. Clearly, the methods effectiveness is related to the robustness of the signature function. Note that a signature database must be distributed and kept up to date very frequently, due to the rapid variation of spam e-mails. To this regard, some proposals are based on *collaborative solutions*, in particular, on Peer-to-Peer (P2P) networks for signature distribution [9, 30]. These approaches are not well suited for developing a personal antispam system.

Other approaches consider spam detection as a *binary classification problem* and several algorithms from the learning theory research field have been used. In these solutions, e-mails are mapped into multidimensional space, each dimension representing the words in the e-mail content; several measures are proposed such as the terms-frequency (tf) or the product between the documents-frequency (df) and terms-frequency, as in [12]. *Statistical filters* based on the the Bayes theory have been also investigated [1, 24].

One of the drawbacks of these methods is that they typically do not consider specific countermeasures for taking into account new spammer tricks, so a complete retraining of the system is needed when these attacks arise. To this regard, it is interesting to remark that a recent study [4] tried to model spam filtering as an adversarial classification problem, in which an intelligent, adaptive adversary modifies patterns to evade the classifier.

Image-based: Image spam has been extensively studied using several techniques primarily developed from the Image Processing and Computer Vision community, using features related to color distribution [2] or textual areas [2, 29]. A classifier is usually trained on such features, trying to discriminate spam images from legitimate ones. In [11], the authors present features that are focused on simple properties of the image, making classification very fast. However, the authors completely disregard the textual part of the emails.

Other approaches basically try the detect textual areas in images following the idea that images which contain text are likely to be spam . In [28] the authors propose an algorithm for text localization. They construct a corner detection algorithm based on a circular template to predict the corner points of the text in an image, which is crucial for text localization. The same idea is presented in [7]. The method proposed there extracts edge features of a binarized image by using higher-order local autocorrelation, and then passes these features to a Support Vector Machine (SVM) for classification . In [18] the authors try instead to extract connected components from the image in order to detect the presence of an embedded text. A quite different approach is followed in Fumera et al. [13], where the authors propose to process each image by using an OCR system for extracting embedded texts.

All these approaches, however, cannot be effectively used when text within images is intentionally distorted and/or obfuscated. As it was noted in [5], in fact, now spammers try to make OCR and text detection techniques ineffective without compromising human readability by placing text on non-uniform background, or by using techniques like the ones exploited in CAPTCHAs[1] (programs that generate and grade tests that humans can pass but current computer programs cannot).

3 System Architecture

As shown in Fig. 1, we design a system that integrates image-based and text-based analysis. The mails, initially, are parsed by a Multipurpose Internet Mail Extensions (MIME) parser that can retrieve the different parts of the e-mails: the text parts, the attached images or text files, the email subject and the headers. The text is thus processed by a *Text Analyzer* module according to the methodology described in Sect. 4 and its output is a classification result obtained using the feature vector of text part of input email. The images are forwarded to the *Image Analyzer* module which gives another classification results for the feature vector that is extracted with the techniques described in Sect. 5 for the image part of the email. We note that the OCR output of the *Image Analyzer* could be also used by *Text Analyzer* in order to build its feature vector. The *Fusion block* has the role to combine two outputs furnishing the final classification of each e-mail using the strategy discussed starting from Sect. 6.

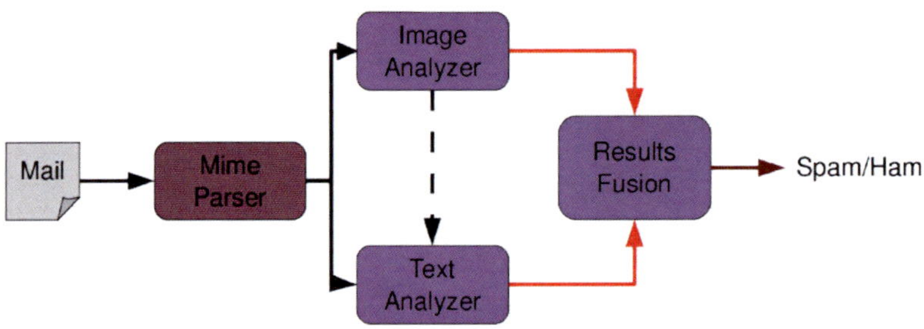

Fig. 1 The proposed system architecture: the dashed line represents the OCR output that is fed to the Text Analyzer

[1] The term CAPTCHA (Completely Automated Turing Test To Tell Computers and Humans Apart) was coined in 2000 by Luis von Ahn, Manuel Blum, Nicholas Hopper and John Langford of Carnegie Mellon University. At that time, they developed the first CAPTCHA to be used by Yahoo – http://www.captcha.net/

Both the Text and the Image analyzer can be implemented by means of different classifiers, each classifier using different features. Next, we will describe in detail the different feature sets used and the combination process.

4 Textual Features

We propose a strategy based on text processing and analysis in order to process both *semantic* and *syntactic* features. Generally speaking, our main idea is to characterize how e-mails belonging to the same class (*ham*[2] or *spam*) do have the same meanings by means of using a set of semantic features in addition to the detection of special characters (syntactic features) that are typically used in spam context.

In particular, at the semantic level we analyze the whole email content taking into account the word localization in a given context, thus measuring the weight of a single word in the document. In this way, we relate the emails content to certain topic by looking at commonly shared words. A topic is described by a region of multidimensional space formed by the vectors of words of different e-mails. In the spam context, examples of e-mail topics are given in Table 1. In Sect. 4.1, we will describe the model used to discover the semantic content of e-mails.

Table 1 The list of contents in spam mails

Spam Topics
Investment/Business
Health/Medicine
Games
Software
Leisure/Travel
Adult
Finance
Product/Service

The use of syntactic features is suitable to detect grammar anomalies in the texts. Typically, the ham e-mails do not have particular occurrences of special characters: these ones can be thus used as signs of low trustworthiness of the received e-mail; the related developed methodology will be described in Sect. 4.2.

4.1 Semantic Features

We propose to use a feature set based on a modified version of Vector Space Model (VSM) [23]. This model is based on the representation of documents as vectors in multidimensional space. The representation of e-mail textual content in the vector space model has a number of advantages, including the uniform treatment of queries

[2] Legitimate (valid) email.

and documents as vectors and the ability to differently weight the different terms; however, it also suffers from its inability to cope with two classic problems arising in natural languages [20], i.e., synonymy and polysemy.

We briefly recall that *synonymy* refers to a case where two different words (say "pupil" and "scholar") have the same meaning, and *polysemy* refers to the case where a term such as "play" has multiple meanings according to different contexts. In fact, as the worst case of the influence of synonymy on similarity measure, we could have two orthogonal vectors with 0 as a result of cosine similarity even if there are two different words that have the same meaning behind those two vectors. The semantic correlation or disambiguation of the these terms can be made by looking at the context in which they are placed, for example, the terms "scholar" can be correlated to "pupil" if the documents, in which they are found, also contain terms like "school", "book", "pen" and so on. In that way the shared terms can increase the value of similarity measures. The idea of looking at the whole email document can be seen also as an overcoming of the independence hypothesis used in a Bayesian filter technique known as bag-of-words model that is one most used approaches to anti-spam filtering. In this model the relationships among a set of words (joint distribution) are simply factorized.

In order to overcome the fault of the vector space model to capture the *synonymy* and *polysemy* relationships, we choose a modified version of VSM, the Latent Semantic Analysis or LSA [23]. Despite LSA being a traditional and well accepted technique used to stick out the semantic contents in text-process community, there are few applications of it in the spam framework. LSA is an application of Singular Value Decomposition (SVD) to a document-by-term $M \times N$ matrix $\mathbf{A}$. In particular, SVD provides a suitable matrix decomposition as $\mathbf{A} = \mathbf{TSD}^T$, where $\mathbf{S}$=diag$(\sigma_1, \ldots, \sigma_r)$ is an $M \times N$ matrix, $\sigma_i = \sqrt{\lambda_i}$ is the ith singular value, $\lambda_i \geq \lambda_{i+1}$ ($\lambda_1, \ldots, \lambda_r$ are the eigenvalues of $\mathbf{AA}^T$), and r is the rank of $\mathbf{A}$. Note that $\mathbf{A}^T\mathbf{A}$ has the same eigenvalues as $\mathbf{AA}^T$.

In the LSA technique, a reduced version of $\mathbf{A}$, $\mathbf{A}_k = \mathbf{T}_k\mathbf{S}_k\mathbf{D}_k^T$, is used, with k being a positive integer that is the maximum rank of $\mathbf{A}$ ($\mathbf{T}_k$ is of $M \times k$, $\mathbf{S}$ is of $k \times k$, and $\mathbf{D}_k$ is of $N \times k$). After that, we can have a representation of documents and terms in the singular value space. We note that this operation does dimensionality reduction since typically $k \ll M$ while documents can be still faithfully represented in the reduced space.

The obtained approximation is computed by taking into account the distance between the two matrices $\mathbf{A}$-$\mathbf{A}_k$ that is minimal according to the Frobenius norm [23]. In other words, we have a reduced space in which the words that have similar co-occurrence patterns are projected (or collapsed) into the same point. The choice of k has been empirically found, with 80 to 100 dimensions being sometimes the optimal choice for collections of about 5,000 terms by 1,000 documents [10]. In order to derive the features for training a classifier during the learning phase, we adopted as text features the projection of the document onto the space obtained by $\mathbf{S}_k\mathbf{D}_k^T$. In the testing phase, in addition to $\mathbf{S}_k\mathbf{D}_k^T$, we use $\mathbf{T}_k$ in order to compute the projection of the input $\mathbf{Q}$ ($\mathbf{Q}$ is of $N \times 1$).

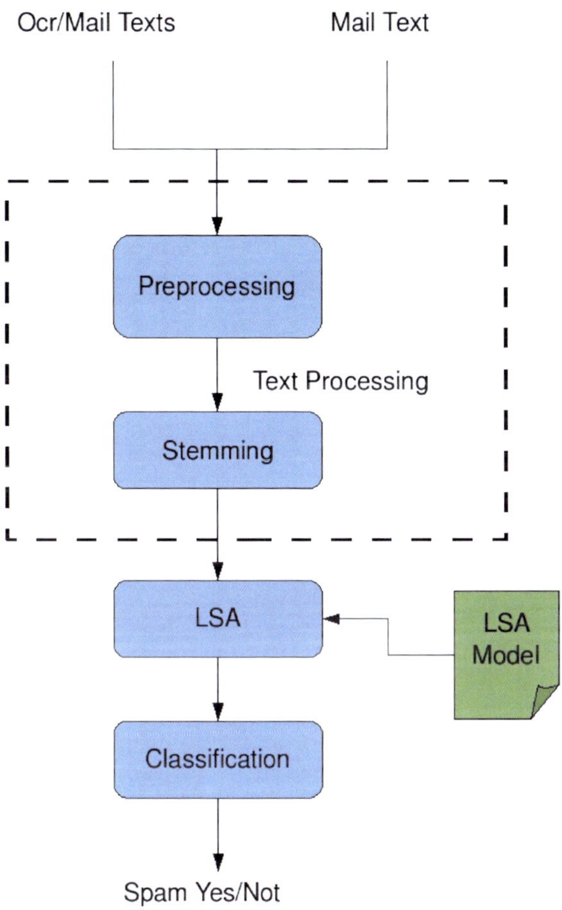

Fig. 2 The different phases of the Text Analyzer

There are different steps used to process the text for generating feature in the training and test phases. The different phases are depicted in Fig. 2 and they consist in:

- The *Preprocessing* module uses a set of intelligent filters that we apply to the email documents with or without the OCR recognized set of words. These filters include:
  - The classical stop word list filter that is used to delete the words that have no particular meaning, e.g., adverbs and pronouns.
  - An intelligent filter that is able to detect and reject the words that are not *human-understandable*, e.g., sequences like "fsdrx", "jkdld". The filter is based on a Support Vector Machine (SVM) classifier trained on several features derived from bigrams and trigrams of English words. We also built a feature vector containing the ratio between the correct bigrams (trigrams) and

all the bigrams (trigrams) for a set of 170000 common English words. Note that the use of this kind of filter has also the aim of enhancing the recognition of the semantic content that can be used in particular spammer attacks, such as the ones which put random words into e-mail texts, thus trying to reduce the effectiveness of current antispam algorithms. This filter can be also employed for rejection of words that are poorly recognized by the Optical Character Recognition algorithms.
  - A Part of Speech filter (POS) module that is able to detects nouns, verbs and adjective; it is used to reject adjectives that typically do not give additional information.

- The *Stemming* module implementing the well known Porter Stemmer algorithms [26] that is used to remove the common morphological and inflexional endings from words in English. The stemming and preprocessing modules together will be called Text Processing (TP).
- The *LSA* module that implements the functionality of the model described above. We computed the following features:
  - Term Frequency (TF)

$$TF_{ij} = \frac{n_{ij}}{\sum_j n_{ij}} \,, \tag{1}$$

where n_{ij} is the number of occurrences of the term t_j in the document d_i, and the denominator is the sum of number of occurrences of all terms in the document d_i.
  - Inverse Document Frequency× Term Frequency

$$IDTF_{ij} = TF_{ij} \log \frac{N_D}{N_D^{t_i}} \,, \tag{2}$$

where N being the number of total documents in the corpus and N^{t_j} the number of documents in the corpus, containing the term t_j.
  - Entropy Weight (WE) [22]:

$$WE_{ij} = TF_{ij}(1 + \sum_j \frac{p_{ij} \log_2(p_{ij})}{log_2(N)}) \,, \tag{3}$$

where $p_{ij} = \frac{TF_{ij}}{TF_j}$ and TF_j is the frequency of the term t_j on the whole document collection.

4.2 Syntactic Features

We propose to use some syntactic features that can be extracted from mail texts in order to distinguish valid and suspected mails.

Spammers, in fact, usually try to obfuscate the textual part of the e-mail body by substituting some characters in order to bypass the effectiveness of antispam filters.

So, we defined another set of features for obtaining a characterization of this kind of obfuscated text. The features we have investigated are mainly based on the presence of special characters, i.e., those characters that should not frequently occur in a legitimate text. The whole set we considered is made up of the following characters: { !, ", #, $, %, &, ', (,), *, +, „ -, . . . , /, @}. Starting from this set we defined six *syntactic features*:

- **text_length:** the number of characters of the whole text
- **words_number:** the number of words in the text
- **ambiguity:** the ratio between the number of special and normal characters
- **correctness:** the ratio between the number of words that do not contain special characters and the number of words that contain special characters
- **special_length:** the maximum length of a continuous sequence of special characters
- **special_distance:** the maximum distance between two special characters belonging to the above considered set.

5 Image Features

In [15], we proposed the approach for the detection of the image spam , in which two different image processing techniques are used. The first technique directly extracts some global features from each image attached to e-mails. Such features should also be able to detect if images were adulterated or not by considering the complexity of the image itself as it is perceived by a human being. The second technique is carried out by means of two steps: first, there is a preprocessing phase with the use of an OCR, then a feature extraction process, starting from the OCR output, tries to characterize this output in order to detect if the embedded text has been voluntarily obfuscated and/or distorted.

5.1 *Visual Features*

The first set of features that we called *visual features*, are directly obtained from the image attached to emails. In order to give an image characterization that should be able to discriminate between normal and adulterated images, we considered features that describe the image texture from a statistic point of view. As said before, in fact, spammers typically try to bypass filters that use an OCR for detecting texts within an image by obfuscating such texts with the addition of some noise or by superimposing a texture (see also Fig. 3). So, texture detection can help in identifying those images that contain spam messages.

For the sake of simplicity, we will consider features extracted from gray-level images, but the same operators can be applied to color images too.

We will use $\{I(x,y), 0 \le x \le N-1, 0 \le y \le M-1\}$ to denote an $N \times M$ image with G gray levels. All the considered statistical texture measures are based on the co-occurrence matrices. Spatial gray level co-occurrence estimates image properties related to second-order statistics. The $G \times G$ gray level co-occurrence matrix $P_{\mathbf{d}}$ for

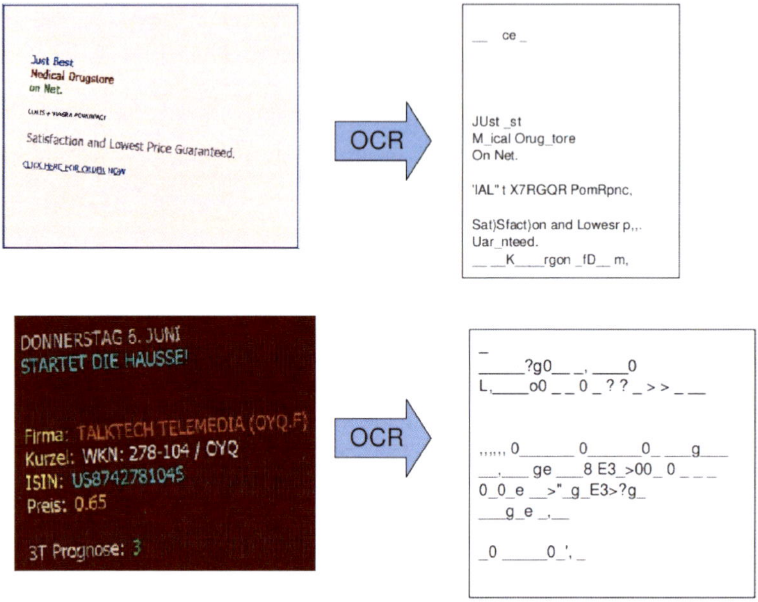

Fig. 3 Outputs obtained by applying *gocr* (available at http://jocr.sourceforge.net) to some spam images

a displacement vector $d = (dx, dy)$ is defined as follows. The entry (i, j) of $P_{\mathbf{d}}$ is the number of occurrences of the pair of gray levels i and j which are a distance $\mathbf{d}$ apart. Formally, it is given as:

$$P_{\mathbf{d}}(i, j) = |\{((r,s),(t,v)) : I(r,s) = i, I(t,v) = j\}|,$$

where $(r,s),(t,v) \in N \times M$, $(t,v) = (r+dx, s+dy)$, and $|\cdot|$ is the cardinality of a set.

Regarding the choice of the displacement vector $\mathbf{d}$, we considered the four direct neighbors of each pixel, i.e., we used four pairs as values of dx and dy for calculating the number of co-occurrences, namely $(0,1)$, $(1,0)$, $(-1,0)$ and $(0,-1)$. We do not perform a normalization of $P_{\mathbf{d}}$ in order to preserve the dependence of the considered features on the image size.

As suggested in [17], from the co-occurrence matrix it is possible to extract features that can be used for detecting a texture within an image. In particular, we calculated the following five features:

- **Contrast**

$$\sum_i \sum_j (i-j)^2 P_{\mathbf{d}}(i, j)$$

is the difference in terms of visual properties that makes an object (or its representation within an image) distinguishable from other objects and the background. In the visual perception of real world, contrast is determined by the difference in the color and brightness of the object and other objects within the same field of view.

- **Entropy:**
$$-\sum_i \sum_j P_{\mathbf{d}}(i,j) log P_{\mathbf{d}}(i,j)$$
is an index of the brightness variation among pixels.
- **Energy:**
$$\sum_i \sum_j P_{\mathbf{d}}^2(i,j)$$
is the spectral content of an image
- **Correlation:**
$$\frac{\sum_i \sum_j (i-\mu_x)(j-\mu_y) P_{\mathbf{d}}(i,j)}{\sigma_x \sigma_y}$$
is an index of the correlation degree among pixels. Here μ_x and μ_y are the means and σ_x and σ_y are the standard deviations of $P_{\mathbf{d}}(x)$ and $P_{\mathbf{d}}(y)$ respectively, where $P_{\mathbf{d}}(x) = \sum_j P_{\mathbf{d}}(x,j)$ and $P_{\mathbf{d}}(y) = \sum_i P_{\mathbf{d}}(i,y)$
- **Homogeneity:**
$$\sum_i \sum_j \frac{P_{\mathbf{d}}(i,j)}{1+|i-j|}$$
is a measure of the brightness variation within the image. If the image is completely black or white, its homogeneity value will be the maximum. On the contrary, if the image contains many brightness variations, this value will be very low.

Another category of features that can be used for characterizing images from a global point of view is based on the complexity of an image for a human reader. We have chosen to consider a feature also proposed in [5]:

- **Perimetric Complexity** is defined as the squared length of the boundary between black and white pixels (the perimeter) in the whole image, divided by the black area.

Note that, differently from [5], we evaluate the perimetric complexity on the whole image after performing binarization with a fixed threshold.

5.2 *OCR-Based Features*

Here we propose to use the same features as considered in Sect. 4.2. In this case, however, special characters are extracted from the output of an OCR that has received an attached image as input.

We have noticed, in fact, that characters embedded into an image are intentionally distorted and/or obfuscated in spam e-mails. Thus, most of the words cannot be correctly detected, as we can see in Fig. 3. Furthermore, several special characters that typically are not present in commonly used words can appear in the OCR output.

6 Combining Text-Based and Image-Based Classifiers

It has been experimentally shown that the combination of an ensemble of classifiers can be of great benefit in many practical pattern recognition applications. Through the appropriate choice of a combination rule, it is possible to suppress the overall effect of the *independent* errors in each observation domain, thus reaching performance better than that of a single classifier.

The combination of classifiers is an important part of our architecture. Anyway, there are some problems that must be taken into account in this case:

- It is necessary to define a method for combining a non-constant number of classifiers, since it is not possible to *a priori* know if there is one or more images attached to the e-mail and/or there is textual information to be processed.
- *Padding attacks* should be avoided. That is, the possibility that an attacker puts a spam message within a *normal* context, for example, by attaching an image containing an embedded spam message to an e-mail that contains *normal* images.

As shown in Fig. 4 we used a two-stage approach. The first stage (denoted as *Classification* in Fig. 4) consists in a simple 3-state *logical OR*, whose behavior is described in Fig. 5. In this way we also consider the case in which a classifier cannot be activated. It happens, for example, when there are no images within the e-mail or when there are no words to be processed by the semantic analysis. In this situation, we assume that the output of the classifier is *undefined*. Note that through this approach we try to address the problem of *padding attacks*, too. Just one correctly classified spam image, in fact, is sufficient so that the block of the visual classifiers declares the email as spam.

Then, at the second stage we adopt a *Behaviour Knowledge Space* (BKS) combining rule [19]. The idea behind this rule is to avoid making unjustified assumption on the classifier ensemble such as classifier independence. In Fig. 6 an example of how it works is shown.

A BKS is a K-dimensional space where each dimension corresponds to the decision of a classifier. Given an e-mail to be assigned to one of 2 possible classes, the ensemble of K classifiers can in theory provide 2^K different decisions.

We must also consider the case in which the output of the 3-state logical OR is *undefined*. In other words, each set of classifiers can attribute a mail to one out of three possible classes, i.e., $\{Spam, Ham, Undefined\}$ and the number of different decisions becomes 3^K.

Each of these decisions constitutes one unit of the BKS. In the learning phase each BKS unit can record 2 different values e_i (say, e_{ham} and e_{spam}), by considering that the actual classes are only *ham* and *spam*. Given a suitably chosen training set, each sample x of this set is classified by all the classifiers and the unit that corresponds to the particular decision of classifiers is activated. It records the actual class of x, say C_j, by adding one to the value of e_j. At the end of this phase, each unit can calculate the best representative class associated to it, defined as the class that exhibits the highest value of e_i. This class corresponds to the most likely class, given a decision of classifiers that activates that unit.

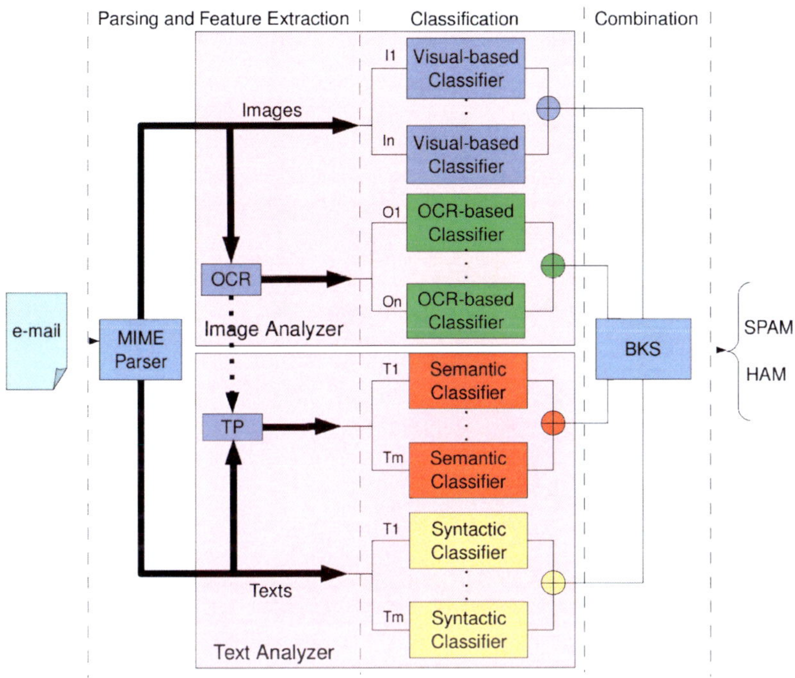

Fig. 4 The proposed approach for combining text-based and image-based classifiers. "TP" stands for *Text Processing*; it is described in Sect. 4.1

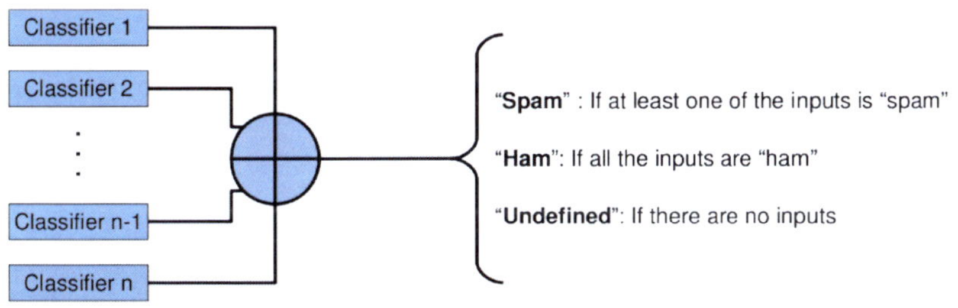

Fig. 5 The 3-state logical OR

In the operating mode, for each e-mail to be classified, the K decisions of the classifiers are collected and the corresponding unit is selected. Then the e-mail is assigned to the best representative class associated to that unit. Since we consider all possible combinations of classifier outputs as the number of available classifiers varies, we are implicitly handling the fact that the number of available classifiers can be different for each e-mail.

It is worth noting that the proposed combining scheme could be also easily extended using different feature sets and different classifiers. This could be required,

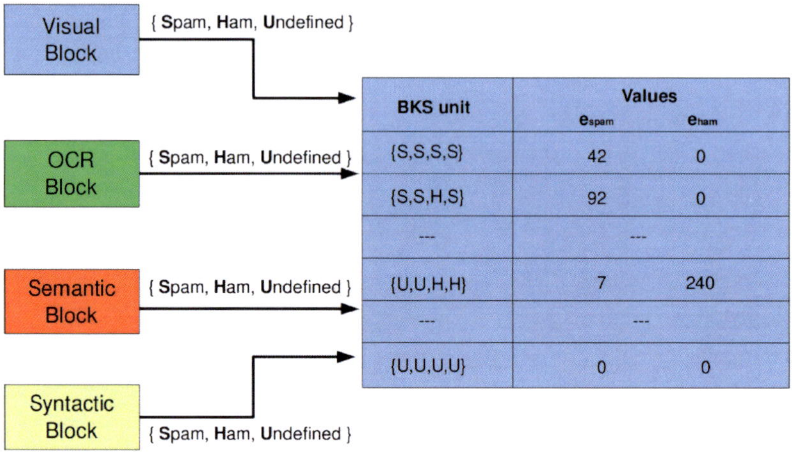

Fig. 6 The Behaviour Knowledge Space for combining classifiers

for example, for addressing new spammers' tricks. In this case, the problem is that the number of BKS unit grows exponentially, so that a larger training set would be needed in order to achieve good classification results. However, as will be shown in the next section, only a subset of all the possible units are typically activated in practice, since some configurations of the decisions are not allowed.

7 Experimental Results

In this section, we will first describe the database used for assessing the effectiveness of the proposed approach, then evaluate if the use of both visual and textual features can improve the performance of the system with respect to the use of a single set of features. Finally, we make a comparison of our approach with a state-of-the-art anti-spam filter, i.e., *SpamAssassin* equipped with two different spam image plug-ins.

Regarding the dataset whose details are given in Table 2, it is composed of 11652 e-mails, 9173 of which contain spam messages. E-mails were collected from the mailboxes of some users of the `studenti.unina.it` mailserver during three years (2005-2007). This mailserver hosts the mailboxes of all the students of the University of Naples Federico II. Among these e-mails, 151 contain *ham* images and 1802 contain *spam* images. Figs. 7 and 8 show some examples of the *ham* and *spam* images from our dataset.

In the first stage of our architecture (*Classification*), we chose a *Decision Tree* for implementing each classifier. In particular, a C4.5 (J48) coming from the open source tool *Weka*[3] was selected.

[3] http://www.cs.waikato.ac.nz/ml/weka/

Table 2 The dataset used in our tests

total # of e-mails		*e-mails with Images*	
Spam	*Ham*	*Spam*	*Ham*
9173	2479	1802	151

Fig. 7 Some examples of *ham* images of the dataset used in our tests

Each single classifier was first trained on a set of 1000 mails (500 for each class) different from those belonging to the dataset in Table 2. After completing the classifiers' training, the whole dataset was split into two distinct subsets. Then, two experiments have been done, each using one subset for training the BKS rule and the other subset for testing. The results of the proposed system have been finally summarized in the value of the classification accuracy, reported on the test set, and compared with those obtained by the single classifiers.

In Fig. 9 the performance of single classifiers and that of the proposed system are reported. It can be noted that the use of the BKS significantly improves the

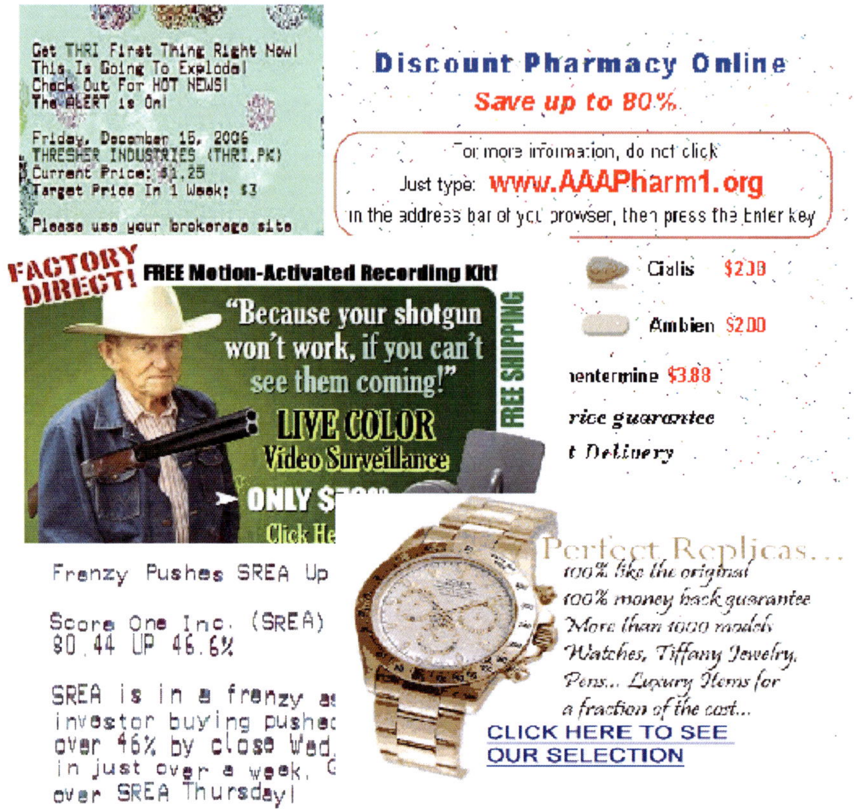

Fig. 8 Some examples of *spam* images of the dataset used in our tests

performance of the single classifiers. It must be remarked, in fact, that the visual-based classifier operates on a subset of the whole dataset (only 1953 mails out of 11652). It is also interesting to note that the number of BKS units that are really activated on the whole dataset is only 18, while their total number is 3^4, i.e., 81. This confirms conclusions made in the previous section.

Finally, in Fig. 10 we report a comparison of the results obtained by our system with those of *SpamAssassin* in its standard configuration and equipped with two plug-ins devised for filtering image spam , namely *Bayes-OCR*[4] and *Fuzzy-OCR*.

It clearly appears that our approach significantly outperforms both *Bayes-OCR* and *Fuzzy-OCR*, by reaching a significantly higher accuracy. Finally, note the time needed for processing the whole dataset by our system is practically the same as the time spent by SpamAssassin with *Fuzzy-OCR*, while it is significantly faster than *SpamAssassin* equipped with *Bayes-OCR*.

[4] This plug-in is available for download at the URL: http://prag.diee.unica.it/n3ws1t0/?q=node/107

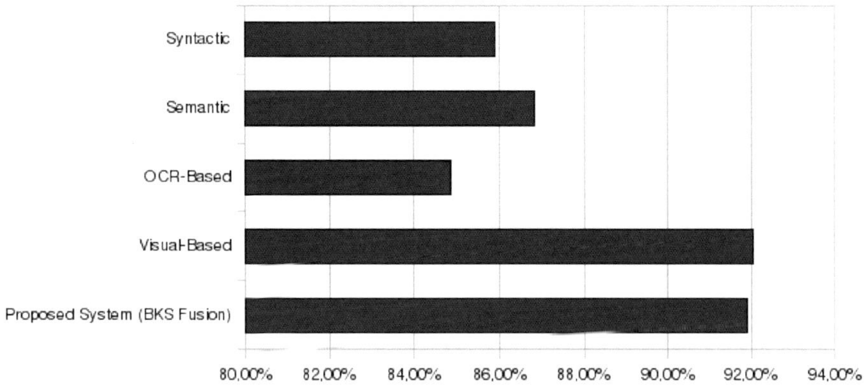

Fig. 9 The accuracy of the four considered single classifiers and of the proposed system. Note that the last two single classifiers - third and fourth rows - processed only e-mails with attached images

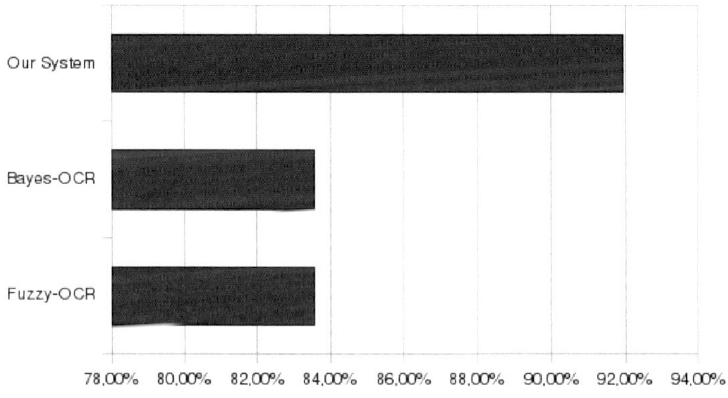

Fig. 10 Comparison between the proposed system and *SpamAssassin* with *Bayes-OCR* and *Fuzzy-OCR* plug-ins

8 Conclusion

In this chapter we presented an approach for addressing the spam e-mail problems, which takes into account some of the recent evolutions of the spammers' tricks as well as the limits of the known methodologies. We proposed to combine visual clues with the semantic information related to the e-mail body by using the Behaviour Knowledge Space rule. This approach allowed us to easily include new modules in our architecture that could be required for addressing new spammers' tricks.

Tests on a dataset of e-mails containing attached images confirmed the effectiveness of the approach and its superiority over the widely used open-source tool such as *SpamAssassin*.

Since the proposed approach has been mainly designed for deploying a personal antispam system, in the future we want to investigate how it is possible to further improve its performance by customizing it for a specific user. This could be done by developing a specific module for taking into account spam images received by the user A that are considered as ham by the user B, such as, for example, those related to a phishing attack against the user A. Moreover, we have planned to better characterize the contribution in terms of CPU time of the various component of our architecture in order to find the best tradeoff between accuracy and computational complexity.

References

1. Androutsopoulos, I., Koutsias, J., Chandrinos, K.V., Paliouras, G., Spyropoulos, C.D.: An evaluation of Naive Bayesian anti-spam filtering. In: Lopez de Mantaras, R., Plaza, E. (eds.) ECML 2000. LNCS (LNAI), vol. 1810, pp. 9–17. Springer, Heidelberg (2000)
2. Aradhye, H.B., Myers, G.K., Herson, J.A.: Image analysis for efficient categorization of image-based spam e-mail. In: Proc. 8th Int. Conf. Document Analysis and Recogn, Seoul, Korea, pp. 914–918. IEEE Comp. Soc., Los Alamitos (2005)
3. Balakumar, M., Vaidehi, V.: Ontology based classification and categorization of email. In: Proc. Int. Conf. Sign. Proc., Communications and Networking, Chennai, India, pp. 199–202. IEEE Comp. Soc., Los Alamitos (2008)
4. Biggio, B., Fumera, G., Roli, F.: Adversarial pattern classification using multiple classifiers and randomisation. In: da Vitoria Lobo, N., Kasparis, T., Roli, F., Kwok, J.T., Georgiopoulos, M., Anagnostopoulos, G.C., Loog, M. (eds.) S+SSPR 2008. LNCS, vol. 5342, pp. 500–509. Springer, Heidelberg (2008)
5. Biggio, B., Fumera, G., Pillai, I., Roli, F.: Image spam filtering using visual information. In: Cucchiara, R. (ed.) Proc. 14th Int. Conf. Image Analysis and Proc., Modena, Italy, pp. 105–110. IEEE Comp. Soc., Los Alamitos (2007)
6. Blanzieri, E., Bryl, A.: A survey of learning-based techniques of email spam filtering. TR DIT-06-056, Informatica e Telecomunicazioni, University of Trento, Italy (2006)
7. Cheng, H., Qin, Z., Liu, Q., Wan, M.: Spam image discrimination using support vector machine based on higher-order local autocorrelation feature extraction. In: Proc. IEEE Conf. Cybern. Intell. Syst., Chendgu, China, pp. 1017–1021. IEEE Comp. Soc., Los Alamitos (2008)
8. Cohen, W.W.: Learning rules that classify e-mail. In: Proc. AAAI Spring Symp. Mach. Learn. in Inf. Access, pp. 18–25. AAAI Press, Menlo Park (1996)
9. Damiani, E., De Capitani di Vimercati, S., Paraboschi, S., Samarati, P.: P2P-based collaborative spam detection and filtering. In: Caronni, G., Weiler, N., Shahmehri, N. (eds.) Proc. 4th Int. Conf. Peer-to-Peer Computing, Zurich, Switzerland, pp. 176–183. IEEE Comp. Soc., Los Alamitos (2004)
10. Deerwester, S., Dumais, S.T., Furnas, G.W., Landauer, T.K., Harshman, R.: Indexing by latent semantic analysis. J. Amer. Soc. Inf. Sci. 41(6), 391–407 (1990)
11. Dredze, M., Gevaryahu, R., Elias-Bachrach, A.: Learning fast classifiers for image spam. In: Proc. 4th Conf. Email and Anti-Spam, Mountain View, CA, pp. 487–493 (2007)
12. Drucker, H., Wu, D., Vapnik, V.N.: Support vector machines for spam categorization. IEEE Trans. Neural Networks 10(5), 1048–1054 (1999)
13. Fumera, G., Pillai, I., Roli, F.: Spam filtering based on the analysis of text information embedded into images. J. Mach. Learn. Research 7, 2699–2720 (2006)

14. Gao, Y., Yang, M., Zhao, X., Pardo, B., Wu, Y., Pappas, T.N., Choudhary, A.: Image spam hunter. In: Proc. IEEE Int. Conf. Acoustics, Speech and Sign. Proc., Las Vegas, NV, pp. 1765–1768. IEEE Comp. Soc., Los Alamitos (2008)
15. Gargiulo, F., Sansone, C.: Visual and OCR-based features for detecting image spam. In: Juan-Císcar, A., Sánchez-Albaladejo, G. (eds.) Proc. 8th Int. Workshop Patt. Recogn. Inf. Syst., Barcelona, Spain, pp. 154–163. INSTICC Press, Setúbal (2008)
16. Han, A., Kim, H.-J., Ha, I., Jo, G.-S.: Semantic analysis of user behaviors for detecting spam mail. In: Proc. IEEE Int. Workshop Semantic Computing and Appl., Incheon, Korea, pp. 91–95. IEEE Comp. Soc., Los Alamitos (2008)
17. Haralick, R.M.: Statistical and structural approaches to texture. Proceedings of IEEE 67(5), 786–804 (1979)
18. Huang, H., Guo, W., Zhang, Y.: A novel method for image spam filtering. In: Proc. 9th Int. Conf. Young Comp. Scientists, Zhang Jia Jie, Hunan, China, pp. 826–830. IEEE Comp. Soc., Los Alamitos (2008)
19. Huang, Y.S., Suen, C.Y.: A method of combining multiple experts for the recognition of unconstrained handwritten numerals. IEEE Trans. Pattern Analysis and Mach. Intell. 17(1), 90–94 (1995)
20. Jurafsky, D., Martin, J.H.: Speech and Language Processing: An Introduction to Natural Language Processing. In: Computational Linguistics and Speech Recognition. Prentice Hall, Upper Saddle River (2009)
21. Liu, W., Fang, W.: Adaptive spam filtering based on fingerprint vectors. In: Proc. ISECS Int. Colloquium Computing, Communication, Control, and Management, Guangzhou, China, pp. 384–388. IEEE Comp. Soc., Los Alamitos (2008)
22. Lochbaum, K.E., Streeter, L.A.: Comparing and combining the effectiveness of latent semantic indexing and the ordinary vector space model for information retrieval. Inf. Proc. and Management 25(6), 665–676 (1989)
23. Manning, C., Schuetze, H.: Foundations of Statistical Natural Language Processing. MIT Press, Cambridge (1999)
24. Metsis, V., Androutsopoulos, I., Paliouras, G.: Spam filtering with Naive Bayes – which Naive Bayes? In: Proc. 3rd Conf. Email and Anti-Spam, Mountain View, CA (2006)
25. Okabe, M., Yamada, S.: Interactive spam filtering with active learning and feature selection. In: Proc. IEEE/WIC/ACM Int. Conf. Web Intell. and Intell. Agent Technology, Sydney, NSW, Australia, pp. 165–168. IEEE Comp. Soc., Los Alamitos (2008)
26. Porter, M.F.: An algorithm for suffix stripping. Program 14(3), 130–137 (1980)
27. Schryen, G.: Anti-Spam Measures: Analysis and Design. Springer, New York (2007)
28. Wan, M., Zhang, F., Cheng, H., Liu, Q.: Text localization in spam image using edge features. In: Proc. Int. Conf. Communications, Circuits and Syst., Fujian, China, pp. 838–842. IEEE Comp. Soc., Los Alamitos (2008)
29. Wu, C.T., Cheng, K.T., Zhu, Q.A., Wu, Y.L.: Using visual features for anti-spam filtering. In: Proc. IEEE Conf. Image Processing, Genoa, Italy, pp. 509–512. IEEE Comp. Soc., Los Alamitos (2005)
30. Zhou, F., Zhuang, L., Zhao, B.Y., Huang, L., Joseph, A.D., Kubiatowicz, J.: Approximate object location and spam filtering on peer-to-peer systems. In: Endler, M., Schmidt, D.C. (eds.) Middleware 2003. LNCS, vol. 2672, pp. 1–20. Springer, Heidelberg (2003)

Weighted Decoding ECOC for Facial Action Unit Classification

Terry Windeatt

Abstract. There are two approaches to automating the task of facial expression recognition, the first concentrating on what meaning is conveyed by facial expression and the second on categorising deformation and motion into visual classes. The latter approach has the advantage that the interpretation of facial expression is decoupled from individual actions as in FACS (Facial Action Coding System). In this chapter, upper face action units (*aus*) are classified using an ensemble of MLP base classifiers with feature ranking based on PCA components. When posed as a multi-class problem using Error-Correcting-Output-Coding (ECOC), experimental results on Cohn-Kanade database demonstrate that error rates comparable to two-class problems (one-versus-rest) may be obtained. The ECOC coding and decoding strategies are discussed in detail, and a novel weighted decoding approach is shown to outperform conventional ECOC decoding. Furthermore, base classifiers are tuned using the ensemble Out-of-Bootstrap estimate, for which purpose, ECOC decoding is modified. The error rates obtained for six upper face *aus* around the eyes are believed to be among the best for this database.

1 Introduction

The topic of this chapter concerns solving a supervised learning problem in face expression recognition using a combination of neural network classifiers. In the case of face recognition, pattern features consist of real numbers representing different aspects of facial features, as described in Sect. 4. In order to design the learning system we follow the well established technique of dividing the example patterns into two sets, a training set to design the classifier and a test set, which is subsequently used to predict the performance when previously unseen examples are applied.

Multiple Classifier Systems (MCS) have become an established method for improving generalisation performance over a single classifier, and the relevant aspects

Terry Windeatt
CVSSP, University of Surrey, University, Guildford, Surrey, UK GU2 7XH
e-mail: t.windeatt@surrey.ac.uk

O. Okun & G. Valentini (Eds.): Appli. of Supervised & Unsuper. Ensemble Meth., SCI 245, pp. 59–77.
springerlink.com
© Springer-Verlag Berlin Heidelberg 2009

are discussed in Sect. 2. The single classifier performance can be quite sensitive to classifier parameters, and it has previously been shown [36] that an ensemble is less sensitive to base classifier complexity. However, even though an ensemble is less likely to over-fit, there is still the difficulty of tuning individual classifier parameters with respect to ensemble performance. Multi-layer perceptrons (MLP) make powerful classifiers that may provide superior performance compared with other classifiers, but are often criticized for the number of free parameters. The common approach to adjusting parameters is to further divide the training set into two to produce a validation set. When the number of examples is in short supply, cross-fold validation may be used. For example, in n-fold cross-validation, the set is randomly split into n equal parts with (n-1) parts used for training and one part used as a validation set to tune parameters. Training is repeated n times with a different partition each time, and the results averaged. However, it is known that these approaches to validation are either inappropriate or very time-consuming. Ideally all the training set should be used for training, so that there is no need for validation. However, this requires that over-fitting be detected by looking at performance on only the training set, which is a difficult problem. In this chapter the OOB estimate (Sect. 2), is used to determine optimal parameters from the training set.

The problem of face expression recognition is difficult because facial expression depends on age, ethnicity, gender, occlusions as well as pose and lighting variation [41]. Facial action unit (*au*) classification is an approach to face expression recognition that decouples the recognition of expression from individual actions. In FACS (facial action coding system) [27] the problem is decomposed into forty-four facial action units, that includes six upper face *aus* around the eyes. This approach has the potential of being applied to a much richer set of applications than an approach that targets facial expression directly. However, the coding process requires skilled practitioners and is time-consuming so that typically there are a limited number of training patterns.

There are various approaches to determining features for discriminating between *aus*. Originally, features were based on geometric measurements of the face that were involved in the *au* of interest [27]. For example, features were extracted based upon whether the eyes were open or closed, the degree of eye opening, and the location and radius of the iris. More recently, holistic approaches based on PCA, Gabor [11] and Haar wavelets represent a more general approach to extracting features [2], and have been shown to give comparable results. The difficulty with these latter approaches is the large number of features. When combined with the limited number of patterns, this can lead to the small sample-size problem, that is when the number of patterns is less than or comparable to the number of features. A method of eliminating irrelevant features is therefore required [3, 26]. In this chapter the Out-of-Bag error estimate is used to optimise the number of features.

In previous work [41, 42] five feature ranking schemes were compared using Gabor features in an MLP ensemble. The schemes were Recursive Feature Elimination (RFE) [14] (Sect. 4) combined with MLP weights and noisy bootstrap, Boosting (single feature selected each round), one-dimensional class-separability measure and Sequential Floating Forward Search (SFFS). A full description of these feature selection techniques may be found in [41]. MLP weights combined with RFE, Sect. 5, perform well for feature selection, even though it is known that MLP weights are not good at selecting most relevant features [32]. It was shown that ensemble performance is relatively insensitive to the feature-ranking method with simple one-dimensional performing at least as well as multi-dimensional schemes. This was a somewhat surprising conclusion, since it is known that sophisticated multi-dimensional schemes out-perform one-dimensional schemes for single classifiers [14]. It was also shown that the ensemble using PCA features with its own inherent ranking outperformed Gabor.

Error-Correcting Output Coding (ECOC) is a well-established method [9, 39] for solving multi-class problems by decomposition into complementary two-class problems, and is fully discussed in Sect. 3. However, the idea behind ECOC is quite simple and so we introduce the main concept here. ECOC is a two-stage process, coding followed by decoding. The coding step is defined by the binary $k \times b$ code word matrix C that has one row (codeword) for each of k classes, with each column defining one of b sub-problems that use a different labeling. Assuming each element of C is a binary variable z a training pattern with target class ω_l for $l = 1...k$ is re-labeled as class Ω_1 if $C_{ij} = z$ and as class Ω_2 if $C_{ij} = \bar{z}$. The two super-classes Ω_1 and Ω_2 represent, for each column, a different decomposition of the original problem. For example, if a column of C is given by $[01001]^T$, this would naturally be interpreted as patterns from class 2 and 5 being assigned to Ω_1 with remaining patterns assigned to Ω_2. This is in contrast to the conventional One-versus-rest code, which can be defined by the diagonal $k \times k$ code matrix. In the decoding step, an unknown pattern is classified according to closest codeword.

In this chapter, features based on Principal Components Analysis (PCA) are used with Error Correcting Output Coding (ECOC) and a weighted decoding strategy based on bootstrapping individual base classifiers is proposed. The principle behind weighted decoding is to reward classifiers that perform well. The weights in this study are fixed in the sense that none change as a function of the particular pattern being classified. Sometimes this is referred to as implicit data-dependence or constant weighting. It is generally recognized that a weighed combination may in principle be superior, but it is not easy to estimate the weights.

Although this chapter employs MLP ensembles, the techniques for OOB, feature selection and ECOC weighted decoding are suitable for any base classifier. The chapter is organised as follows. Section 2 discusses ensemble techniques and Bootstrapping , Sect. 3 the ECOC method including weighted decoding, Sect. 4 describes the database and design decisions for *au* classification , and Sect. 5 compares 2-class classification with weighted and conventional ECOC decoding.

2 Ensembles and Bootstrapping

For some classification problems, both two class and multiclass, it is known that the lowest error rate is not always reliably achieved by trying to design a single best classifier. An alternative approach is to employ a set of relatively simple sub-optimal classifiers and to determine a combining strategy that pools together the results. Although various systems of multiple classifiers have been proposed, most use similar constituent classifiers, which are often called base classifiers . A necessary condition for improvement by combining is that the results of the base classifiers are not too well correlated, as discussed in [34]. There are some popular approaches for reducing correlation that are based on perturbing feature sets, perturbing training sets or injecting randomness [10]. For example two well-known training set perturbation methods are Bagging [4] and Boosting [13] . All these perturbation techniques have in common that each base classifier handles the same problem in the sense that the class labelling is identical. There is another type of correlation reduction technique, aimed solely at multiclass problems, that perturbs class labels. In a method like Error Correcting Output Coding (ECOC) each base classifier solves a sub-problem that uses a different class labelling. Techniques like binary decision clustering [33] and pairwise coupling [16] may also be considered in this category.

The architecture envisaged is a simple MCS framework in which there are parallel MLP base classifiers , as shown in Fig. 1. For realistic problems, slow convergence and lack of guarantee of global minima are drawbacks of MLP training [37]. An MLP Ensemble offers a way of solving some of these problems [15]. The rationale is that it may be easier to optimise the design of a combination of relatively simple MLP classifiers than to optimise the design of a single complex MLP classifier. An MLP with random starting weights is a suitable base classifier since randomisation is known to be beneficial in the MCS context. Problems of local minima and computational slowness may be alleviated by the MCS approach of pooling together the decisions obtained from locally optimal classifiers. However, there is still the problem of tuning base classifiers.

Although it is known that diversity among base classifiers is a necessary condition for improvement in ensemble performance, there is no general agreement about how to quantify the notion of diversity among a set of classifiers. Experimental evidence in [21] casts doubt on the usefulness of diversity measures for predicting ensemble accuracy. Diversity measures can be categorised into pair-wise and non-pair-wise, but to apply pair-wise measures to finding overall diversity it is necessary to average over the classifier set. These pair-wise diversity measures are normally computed between pairs of classifiers and take no account explicitly of the target labels. As explained in [35], the accuracy-diversity dilemma arises because when base classifiers become very accurate their diversity must decrease, so that it is expected that there will be a trade-off. A class separability measure that combines accuracy and diversity for two-class problems is described in [36]. For two-class problems, over-fitting may be detected by observing the class separability measure computed on the training set as it varies with base classifier complexity. In this chapter a modified version of the class separability measure is proposed in Sect. 3.4 for the weighted decoding strategy.

Bootstrapping is an ensemble technique which implies that if μ training patterns are randomly sampled with replacement, $(1-1/\mu)\mu \approx 37\%$ of them are removed

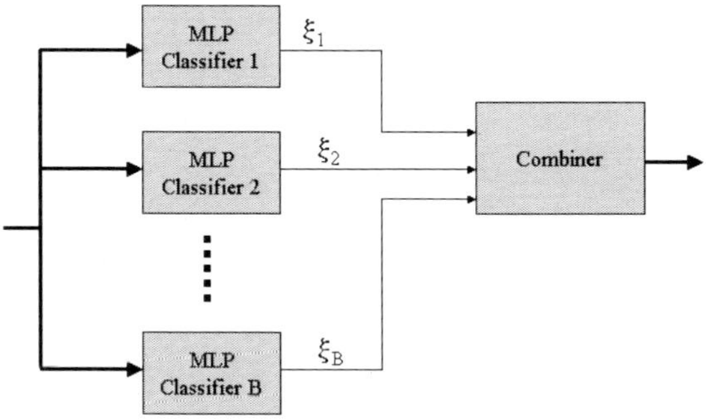

Fig. 1 Ensemble MLP architecture

with remaining patterns occurring one or more times. An advantage of Bootstrapping is that the Out-of-Bootstrap (OOB) error estimate may be used to tune base classifier parameters, and furthermore, the OOB is a good estimator of when to stop eliminating features [40]. Normally, deciding when to stop eliminating irrelevant features is difficult and requires a validation set or cross-validation techniques. The base classifier OOB estimate uses the patterns left out of training, and should be distinguished from the ensemble OOB. For the ensemble OOB, all training patterns contribute to the estimate, but the only participating classifiers for each pattern are those that have not been used with that pattern for training (that is, approximately thirty-seven percent of classifiers). Note that OOB gives a biased estimate of the absolute value of generalisation error [5], but for tuning purposes the estimate of the absolute value is not important. The ensemble OOB estimate is incorporated into the ECOC decoding strategy in Sect. 3.2.

3 Error-Correcting Output Coding ECOC

There are several reasons for decomposing the original multiclass problem into separate and complementary two-class problems. Firstly, some accurate and efficient two-class classifiers do not naturally scale up to multiclass. Attention can then be focused on developing an effective technique for the two-class case, without having to consider explicitly the design and automation of the multiclass classifier. It is also hoped that the parameters of a simple classifier run several times are easier to determine than a complex classifier run once and may facilitate more efficient solutions. Finally, solving different 2-class sub-problems, perhaps repeatedly with random perturbation, may help to reduce error in the original problem.

It needs to be remembered however, that even if ECOC successfully produces accurate and diverse classifiers, there is still the need to choose or design a suitable combining strategy. Bagging and Boosting originally used respectively the majority and weighted vote, which are both hard-level combining strategies. By hard-level we mean that a single-hypothesis decision is taken for each base classifier, in contrast with soft-level which implies a measure of confidence associated with the decision. The ECOC method was originally motivated by error-correcting principles, as discussed in Sect. 3.1 and used a Hamming Distance-based hard-level combining strategy. When it could be shown that ECOC produced reliable probability estimates [20], the decision-making strategy was changed to soft-level (L^1 norm Eq. (2)).

3.1 Motivation

First let us motivate the need for a suitable output coding by discussing the case of Multi-layer Perceptron (MLP) network. A single multiple output MLP can handle a multiclass problem directly. The standard technique is to use a k-dimensional binary target vector that represents each one of k classes using a single binary value at the corresponding position, for example $[0,...0,1,0,...0]$ which is sometimes referred to as one-per-class (OPC) encoding. The reason that a single multiclass MLP is not a suitable candidate for use as a base classifier is that all nodes share in the same training, so errors are far from independent and there is not much benefit to be gained from combining. However a 2-class MLP is a suitable base classifier, and independence among classifiers is achieved by the problem decomposition defined by the coding method, as well as by injection of randomness through the starting weights. Of course, no guarantee can be given that a single MLP with superior performance will not be found, but the assumption is that even if one exists its parameters would be more difficult to determine.

An alternative to OPC is distributed output coding [25], in which k binary vectors are assigned to the k classes on the basis of meaningful features corresponding to each bit position. For this to provide a suitable decomposition some domain knowledge is required so that each classifier output can be interpreted as a binary feature which indicates the presence or otherwise of a useful feature of the problem at hand. The vectors are treated as code words so that a test pattern is assigned to the class that is closest to the corresponding code word. It is this method of assigning, which is analogous to the assignment stage of error-correcting coding, that provides the motivation for employing ECOC in classification .

The first stage of the ECOC method, as described in Sect. 3.2, gives a strategy to decompose a multiclass problem into complementary two-class sub-problems. The second stage of the ECOC method is the decoding step, which was originally based on error-correcting principles under the assumption that the learning task can be modelled as a communication problem, in which class information is transmitted over a channel [8]. In this model, errors introduced into the process arise from various sources including the learning algorithm, features and finite training sample. The motivation for encoding multiple classifiers using an error-correcting code with Hamming Distance-based decoding was to provide error insensitivity with respect to individual classification errors. From the transmission channel viewpoint, we would expect that the one-per-class and distributed output coding matrices would not perform as well as the ECOC matrix, because of inferior error-correcting capability.

3.2 ECOC Algorithm and OOB Estimate

In the ECOC method, a $k \times b$ binary code word matrix C has one row (code word) for each of k classes, with each column defining one of b sub-problems that use a different labelling. Specifically, for the jth sub-problem, a training pattern with target class w_i $(i = 1...k)$ is re-labelled either as class Ω_1 or as class Ω_2 depending on the value of C_{ij} (typically zero or one). One way of looking at the re-labelling is to consider that for each column the k classes are arranged into two super-classes Ω_1 and Ω_2.

A test pattern is applied to the b trained classifiers forming vector

$$\mathbf{y} = [y_1, y_2, ...y_b]^T , \tag{1}$$

in which y_j is the real-valued output of jth base classifier.

The distance between output vector and code word for each class is given by

$$L_i^1 = \sum_{j=1}^{b} |C_{ij} - y_j| . \tag{2}$$

Equation (2) represents the L^1 norm or Minkowski distance, but if y_j in Eq. (2) is taken as binary decision, this reduces to Hamming Distance. The decoding rule is to assign a test pattern to the class corresponding to closest code word $\arg\min_i(L_i^1)$.

A diagrammatic representation of the decoding step for a three class problem is given in Fig. 2, in which the test pattern is assigned to the code word that has minimum Hamming Distance compared with ECOC ensemble outputs.

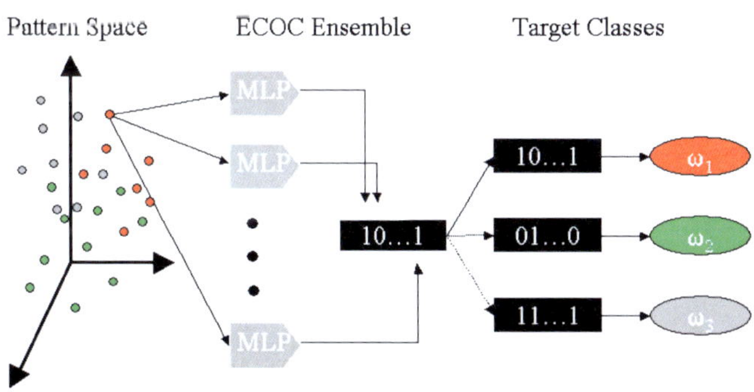

Fig. 2 Representation of the Hamming-based decoding step for a three class problem

To obtain the ensemble OOB estimate, the pth pattern is classified using only those classifiers that are in the set OOB_m, defined as the set of classifiers for which the pth pattern is out-of-bootstrap. For the OOB estimate, the summation in Eq. (2) is therefore modified to

$$L_i^1 = \sum_{j \in OOB_m} |C_{ij} - y_j| \, . \tag{3}$$

In other words it is necessary, for each pattern, to remember which classifier used that pattern for training. In the decoding step the columns of ECOC matrix C are removed if they correspond to classifiers that used the pth pattern for training. Therefore, on average, the column size of the C is about one third of the total number of classifiers.

3.3 Coding Strategies and Errors

When the ECOC technique was first developed it was believed that the ECOC code matrix should be designed to have certain properties to enable it to generalise well [9]. Various coding strategies have been proposed, but most ECOC code matrices that have been investigated previously are binary and problem-independent, that is pre-designed. Random codes have received much attention, and were first mentioned in [8] as performing well in comparison with error-correcting codes. In [9] random, exhaustive, hill-climbing search and BCH (Bose-Ray-Chaudhuri-Hocquenghem) coding methods were used to produce ECOC code matrices for different column lengths. Random codes were investigated in [24] for combining Boosting with ECOC, and it was shown that a random code with a near equal column split of labels was theoretically better. Random codes were also shown in [18] to give Bayesian performance if pairs of code words were equidistant, and it was claimed that a long enough random code would not be outperformed by a pre-defined code. In [38] a random assignment of class to codeword was suggested in order to reduce sensitivity to code word selection.

According to error-correcting theory, an ECOC matrix designed to have d bits error-correcting capability will have a minimum Hamming Distance $2d+1$ between any pair of code words. Assuming each bit is transmitted independently, it is then possible to correct a received pattern having d or fewer bits in error, by assigning the pattern to the code word closest in Hamming distance.

While in practice errors are not independent, the experimental evidence is that application of the ECOC method does lead to reduced test error rate. From the perspective of error-correcting theory, it is therefore desirable to use a matrix C containing code words having high minimum Hamming Distance between any pair. Besides the intuitive reason based on error-correcting theory, this distance property has been confirmed from other perspectives. In [1] it was shown that a high minimum distance between any pair implies a reduced upper bound on the generalisation error, and in [18] it was shown for a random matrix that if the code is equidistant, then decision-making is optimum.

Maximising Hamming Distance between any pair of code words is intended to remove individual classification errors on the re-labelled training sets, but even if classifiers are perfect (Bayesian) there will still be errors due to decoding. The decoding errors can be categorised into those due to inability of sub-problems to represent the main problem, and those due to the distance-based decision rule.

Sub-problems are more independent and likely to benefit from combining if Hamming distance between columns is maximised, remembering that a column and its complement represent identical classification problems [9]. The distance-based effect on decoding error can be understood by analysing the relationship between decoding strategy and Bayes decision rule. Consider that the decomposition of a multiclass classification problem into binary sub-problems in ECOC can be interpreted as a transformation between spaces from the original output **q** to **p**, given in matrix form by

$$\mathbf{p} = C^T \mathbf{q} \,, \tag{4}$$

where **q** are individual class probabilities.

Using the distance-based decision rule from (Eq. (2)) and Eq. (4)

$$L_i^1 = \sum_{j=1}^{b} |(\sum_{l=1}^{k} q_l C_{lj}) - C_{ij}| \tag{5}$$

and knowing that $\sum_{l=1}^{k} q_l = 1$, we have

$$L_i^1 = (1 - q_i) \sum_{j=1}^{b} |C_{ij} - C_{lj})| \,. \tag{6}$$

From Eq. (6), we see that L_i^1 is the product of $1 - q_i$ and Hamming Distance between code words. When all pairs of code words are equidistant, minimising L^1 implies maximising posterior probability which is equivalent to Bayes rule

$$\arg\max_i(q_i) = \arg\min_i(L_i^1) \,. \tag{7}$$

From the foregoing discussion, the main considerations in designing ECOC matrices are as follows

- minimum Hamming Distance between rows (error-correcting capability),
- variation of Hamming Distance between rows (effectiveness of decoding),
- number of columns (repetition of different parts of sub-problems),
- Hamming Distance between columns and complement of columns (independence of base classifiers).

From the theory of error-correcting codes [23] we know that finding a matrix with long code words, and having maximum and equal distance between all pairs of rows is complex. In [39] we compare random, equidistant and non-equidistant code matrices as number of columns is varied, but do not address explicitly the distance requirement between columns. Lack of experimental results on equidistant codes in previous work can be attributed to the difficulty in producing them. In [39] we produced equidistant codes by using the BCH method [23], which employs algebraic techniques from Galois field theory. Although BCH has been used before for ECOC, our implementation was different in that we first over-produced the number of rows (BCH requires number to be power of 2), before selecting a subset of rows.

Although various heuristics have been employed to produce better binary problem independent codes there appears to be little evidence to suggest that performance significantly improves by a clever choice of code [8, 9]. A three-valued code [1] was suggested which allows specified classes to be omitted from consideration (don't care for third value), thereby permitting integrated representation of methods such as all-pairs-of-classes [16]. Theoretical and experimental evidence indicates that, providing a problem-independent code is long enough and base classifier is powerful enough, performance is not much affected [18]. In this chapter, a random code with near equal split of labels in each column is used with $b = 200$ and $k = 12$.

In [7] problem-dependent codes were investigated and it is claimed that designed continuous codes show more promise than designed discrete codes. A sub-class problem-dependent code design is suggested in [12], in which SFFS is used to split classes based on maximising mutual information between data and respective class labels. In this chapter, it is proposed that a useful way to consider problem-dependence is to consider it as a generate-and-test search in the coding-decoding strategy. The question then is to decide how much intelligence is put into the coding or the decoding step. In Section 3.4, we discuss a method of problem-dependent decoding, that uses a random code with weighted decoding.

3.4 Weighted Decoding

One way to introduce problem-dependence is through the decoding scheme. First, consider a modification of the decoding step in which each column of the ECOC matrix is weighted. In the test phase, if the jth classifier produces an estimated probability $\hat{q}_j$ that a test pattern comes from the super-class defined by the jth decomposition. The pth test pattern is assigned to the closest code word, for which weighted distance of the pth pattern to the ith code word is defined as

$$D_{pi} = \sum_{j=1}^{B} \alpha_{jl} |C_{ij} - \hat{q}_{pj}| \,, \tag{8}$$

where $l = 1, \ldots, k$ and $\alpha_j l$ in Eq. (8) allows for lth class and jth classifier to be assigned a different weight.

Although this appears to be an obvious way to introduce weighted decoding, there is a difficulty in estimation of the values of the weights. In this chapter we propose a different weighted decoding scheme, that treats the outputs of the base classifiers as binary features [43]. By using the diagonal matrix $C_{ij} = 1$ if and only if $i = j$ the problem is recoded as k 2-class problems where each problem is defined by a different binary-to-binary mapping. There are many strategies that may be used to learn this mapping, but we use a weighted vote with weights set by class-separability measure applied to the training data, which was defined in [36].

Let z_{mj} indicate the binary output of the jth classifier applied to the mth training pattern, so that the output of base classifiers for the mth pattern is given by

$$\mathbf{z_m} = \left[z_{m1}, z_{m2}, \ldots z_{mb}\right]^T \,. \tag{9}$$

Assuming in Eq. (9) that a value of 1 indicates agreement of the output with target label and 0 disagreement, we can define counts for jth classifier as follows

$$N_j^{11} = z_{mj} \wedge z_{nj}$$
$$N_j^{00} = \bar{z}_{mj} \wedge \bar{z}_{nj}$$

where the mth and nth pattern are chosen from different classes.

The weight for the jth output is then defined as

$$w_j = \frac{1}{K}\left(\sum_{allpairs} N_j^{11} - \sum_{allpairs} N_j^{00}\right), \tag{10}$$

where K is a normalization constant and the summation is over all pairs of patterns from different class.

The motivation behind Eq. (10) is that the weight is computed as the difference between positive and negative correlation with respect to target class. In [36] this is shown to be a measure of class separability.

4 Dataset and Feature Extraction

The Cohn-Kanade database [19] contains posed expression sequences from a frontal camera from 97 university students. Each sequence goes from neutral to target display but only the last image is *au* coded. Facial expressions in general contain combinations of action units (*aus*), and in some cases *aus* are non-additive (one action unit is dependent on another). To automate the task of *au* classification, a number of design decisions need to be made, which relate to the following 1) subset of image sequences chosen from the database, 2) whether or not the neutral image is included in training, 3)image resolution, 4)normalisation procedure, 5)size of window extracted from the image, if at all, 6) features chosen for discrimination. Furthermore classifier type/parameters, and training/testing protocol need to be chosen. Researchers choose different decisions in these areas, and in some cases are not explicit about which choice has been made. Therefore it is difficult to make a fair comparison with previous results.

We concentrate on the upper face around the eyes, involving *au1(inner brow raised), au2(outer brow raised), au4(brow lowered), au5(upper eyelid raised), au6(cheek raised),* and *au7(lower eyelid tightened).* We use the MLP ensemble, given in Fig. 1 and random training/test split of 90/10 repeated twenty times and averaged. Other decisions we made were:

1. All image sequences of size 640 x 480 chosen
2. Last image in sequence (no neutral) chosen giving 424 images, 115 containing *au1*
3. Full image resolution, no compression
4. Manually located eye centres plus rotation/scaling into 2 common eye coordinates

5. Window extracted of size 150 x 75 pixels centred on eye coordinates
6. Principal Components Analysis (PCA) applied to raw image with PCA ordering

With reference to decision 2, some studies use only the last image in the sequence but others use the neutral image to increase the numbers of *non-aus*. Furthermore, some researchers consider only images with single *au*, while others use combinations of *aus*. We consider the more difficult problem, in which neutral images are excluded and images contain combinations of *aus*. With reference to decision 4 there are different approaches to normalisation and extraction of the relevant facial region. To ensure that our results are independent of any eye detection software, we manually annotate the eye centres of all images, and subsequently rotate and scale the images to align the eye centres horizontally. A further problem is that some papers only report overall error rate. This may be misleading since class distributions are unequal, and it is possible to get an apparently low error rate by a simplistic classifier that classifies all images as *non-au*. For the reason we report area under ROC curve, similar to [3].

With reference to decision 6, PCA, or Karhunen-Loeve expansion [29], is a well-known statistical method that was applied to the coding and decoding of images in [30]. PCA minimises mean-squared error when a finite number of basis functions are used in the expansion. Furthermore the entropy, defined in terms of average squared coefficients used in the expansion, is also minimised. The latter property is desirable for pattern recognition, in that features are clustered in the dimensionality reduction process. In the context of face recognition, the principal components of the distribution of faces are found, which is equivalent to finding the eigenvectors of the set of face images. Each face image in the training set may be represented by a linear combination of the 'eigenfaces', which is the name given to each eigenvector in the context of facial decomposition. The corresponding eigenvalues give a numerical value of the importance of each eigenface for reconstruction of the original images. Our purpose is not reconstruction, but we can characterise each image by the highest eigenvalues thereby reducing dimensionality.

A summary of the method follows, but for full details see reference [30]. First each 2-dim array of pixels of the window defined in decision 5 is represented by a 1-dimensional vector of size 150 x 75 = 11250. Now it is desired to find the μ orthonormal vectors u_j with associated eigenvalues λ_j of the covariance matrix W of the training set. Given the training set of vectors $x_i, i = 1,\ldots,\mu,\ x_i \in R^D$,each belonging to one of k classes $\{\omega_1, \omega_2, \ldots, \omega_k\}$, we compute the mean face image given by

$$x_{mean} = 1/\mu \sum_{i=1}^{\mu} x_i \,. \tag{11}$$

The mean image in Eq. (11) is subtracted from each training set image to give

$$t_i = x_i - x_{mean} \,. \tag{12}$$

Now the covariance matrix is given by

$$W = 1/\mu \sum_{i=1}^{\mu} t_i t_i^{\;T} = BB^T \tag{13}$$

where $B = [t_1 t_2 \dots t_\mu]$ and W is size $D \times D$. Following [30], we solve the simpler problem $B^T B$, which is $\mu \times \mu$, for obtaining the eigenvectors and eigenvalues.

Now the eigenvalues may be sorted to indicate the order of significance of the eigenvectors. Thus each face image is represented by the set of real numbers, or weights, corresponding to the P most significant eigenvalues, where P is to be determined experimentally (using OOB). The low-dimensional representation of each training pattern t_i, given by $u_k^T(t_i - x_{mean})$ for $k = 1 \dots P$ is used to train the network. An unknown test pattern t_T is projected using $u_k^T(t_T - x_{mean})$ for $k = 1 \dots P$ and input to the trained network for classification.

The ultimate goal in *au* classification is to detect combination of *aus*. In the ECOC approach, a random 200×12 code matrix is used to treat each *au* combination as a different class. After removing classes with less than four patterns this gives a 12-class problem with *au* combinations as shown in Table 1. In Sect. 5, to compare the ECOC results with 2-class classification, we compute test error by interpreting super-classes as 2-class problems, defined as either containing or not containing respective *au*. For example, *sc2, sc3, sc6, sc11, sc12* in Table 1 are interpreted as *au1*, and remaining super-classes as *non-au1*.

Table 1 ECOC super-classes of action units and number of patterns

ID	sc1	sc2	sc3	sc4	sc5	sc6	sc7	sc8	sc9	sc10	sc11	sc12
au	{}	1,2	1,2,5	4	6	1,4	1,4,7	4,7	4,6,7	6,7	1	1,2,4
#pat	149	21	44	26	64	18	10	39	16	7	6	4

5 Experiments on Cohn-Kanade Database

This section contains three sets of example experiments aimed at 2-class and multi-class formulations of *au* classification, for the Cohn-Kanade database described in Sect. 4. The goal is to demonstrate that weighted decoding ECOC outperforms conventional ECOC decoding, when base classifiers are tuned using OOB estimate. For experiments on UCI benchmark data [22] that demonstrate the use of OOB for ECOC ensemble design and provide an experimental comparison of feature selection schemes for ECOC ensembles, the reader is referred to [40, 42].

In the experiments in this section, the MLP ensemble uses two hundred single hidden-layer MLP base classifiers , with Levenberg-Marquardt training algorithm [17] and default parameters. Random perturbation of the MLP base classifiers is caused by different starting weights on each run, combined with bootstrapped training patterns. In our framework, we vary the number of hidden nodes, with a single

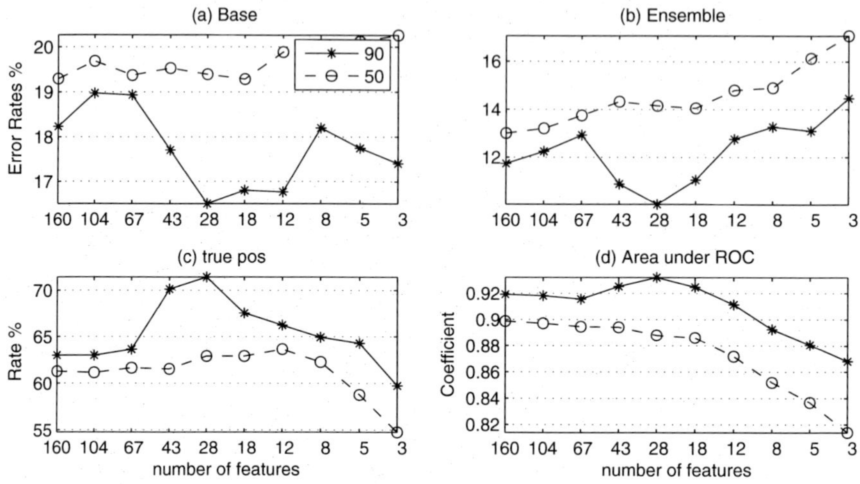

Fig. 3 Mean test error rates, True Positive and area under ROC for RFE MLP ensemble au1 classification for 90/10 and 50/50 train/test splits

node for linear perceptron, and keep the number of training epochs fixed at 20. For a comparison of feature extraction, the first experiment uses Gabor features [11], which have generally been found to give better performance than PCA for single classifiers [28]. The second and third experiments use PCA as described in Sect. 4.

In the first experiment, which comes from [41], we use RFE with MLP weights to rank Gabor features. RFE is a simple algorithm [14], and operates recursively as follows:

1. Rank the features according to a suitable feature-ranking method
2. Identify and remove the r least ranked features

If $r \geq 2$, which is usually desirable from an efficiency viewpoint, this produces a feature subset ranking. The main advantage of RFE is that the only requirement to be successful is that at each recursion the least ranked subset does not contain a

Table 2 Mean best test error rates for 2-class problems and area under ROC showing nodes/features for au classification with optimized PCA features and MLP ensemble

	Test error %	*Area under ROC*
au1	9.4/16/28	0.97/16/36
au2	3.5/4/36	0.99/16/22
au4	9.1/16/36	0.95/16/46
au5	5.5/1/46	0.97/1/46
au6	10.5/1/36	0.94/4/28
au7	10.3/1/28	0.92/16/60
mean	8.1	0.96

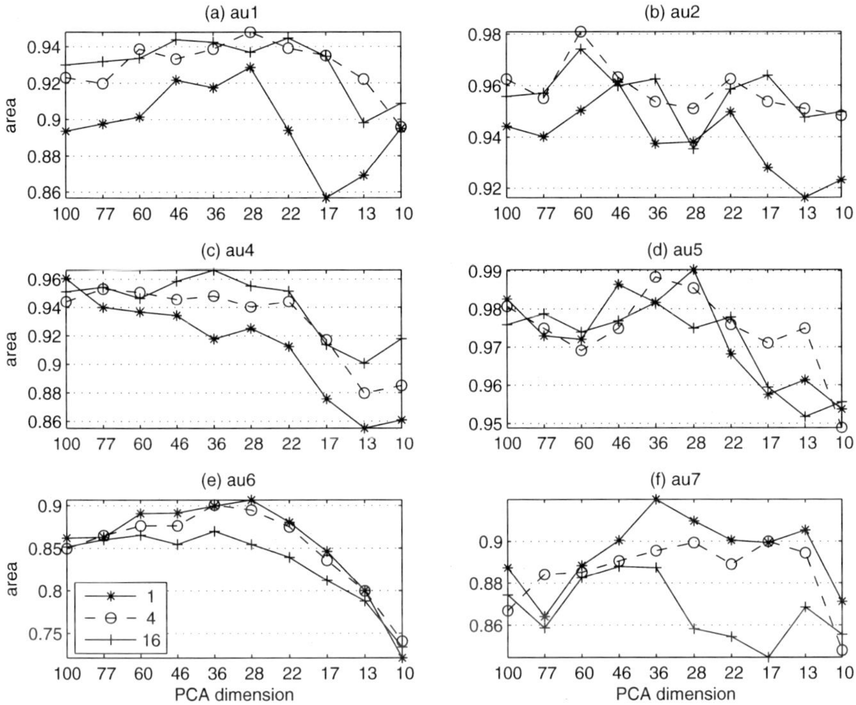

Fig. 4 Area under ROC at logarithmic scale for weighted decoding ECOC MLP ensemble (the number of hidden nodes tried was 1, 4, and 16), trained for 20 epochs versus the number of PCA features extracted

Table 3 Mean best test error rates and area under ROC for ECOC L1 norm decoding showing nodes/features for au classification with optimized PCA features and MLP ensemble

	Test error %	*Area under ROC*
au1	10.3/1/10	0.92/16/46
au2	3.4/1/36	0.96/16/28
au4	12.0/16/28	0.92/4/28
au5	3.6/16/36	0.99/1/36
au6	13.1/1/77	0.88/1/77
au7	11.6/1/28	0.89/4/46
mean	9.0	0.93

strongly relevant feature [44]. It was found that lower test error was obtained with non-linear base classifier and Fig. 3 shows test error rates, using an MLP ensemble with 16 nodes. The minimum base error rate for 90/10 split is 16.5 percent achieved for 28 features, while the ensemble is 10.0 percent at 28 features. Note that for 50/50 split there are too few training patterns for feature selection to have much effect.

Table 4 Mean best test error rates and area under ROC for ECOC weighted decoding showing nodes/features for au classification with optimized PCA features and MLP ensemble

	Weighted error %	*Weighted ROC*
au1	9.2/4/36	0.94/16/36
au2	2.8/16/22	0.98/1/46
au4	9.5/1/28	0.94/4/28
au5	3.2/1/36	0.99/1/36
au6	12.8/1/77	0.90/1/28
au7	10.9/4/46	0.92/1/36
mean	8.1	0.95

Since class distributions are unbalanced, the overall error rate may be mis-leading, as explained in Sect. 4. Therefore, we show the true positive rate in Fig. 3 c) and area under ROC in Fig. 3 d). Note that only 71 percent of au1s are correctly recognised. However, by changing the threshold for calculating the ROC, it is clearly possible to increase the true positive rate at the expense of false negatives.

The second set of experiments detects *au1, au2, au4, au5, au6, au7* using six different 2-class classification problems, where the second class contains all patterns not containing respective *au*. The MLP ensemble uses majority vote combining rule and PCA features are used to train the base classifiers . The best error rate of 9.4 percent for *au1* was obtained with 16 nodes and 28 features. The 9.4 percent error rate for *au1* is equivalent to 73 percent of *au1s* correctly recognised. The best ensemble error rate, area under ROC with number of features and number of nodes for all upper face *aus* are shown in Table 2. Note that number of nodes for best area under ROC is generally higher than for best error rate, indicating that error rate is more likely to be susceptible to over-fitting.

The third set of experiments uses ECOC method described in Sect. 3, and Fig. 4 shows area under ROC for the six *aus*, as number of PCA features is reduced. Table 3 shows best $L1$ norm decoding classification error and area under ROC, while Table 4 shows respective weighted decoding. It may be seen that weighted consistently outperforms $L1$ norm decoding. Also it may be seen from Table 2 that 2-class classification with optimized PCA features on average slightly outperforms ECOC. However, the advantage of ECOC is that all problems are solved simultaneously, and furthermore the combination of *aus* is recognized. As a 12-class problem, the mean best error rate over the twelve classes defined in Table 1 is 38.2 percent, showing that recognition of combination of *aus* is a difficult problem.

6 Discussion

The results for upper face *aus*, shown in Table 2 and Table 4, are believed to be among the best on this database (recognising the difficulty of making fair comparison as explained in Sect. 3).There are two possible reasons why the ECOC decoding strategy works well. Firstly, the data is projected into a high-dimensional space and therefore more likely to be linearly separable [6]. Secondly, although the full training set is used to estimate the weights, each base classifier is bootstrapped and therefore is trained on a subset of the data, which guards against overfitting. As indicated in Sect. 2, bootstrapping also facilitates the OOB estimate for removing irrelevant features without validation.

7 Conclusion

In this chapter, an information theoretic approach of coding and decoding has been applied to both feature extraction and multi-class classification. For upper face *au* classification, weighted decoding ECOC achieves comparable performance to optimized 2-class classifiers. However, ECOC has the advantage that all *aus* are detected simultaneously, and further work is aimed at determining whether problem-dependent rather than random codes can improve results. Furthermore, the ultimate aim of this work is to apply the technique to improve robustness of face verification systems, and to better recognise driver fatigue.

References

1. Allwein, E.L., Schapire, R.E., Singer, Y.: Reducing multi-class to binary: a unifying approach for margin classifiers. J. Mach. Learn. Res. 1, 113–141 (2000)
2. Bartlett, M.S., Littlewort, G., Lainscsek, C., Fasel, I., Movellan, J.: Machine learning methods for fully automatic recognition of facial expressions and facial actions. In: Proc. IEEE Conf. Syst., Man and Cybernetics, The Hague, The Netherlands, pp. 592–597. IEEE Comp. Soc., Los Alamitos (2004)
3. Bartlett, M.S., Littlewort, G., Frank, M., Lainscsek, C., Fasel, I., Movellan, J.: Fully automatic facial action recognition in spontaneous behavior. In: Proc. 7th IEEE Conf. Automatic Face and Gesture Recogn., Southampton, UK, pp. 223–238. IEEE Comp. Soc., Los Alamitos (2006)
4. Breiman, L.: Bagging predictors. Mach. Learn. 24(2), 123–140 (1997)
5. Bylander, T.: Estimating generalisation error two-class datasets using out-of-bag estimate. Mach. Learn. 48(1-3), 287–297 (2002)
6. Cover, T.M.: Geometrical and statistical properties of systems of linear inequalities with applications in pattern recognition. IEEE Trans. Electronic Comp. 14(3), 326–334 (1965)
7. Crammer, K., Singer, Y.: On the learnability and design of output codes for multiclass problems. Mach. Learn. 47(2-3), 201–233 (2002)
8. Dietterich, T.G., Bakiri, G.: Error-correcting output codes: a general method for improving multiclass inductive learning programs. In: Proc. 9th Natl. Conf. Artif. Intell., Anaheim, CA, pp. 572–577. AAAI/MIT Press, Cambridge (1991)
9. Dietterich, T.G., Bakiri, G.: Solving multiclass learning problems via error-correcting output codes. J. Artif. Intell. Research 2, 263–286 (1995)
10. Dietterich, T.G.: Ensemble methods in machine learning. In: Kittler, J., Roli, F. (eds.) MCS 2000. LNCS, vol. 1857, pp. 1–15. Springer, Heidelberg (2000)
11. Donato, G., Bartlett, M.S., Hager, J.C., Ekman, P., Sejnowski, T.J.: Classifying facial actions. IEEE Trans. Patt. Analysis Mach. Intell. 21(10), 974–989 (1999)

12. Escalara, S., Tax, D.M.J., Pujol, O., Radeva, P., Duin, R.W.: Subclass problem-dependent design for error-correcting output codes. IEEE Trans. Patt. Analysis Mach. Intell. 30(8), 1041–1054 (2008)
13. Freund, Y., Schapire, R.E.: A decision-theoretic generalisation of on-line learning and application to boosting. J. Comp. Syst. Sci. 55(1), 119–139 (1997)
14. Guyon, I., Weston, J., Barnhill, S., Vapnik, V.: Gene selection for cancer classification using support vector machines. Mach. Learn. 46(1-3), 389–422 (2002)
15. Hansen, L.K., Salamon, P.: Neural network ensembles. IEEE Trans. Patt. Analysis Mach. Intell. 12(10), 993–1001 (1990)
16. Hastie, T., Tibshirani, R.: Classification by pairwise coupling. The Annals of Stat. 26(2), 451–471 (1998)
17. Haykin, S.: Neural Networks: a Comprehensive Foundation. Prentice Hall, Upper Saddle River (1999)
18. James, G.M., Hastie, T.: The error coding method and PiCT. Computational and Graphical Stat. 7(3), 377–387 (1998)
19. Kanade, T., Cohn, J.F., Tian, Y.: Comprehensive database for facial expression analysis. In: Proc. 4th Int. Conf. Automatic Face and Gesture Recogn., Grenoble, France, pp. 46–53. IEEE Comp. Soc., Los Alamitos (2000)
20. Kong, E.B., Diettrich, T.G.: Error-correcting output coding corrects bias and variance. In: Prieditis, A., Russell, S.J. (eds.) Proc. 12th Int. Conf. Mach. Learn., Tahoe City, CA, pp. 313–321. Morgan Kaufmann, San Francisco (1995)
21. Kuncheva, L.I., Whitaker, C.J.: Measures of diversity in classifier ensembles. Mach. Learn. 51(2), 181–207 (2003)
22. Merz, C.J., Murphy, P.M.: UCI repository of machine learning databases (1998), `http://www.ics.uci.edu/~mlearn/MLRepository.html`
23. Peterson, W.W., Weldon, J.R.: Error-Correcting Codes. MIT Press, Cambridge (1972)
24. Schapire, R.E.: Using output codes to boost multiclass learning problems. In: Fisher, D.H. (ed.) Proc. 14th Int. Conf. Mach. Learn., Learn., Nashville, TN, pp. 313–321. Morgan Kaufmann, San Francisco (1997)
25. Sejnowski, T.J., Rosenberg, C.R.: Parallel networks that learn to pronounce English text. Complex Systems 1(1), 145–168 (1987)
26. Silapachote, P., Karuppiah, D.R., Hanson, A.R.: Feature selection using Adaboost for face expression recognition. In: Villanueva, J.J. (ed.) Proc. 4th IASTEAD Int. Conf. Visualization, Imaging and Image Proc., Marbella, Spain, pp. 84–89. ACTA Press, Calgary (2004)
27. Tian, Y., Kanade, T., Cohn, J.F.: Recognising action units for facial expression analysis. IEEE Trans. Patt. Analysis Mach. Intell. 23(2), 97–115 (2001)
28. Tian, Y., Kanade, T., Cohn, J.F.: Evaluation of Gabor-based facial action unit recognition in image sequences of increasing complexity. In: Proc. 5th Int. Conf. Automatic Face and Gesture Recogn., Washington, DC, pp. 229–234. IEEE Comp. Soc., Los Alamitos (2002)
29. Tou, J.T., Gonzales, R.C.: Pattern Recognition Principles. Addison-Wesley, Reading (1974)
30. Turk, M.A., Pentland, A.P.: Face recognition using eigenfaces. In: Proc. IEEE Int. Conf. Comp. Vision Patt. Recogn., Maui, HW, pp. 586–591. IEEE Comp. Soc., Los Alamitos (1991)
31. Valentini, G., Dietterich, T.G.: Bias-variance analysis of support vector machines for the development of SVM-based ensemble methods. J. Mach. Learn. Res. 5, 725–775 (2004)
32. Wang, W., Jones, P., Partridge, D.: Assessing the impact of input features in a feedforward neural network. Neural Computing Appl. 9(2), 101–112 (2000)

33. Wilson, C.L., Grother, P.J., Barnes, C.S.: Binary decision clustering for neural network-based optical character recognition. Patt. Recogn. 29(3), 425–437 (1996)
34. Windeatt, T.: Diversity measures for multiple classifier system analysis and design. Inf. Fusion 6(1), 21–36 (2004)
35. Windeatt, T.: Spectral measure for multi-class problems. In: Roli, F., Kittler, J., Windeatt, T. (eds.) MCS 2004. LNCS, vol. 3077, pp. 184–193. Springer, Heidelberg (2004)
36. Windeatt, T.: Accuracy/diversity and ensemble classifier design. IEEE Trans. Neural Networks 17(5), 287–297 (2006)
37. Windeatt, T.: Ensemble MLP classifier design. In: Jain, L.C., Sato-Ilic, M., Virvou, M., Tsihrintzis, G.A., Balas, V.E., Abeynayake, C. (eds.) Computational Intelligence Paradigms. Studies in Computational Intelligence, vol. 137, pp. 133–147. Springer, Heidelberg (2008)
38. Windeatt, T., Ghaderi, R.: Multi-class learning and error-correcting code sensitivity. Electronics Letters 36(19), 1630–1632 (2000)
39. Windeatt, T., Ghaderi, R.: Coding and decoding strategies for multiclass learning problems. Inf. Fusion 4(1), 11–21 (2003)
40. Windeatt, T., Prior, M.: Stopping criteria for ensemble-based feature selection. In: Haindl, M., Kittler, J., Roli, F. (eds.) MCS 2007. LNCS, vol. 4472, pp. 271–281. Springer, Heidelberg (2007)
41. Windeatt, T., Dias, K.: Feature-ranking ensembles for facial action unit classification. In: Prevost, L., Marinai, S., Schwenker, F. (eds.) ANNPR 2008. LNCS (LNAI), vol. 5064, pp. 267–279. Springer, Heidelberg (2008)
42. Windeatt, T., Prior, M., Effron, N., Intrator, N.: Ensemble-based feature selection criteria. In: Perner, P. (ed.) Poster Proc. 5th Int. Conf. Mach. Learn. Data Mining in Patt. Recogn., Leipzig, Germany, pp. 168–182. IBal Publishing, Leipzig (2007)
43. Windeatt, T., Smith, R.S., Dias, K.: Weighted decoding ECOC for facial action unit classification. In: Okun, O., Valentini, G. (eds.) Proc. 2nd Workshop Supervised and Unsupervised Ensemble Methods and Their Appl., Patras, Greece, pp. 26–30 (2007)
44. Yu, L., Liu, H.: Efficient feature selection via analysis of relevance and redundancy. J. Mach. Learn. Res. 5, 1205–1224 (2004)

Prediction of Gene Function Using Ensembles of SVMs and Heterogeneous Data Sources

Matteo Re and Giorgio Valentini

Abstract. The ever increasing amount of biomolecular data available in public domain databases for a broad range of organisms coupled with recent advances in machine learning research has stimulated interest in computational approaches on gene function prediction. In this context data integration from heterogeneous biomolecular data sources plays a key role. In this contribution we test the performance of several ensembles of SVM classifiers, in which each component learner has been trained on different types of data, and then combined using different aggregation techniques. The compared combination methods are the widely adopted linear weighted combination, the logarithmic weighted combination and the similarity based decision template approach. The results show that heterogeneous data integration through ensemble methods represents a valuable research line in gene function prediction .

1 Introduction

Functional classification of unannotated genes and the improvement of the existing gene functional annotation catalogs, are of capital importance in modern functional genomics and bioinformatics. Gene functional classification may provide useful insights in pharmacogenomics, being able to provide indications for the development of target specific drugs. More in general, it plays a key role in molecular biology, given its ability to detect previously unknown role of genes and their products in physiological and pathological processes.

Nevertheless, the application of automated systems in this research area is strongly limited by the intrinsic difficulty of this task, which is mainly caused by the natural heterogeneity of the involved data. Different types of biomolecular data, ranging from expression profiles to phylogenetic gene-specific evolution

Matteo Re · Giorgio Valentini
Dipartimento di Scienze dell'Informazione, Università degli Studi di Milano, Italy
e-mail: {re, valentini}@dsi.unimi.it

O. Okun & G. Valentini (Eds.): Appli. of Supervised & Unsuper. Ensemble Meth., SCI 245, pp. 79–91.
springerlink.com © Springer-Verlag Berlin Heidelberg 2009

rates and many others can in principle provide useful information for the automated assessment of the functional role of genes. The extent of the degree to which the presence of a specific type of experimental data could result in the improvement of classification performance is expected to vary according to the specific gene and the particular bio molecular process under investigation.

Several approaches for heterogeneous biomolecular data integration have been proposed in the literature. The first one corresponds to the "early integration" technique, by which different vectorial data are concatenated [9]. Other methods are based on modeling networks of functional relationships between proteins; in this context graphical models provide a probabilistic framework for data integration [16]. Kernel methods and techniques based on kernel fusion methods represent another important research area with significant applications in the integration of different bio-molecular data sources for gene function prediction [8] .

In the aforementioned scenario the application of methods able to deal with both different data sources and the problem to integrate the prediction obtained from different learners is clearly appealing. It is widely accepted that combining multiple classifiers can provide advantages over the monolithic approach to pattern classifier design [12], but a systematic evaluation of the impact on classification performances of different combination rules suitable to merge the output of gene function classifiers trained on different data sources, as today, has not been explored. To our knowledge, only some works have been proposed, such as the "late integration" of kernels trained on different sources of data [9], or the Naive-Bayes integration of the outputs of SVMs in the context of the hierarchical classification of genes [4]. In this work we investigate the effectiveness of three classifier fusion strategies using ensembles of Support Vector Machines [17] each of which trained to produce a probabilistic-like classification output [10].

In the next section we present the ensemble methods we used in our experiments. In Sect. 3 we describe the different types of high-throughput bio-molecular data and the experimental setting we adopted to classify yeast genes according the highest level classes of the FunCat taxonomy [11]. Section 4 presents the main results obtained by comparing performances of single SVMs with respect to to ensembles that merge 6 different sources of biomolecular data. The conclusions summarize the main achievements and drawbacks of the proposed data fusion ensemble approach.

2 Methods

Ensembles of classifiers have enjoyed great attention because of their excellent generalization performance in a wide spectrum of applications. One of the main ideas behind the effectiveness of ensemble systems is that if the single classifiers composing the ensemble are diverse, then they are expected to make different errors, and combining the output produced by these classifiers can in principle reduce the error through averaging [6].

Diversity can be achieved using different sources data, thus obtaining different "views" of the same phenomenon. In particular the objective of data fusion is to extract complementary pieces of information from different data sources and then merge them achieving a more informed decision about the phenomenon under analysis. Working with heterogeneous data sources, data fusion can be realized by means of an ensemble system composed by learners trained on different datasets and then combining the outputs of the component learners.

The continuous output assigned to an instance vector **x** by a binary classifier can be interpreted as the support given by the membership of **x** to a specific functional class.

In particular, with SVMs , a probabilistic output can bc obtained by applying a sigmoid fitting to their output [10]. As a consequence, a trained classifier computes a function $d_j : X \rightarrow [0,1]$ that estimates the probability that a given example $\mathbf{x} \in X$ belongs to a specific class ω_j. An ensemble combines the outputs of T base learners using a suitable combining function g to compute the overall support (e.g., the probability) μ_j for a given class ω_j:

$$\mu_j(\mathbf{x}) = g(d_{1,j}(\mathbf{x}), \ldots, d_{T,j}(\mathbf{x})) \,. \tag{1}$$

2.1 Linear Weighted Combination with Linear and Logarithmic Weights

Among the algebraic combiners, the simplest is the mean rule, which calculates the support μ_j for the membership of a current instance **x** to the ω_j class as the average of all classifiers outputs:

$$\mu_j(\mathbf{x}) = \frac{1}{T} \sum_{t=1}^{T} d_{t,j}(\mathbf{x}) \,. \tag{2}$$

In our experiments we used the weighted average rule, in order to take into account the reliability of each base learner in the computation of the support μ_j:

$$\mu_j(\mathbf{x}) = \sum_{t=1}^{T} w_t d_{t,j}(\mathbf{x}) \,. \tag{3}$$

The weights w_t are usually computed using an estimate of the overall accuracy of the base learners, but in our experimental setting, where the gene functional classes are largely unbalanced (positive examples are fewer than negative ones), we chose the F-measure (the harmonic mean between precision and recall) to compute the weights:

$$w_t = \frac{F_t}{\sum_{t=1}^{T} F_t} \,. \tag{4}$$

The F-measure F_t of the tth base learner can be estimated by "internal" cross-validation on the training set.

It can be shown [5] that if we have T independent classifiers, each of which associated with some performance measure (such as the accuracy or the F-measure), the accuracy of an ensemble produced by combining the learners outputs by weighted majority voting is maximized if the output weights satisfy the proportionality

$$w_t \propto log \frac{p_t}{1-p_t} , \tag{5}$$

where p_t is an estimate of the reliability of the tth base learner (e.g., accuracy or F-measure).

In our experiments we implemented the weighted logarithmic combination by adding a small ε in order to avoid division by zero in Eq. 5, and then by normalizing in order to obtain positive weights that sum to 1:

$$\hat{w}_t = ln \frac{F_t + \varepsilon}{1 - F_t + \varepsilon} , \tag{6}$$

$$w_t = \frac{\hat{w}_t - ln\varepsilon}{\sum_{t=1}^{T}(\hat{w}_t - ln\varepsilon)} . \tag{7}$$

Once computed the weights for each classifier, according to Eq. 4 for the linear weighted combination or to Eq. 7 for logarithmic weighted combination, the final decision $D_j : X \rightarrow \{0, 1\}$ of the ensemble is taken using the probability μ_j for the class ω_j (Eq. 3):

$$D_j(\mathbf{x}) = \begin{cases} 1, \text{ if } \mu_j(\mathbf{x}) > 0.5 , \\ 0, \text{ otherwise} , \end{cases} \tag{8}$$

where output 1 corresponds to positive and 0 to negative predictions for ω_j.

2.2 Decision Templates

The main idea behind decision templates is to compare a"prototypical answer" of the ensemble for the examples of a given class (the template), to the current answer (the decision profile) of the ensemble to a specific example whose class needs to be predicted [7].

The decision profile DP($\mathbf{x}$) for a instance $\mathbf{x}$ is a matrix composed by the $d_{t,j} \in [0,1]$ elements representing the support given by the tth classifier to class ω_j. The decision profiles matrices are effective tools that allows us to effectively summarize the information produced by all the members of an ensemble system and also provide conceptual blocks at the basis of the decision template technique.

Decision templates DT_j are the averaged decision profiles obtained from $\mathbf{X}_j$, the set of training instances belonging to the class ω_j:

$$DT_j = \frac{1}{|\mathbf{X}_j|} \sum_{\mathbf{x} \in \mathbf{X}_j} DP(\mathbf{x}) . \tag{9}$$

Note that the sum in Eq. 9 refers to matrices, and hence decision templates are matrices with a number of rows equal to the number of the base learners and a number of columns equal to the number of the classes.

Given a test instance we first compute its decision profile and then we calculate the similarity S_j between $DP(\mathbf{x})$ and the decision template DT_j for each class ω_j. As similarity measure the Euclidean distance is usually applied:

$$S_j(\mathbf{x}) = 1 - \frac{1}{TC} \sum_{t=1}^{T} \sum_{k=1}^{C} [DT_j(t,k) - d_{t,k}(\mathbf{x})]^2 . \quad (10)$$

The final decision of the ensemble is taken by assigning a test instance as:

$$D(\mathbf{x}) = \arg\max_j S_j(\mathbf{x}) . \quad (11)$$

In our experimental setting we consider dichotomic problems, thus obtaining two-columns decision template matrices. Note that for each gene functional class we have two decision templates, one for the positive examples for that class (DT_P) and another one for the negatives (DT_N).

It is easy to see that with dichotomic problems the similarity measure (Eq. 10) for the positive (S_P) and negative (S_N) class becomes:

$$S_P(\mathbf{x}) = 1 - \frac{1}{T} \sum_{t=1}^{T} [DT_P(t,1) - d_{t,1}(\mathbf{x})]^2 , \quad (12)$$

$$S_N(\mathbf{x}) = 1 - \frac{1}{T} \sum_{t=1}^{T} [DT_N(t,2) - d_{t,2}(\mathbf{x})]^2 \quad (13)$$

and the final decision of the ensemble is

$$D(\mathbf{x}) = \arg\max_{\{P,N\}} (S_P(\mathbf{x}), S_N(\mathbf{x})) . \quad (14)$$

3 Experimental Setup

3.1 Heterogeneous Biomolecular Datasets

In order to test the effectiveness of fusion methods for various continuous-valued predictions, we collected a set of data sources used in bioinformatics experiments, published in literature or obtained from public databases. We chose to perform our experiments using data collected on *S. cerevisiae* because it is among the most studied and well characterized model organisms and because of the great amount of biomolecular data available for this species.

Biological functions are mediated in cell by several types of molecules but, in the large majority of biological processes, the final effectors are proteins. Despite the complexity of single proteins, the realization of a single biomolecular process usually requires the coordinated action of more than a single molecule, and the composition of the set of molecules is expected to be highly informative. We thus decided to use protein-protein interaction data collected from BioGrid [15], a database of protein and genetic interactions from STRING [18], a collection of protein functional interactions inferred from heterogeneous data sources, comprising, among the others, experimental data and information found in literature.

The evolutionary pressures acting during evolution on genes lead progressively to a saturation of the number of mutations that can be tolerated without disrupting the functionality of gene products and this reduces the ability to evolve new biomolecular functions. The conservative action of evolutionary pressures is particularly strong for single-copy genes but act in a more relaxed fashion on members of clusters of genes derived from duplication events in entire genomic regions. Provided the presence of multiple copies of a particular gene, the organism is allowed to explore evolutionary paths that would be otherwise precluded and that can lead to the evolution of novel functions by means of the changes occurring at nucleotide level in DNA sequences encoding protein products. As the evolutionary time increases, the genes belonging to the gene cluster will get even different at both nucleotide and aminoacidic sequence level, but the relations between members of the same gene families, which often share a similar biological role, can be detected using classical alignment approaches such as those proposed in [13] and [1]. The use of this type of experimental data enables the detection and quantification (using opportune similarity measures) of homology relationships occurring between genes through a simple nucleotide sequences comparison. In the aim to catch homology and, hopefully, functional relations existing between genes belonging to the same functional classes, we included as data source into our experimental framework the Pairwise Similarity Smith-Waterman dataset published in [8] (data kindly provided by the authors).

Even if protein-protein interactions and evolutionary signatures can provide useful information for functional classification of genes, a potential source of information about the functional role of a gene could be provided by the tight connection between the structure of a protein and its ability to perform a particular biological task. Proteins are constituted by structured regions usually referred as domains joined by unstructured regions named loops. Each specific domain constituting a protein performs a specific task (either structural or biochemical) and thus the presence of particular kinds of domains into the protein structure could be of capital importance for the prediction of its function. In order to account for this source of information we included data published in [2]. This dataset has been processed in order to provide two types of information: the presence/absence of a particular protein domain in the proteins encoded by genes comprised in the dataset and the E-value assigned to each gene product to a collection of profile-HMMs, each of which trained on a specific domain family. The E-values have been obtained by the HMMR software toolkit (`http://hmmer.janelia.org`).

The activation of a gene (and its functional products) is strictly regulated in cell in order to avoid interference between molecular processes, and this regulation is in part realized by modulating the transcriptional state of the gene. Genes involved in the realization of the same biological process are expected to show some similarities in their expression profiles. We thus included into our experiment a dataset obtained by the integration of microarray hybridization experiments published in [3, 14]. The main data sets used in the experiments are summarized in Table 1.

Table 1 Datasets

Data code	Dataset	examples	features
L_1	Protein domain binary	3529	4950
L_2	Protein domain log-E	3529	5724
L_3	Gene expression	4532	250
L_4	PPI - BioGRID	4531	5367
L_5	PPI - vonMering	2338	2559
L_6	Pairwise similarity	3527	6349

*3.2 The Functional Catalogue (*FunCat*)*

In order to associate each of the genes constituting the aforementioned datasets, we used functional annotations collected in the Functional Catalogue (FunCat index-FunCat) database [11], version (2.1), initially developed at MIPS during the early stages of sequencing of the yeast genome. The Functional Catalogue is constituted by hierarchically structured controlled vocabulary of functional categories. FunCat is the natural choice for our experiments, since it was originally developed to describe yeast functional processes.

In order to reduce the number of classification tasks required by the experimental setting, we choose to consider only the first level FunCat classes. In other words, we selected the roots of the trees of the FunCat forest (that is the most general and wide functional classes of the overall taxonomy). We also removed in the list of the target functional classes all the classes represented by less than 20 genes. This restricted our classifications to only 16 functional classes:

01:METABOLISM
02:ENERGY
10:CELL CYCLE AND DNA PROCESSING
11:TRANSCRIPTION
12:PROTEIN SYNTHESIS
14:PROTEIN FATE (folding,modification,destination)
16:PROTEIN WITH BINDING FUNCTION OR COFACTOR REQUIREMENT (structural or catalytic)
18:REGULATION OF METABOLISM AND PROTEIN FUNCTION
20:CELLULAR TRANSPORT AND TRANSPORT ROUTES
30:CELLULAR COMMUNICATION/SIGNAL TRANSDUCTION MECHANISM
32:CELL RESCUE, DEFENSE AND VIRULENCE
34:INTERACTION WITH THE ENVIRONMENT
40:CELL FATE
41:DEVELOPMENT(Systemic)
42:BIOGENESYS OF CELLULAR COMPONENTS
43:CELL TYPE DIFFERENTIATION

3.3 Base Learners Tuning and Generation of Optimized Classifiers

The construction of an ensemble of classifiers trained to perform functional classification using heterogeneous data (as for any ensemble of classifiers) requires the definition of two key points: the way the base learners have to be trained and a strategy to combine the output of the different learner components. This section is dedicated to the former question while the second one has just be treated in Sect. 2.

Being the main objective of this experiment, the evaluation of different strategies of fusion of the continuous output, produced by different classifiers on heterogeneous datasets, is done in the simplest way of adapting the single sources of data: through the intersection between all the datasets. This led to the definition of a common set constituted by 1901 genes.

For each of the 16 target functional classes we tuned the base learners as binary classifiers (thus labeling samples as belonging to the "target functional class" or to "other functional classes") using a classical inner cross-validation tuning scheme.

More precisely, each dataset was split into a training set and a test set (composed, respectively, by the 70% and 30% of the available samples) and the resulting training set was further split into 3 balanced folds, meaning that the proportion of positive and negative samples constituting each fold was kept equal for each fold.

The balanced folds have been used to perform a 3-folds cross validation for model selection. The averaged accuracy, precision, recall and F-measure across folds were collected for each combination of a list of tuning parameters. We chose RBF gaussian kernels for all the training tasks involved in the experiment, tuning each SVM for a cost ranging from 10^{-2} to 10^{2} and a value of sigma varying in the same range. During the tuning stage we experienced problems in tuning the learners dedicated to the classification of the Pairwise similarity dataset, for which only negative classifications were produced in 11 out of 16 learning tasks. We thus changed the tuning setting for this dataset by using a polynomial kernel and varying the degree hyper-parameter from 2 to 5 while keeping the cost varying from 10^{-2} to 10^{2}.

Among the commonly used performance metrics suitable to drive the optimization process, considering that negative examples are fewer than positive examples, we decided to tune the base learners by choosing the set of parameters producing the maximum averaged F-measure during the tuning stage. Once defined the best set of parameters associated to each learner in each learning task, we used them to train the optimal model on the whole training set.

The generalization performance has been estimated on the separated test set.

For the experiments we used the *Lagrange* cluster composed by 208 nodes equipped with Intel Xeon 3.16 GHz QuadCore processors and 16 GB of RAM memory at each node (`http://www.cilea.it`).

4 Results

The performance characteristics obtained in the learning tasks associated to the prediction of the FunCat functional classes are reported in Table 2.

The table reports the F-measures obtained through the evaluation of the test set (570 genes) using the best models selected by internal 3-folds cross-validation. The first column refers to the FunCat identifiers of first-level functional classes. The next 6 columns (from L_1 to L_6) correspond to single SVMs trained respectively on the six datasets described in Table 1. L_{avg} represents the averaged results of the single SVMs across the six datasets, and the three last columns refers respectively to the weighted linear, logarithmic and decision template ensembles. The performance of the best performing single learner and the best performing ensemble are highlighted in boldface.

Note that among the first level FunCat functional classes, the "DEVELOPMENT(Systemic)" class, (class ID: 41) is not reported because, after the intersection of the data sources described in Table 1, it failed to reach the minimum amount of positive samples (20 genes belonging to the target functional class) required by our experimental protocol.

Looking at the last row of Table 2, we see that, on average, data integration methods through ensembles provide better results than single SVMs trained on homogeneous bio-molecular data, independently of the applied combination rule. In particular, decision templates largely outperform both the single SVMs and the other ensembles. Performance of the weighted and logarithmic ensembles is quite comparable, better than the average single SVM and in most cases better than the single SVM trained on a single data set.

F-measures are summarized in Fig. 1: all ensemble methods outperform the F-measure obtained, on average, by the single SVMs . The best single SVM for each

Table 2 F-measures computed on the test sets (see text for more details)

FunCat class	L_1	L_2	L_3	L_4	L_5	L_6	L_{avg}	E_{lin}	E_{log}	E_{DT}
01	0.6240	0.6486	0.4854	0.6461	0.5283	**0.7576**	0.6150	0.7835	**0.7860**	0.7845
02	0.2258	0.3478	0.2941	0.2318	0.3125	**0.4000**	0.3020	0.2857	0.3125	**0.4324**
10	0.5240	**0.6819**	0.1916	0.4059	0.3800	0.5963	0.4632	0.5887	0.5887	**0.6666**
11	0.5607	**0.7213**	0.2395	0.4397	0.4524	0.5693	0.4971	0.5673	0.5673	**0.6722**
12	0.6060	0.6616	0.3207	0.4793	**0.7361**	0.5181	0.5536	**0.6814**	0.6412	0.6715
14	0.4331	0.5622	0.4221	**0.6234**	0.2772	0.6191	0.4895	0.6776	0.6581	**0.6846**
16	0.3771	**0.4661**	0.2561	0.4086	0.2040	0.5146	0.3710	0.5217	0.4978	**0.5543**
18	0.0000	0.0526	0.1764	0.2352	0.0000	**0.2857**	0.1249	0.2424	0.2424	**0.3333**
20	0.4457	**0.6461**	0.2733	0.4588	0.1492	0.4802	0.4088	**0.5828**	0.5212	0.5465
30	0.0975	0.3913	0.1818	0.2702	0.0000	**0.4266**	0.2279	0.2285	0.2352	**0.5769**
32	0.2278	0.2650	0.2025	**0.3146**	0.0000	0.2684	0.2130	0.1842	0.1351	**0.2500**
34	0.3023	**0.3544**	0.0909	0.2133	0.0000	0.1834	0.3023	0.1764	0.1764	**0.4509**
40	0.2307	0.2745	0.1250	0.1250	0.0000	**0.3000**	0.1758	0.1304	0.1304	**0.3409**
42	0.4129	**0.5524**	0.0847	0.0344	0.1068	0.4052	0.2660	0.4736	0.3333	**0.5279**
43	0.4150	**0.6016**	0.2000	0.2000	0.0000	0.4153	0.3053	0.3956	0.3414	**0.4600**
AVG	0.3655	**0.4818**	0.2363	0.3391	0.2098	0.4493	0.3544	0.4347	0.4111	**0.5302**

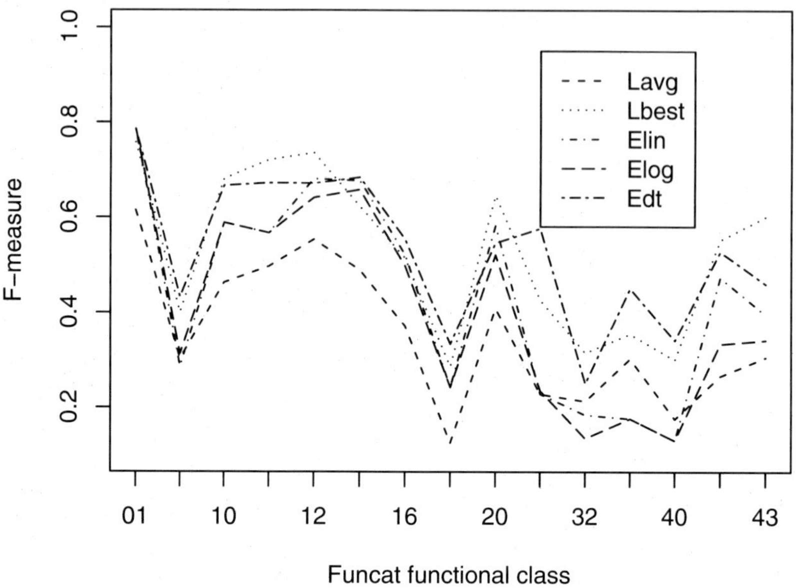

Fig. 1 Comparison of the F-measures achieved in gene prediction: Lavg stands for the average across SVM base learners; Lbest for the best single SVM; Elin, Elog, Edt for weighted linear, logarithmic and decision template ensembles

task outperforms weighted linear and logarithmic ensembles, but decision templates are in most cases better than the best single SVM .

5 Discussion

The prediction of the functional class of genes using heterogeneous data sources is among the most difficult problems in bioinformatics. The difficulty of the task comes mainly from the not very definition of the "biological process" entity, due to the high number of interconnections linking biological processes, and due to the different relevance of diverse biomolecular datasets with respect to different functional classes. A dataset providing critical information for the prediction of a particular functional class could merely represent noise in a classification task targeting a different functional class.

As shown by data reported in Table 2, the best performing ensemble system outperforms, on average, the base learners in all the functional classification tasks. The winner combination method in 13 out of 15 cases is the ensemble based on decision templates . If compared with results obtained by the best performing base learner, the best performing ensemble systems win in 8 out of 15 cases. This comes as a surprise because of the presence (in these ensembles) of the base learners of very poor performance (as reported in the cases of tasks aimed to predict the 20 (CELLULAR TRANSPORT AND TRANSPORT ROUTES), 32 (CELL RESCUE

DEFENSE AND VIRULENCE) and 42 (BIOGENESIS OF CELLULAR COMPONENTS) FunCat classes. In this work we did not explore the effects on ensembles performance of methods that selects subsets of base learners. Indeed, in this preliminary test, we were mainly interested in the evaluation of the potential benefits introduced by the use of data fusion methods in complex functional prediction tasks, but base learner selection approaches will be the object of future investigations.

Despite the expected negative effects of poor performing base learners on the ensemble performance, the ensemble systems are surprisingly robust as demonstrated by the results obtained in the test prediction of the class 34 (INTERACTION WITH THE ENVIRONMENT). In this classification task, despite the presence of 4 out of 6 base learners with accuracy around 20% or less, and the remaining learners with an F-measure of 0.3023 and 0.3544 (L_1 and L_2, respectively), the decision templates based ensemble system outperforms the best base learner by about 10%. This could be interpreted as the ability of the different data sources to provide diverse pieces of information.

The proteins involved in the interaction with the environment are characterized by very peculiar chemical and physical properties, as in the case of cell membrane spanning proteins (expected to be over represented in this functional class), which are composed by hydrophilic and hydrophobic alternate regions, making them easily detectable by protein-domain detection methods and sharing features making them the objective of evolutionary pressures easily detectable by local alignment methods. The ability of ensemble methods to correctly predict this FunCat functional class thus comes as a little surprise, being the aforementioned source of information well represented by 3 out of 6 data sources (the Protein domain binary, Protein domain Log-E and Pairwise similarity, Table 1).

It should be also noted that the realization of the complex pathways enabling the cell to correctly interact with its environment requires the interaction between a high number of proteins, making the proteins interaction data sources potentially informative. This indicates that in some cases the poor performance (either of the base learners or of the ensemble systems) could be explained by the absence of datasets containing relevant information with respect to the specific functional prediction task. We thus plan to extend the number of the datasets to be included in the future analysis.

6 Conclusions

In this work we investigated the impact on yeast genes functional classification performance. Our experiments consisted in the integration (by mean of a simple intersection procedure) of 6 different data sources and in the training (using standard tuning protocols) of 6 SVMs. We then tested linear weighted average, logarithmic weighted average and the decision template techniques to combine the output of the 6 base learners and we evaluated the performance of the single learners and of the ensemble systems. The aforementioned protocol was repeated (in separated binary classification tasks) for each FunCat functional class performed on the common set of genes shared by all the data sources to be integrated.

Our experiments demonstrated the potential benefits introduced by the use of ensemble-based prediction systems in functional classification of genes. The ensemble systems were able to outperform the best performing base learner in 8 out of 15 classification tasks. Considering that poor performance of base learners is functional class specific, by appropriately combining subsets of base learners for each specific FunCat class we could, in principle, improve the performances of SVM ensembles. In conclusion, the results obtained with simple combination strategies show that heterogeneous data integration through ensemble methods represents a valuable research line in gene function prediction indexgene function prediction.

Acknowledgements. The authors gratefully acknowledge partial support by the PASCAL2 Network of Excellence under EC grant no. 216886. This publication only reflects the authors' views.

References

1. Altschul, S.F., Gish, W., Miller, W., Myers, E.W., Lipman, D.J.: Basic local alignment search tool. J. Molecular Biol. 215(3), 403–410 (1990)
2. Finn, R.D., Tate, J., Mistry, J., Coggill, P.C., Sammut, S.J., Hotz, H.R., Ceric, G., Forslund, K., Eddy, S.R., Sonnhammer, E.L., Bateman, A.: The Pfam protein families database. Nucleic Acids Res. 36(database issue), 281–288 (2008)
3. Gasch, P., Spellman, P.T., Kao, C.M., Carmel-Harel, O., Eisen, M.B., Storz, G., Botstein, D., Brown, P.O.: Genomic expression programs in the response of yeast cells to environmental changes. Molecular Biol. Cell 11(12), 4241–4257 (2000)
4. Guan, Y., Myers, C., Hess, D., Barutcuoglu, Z., Caudy, A., Troyanskaya, O.: Predicting gene function in a hierarchical context with an ensemble of classifiers. Genome Biol. 9(S1) (2008)
5. Kuncheva, L.: Combining Pattern Classifiers: Methods and Algorithms. John Wiley & Sons, Hoboken (2004)
6. Kuncheva, L., Whitaker, C.: Measures of diversity in classifier ensembles. Machine Learning 51(2), 181–207 (2003)
7. Kuncheva, L., Bezdek, J., Duin, R.: Decision templates for multiple classifier fusion: an experimental comparison. Patt. Recogn. 34(2), 299–314 (2001)
8. Lanckriet, G., De Bie, T., Cristianini, N., Jordan, M., Noble, W.: A statistical framework for genomic data fusion. Bioinformatics 20(16), 2626–2635 (2004)
9. Pavlidis, P., Weston, J., Cai, J., Noble, W.: Learning gene functional classification from multiple data. J. Comp. Biol. 9(2), 401–411 (2002)
10. Platt, J.: Probabilistic outputs for support vector machines and comparisons to regularized likelihood methods. In: Smola, A., Bartlett, P., Schölkopf, B., Schuurmans, D. (eds.) Advances in Large Margin Classifiers, pp. 61–74. MIT Press, Cambridge (1999)
11. Ruepp, A., Zollner, A., Maier, D., Albermann, K., Hani, J., Mokrejs, M., Tetko, I., Guldener, U., Mannhaupt, G., Munsterkotter, M., Mewes, H.: The FunCat, a functional annotation scheme for systematic classification of proteins from whole genomes. Nucleic Acids Res. 32(18), 5539–5545 (2004)

12. Roli, F., Kittler, J., Windeatt, T. (eds.): MCS 2004. LNCS, vol. 3077. Springer, Heidelberg (2004)
13. Smith, T., Waterman, M.: Identification of common molecular subsequences. J. Molecular Biol. 147(1), 195–197 (1981)
14. Spellman, P., Sherlock, G., Zhang, M.Q., Iyer, V.R., Anders, K., Eisen, M.B., Brown, P.O., Botstein, D., Futcher, B.: Comprehensive identification of cell cycle-regulated genes of the yeast saccharomices cerevisiae by microarray hybridization. Molecular Biol. Cell 9(12), 3273–3297 (1998)
15. Stark, C., Breitkreutz, B., Reguly, T., Boucher, L., Breitkreutz, A., Tyers, M.: BioGRID: a general repository for interaction datasets. Nucleic Acids Res. 34(database issue), D535–D539 (2006)
16. Troyanskaya, O.G., Dolinski, K., Owen, A.B., Altman, R.B., Botstein, D.: A Bayesian framework for combining heterogeneous data sources for gene function prediction (in saccharomices cerevisiae). Proc. Natl. Acad. Sci. USA 100(14), 8348–8353 (2003)
17. Vapnik, V.N.: The Nature of Statistical Learning Theory. Springer, New York (1995)
18. von Mering, C., Huynen, M., Jaeggi, D., Schmidt, S., Bork, P., Snel, B.: STRING: a database of predicted functional associations between proteins. Nucleic Acids Res. 31(1), 258–261 (2003)

Partitioner Trees for Classification: A New Ensemble Method

Georg Krempl and Vera Hofer

Abstract. Two major types of ensemble methods exist: while the first type is based on the idea of varying the training data (Boosting, Bagging), the second type tries to exploit information on each classifier's area of expertise (Grading, Delegating, Arbitrating). This paper presents a new ensemble method called partitioner trees that combines both approaches. Information on misclassifications is used to train meta classifiers called partitioners and to split training data into disjoint subsets. On these subsets succeeding specialised classifiers are constructed. This process yields a binary decision tree with partitioners on the inner nodes to perform the splitting and specialised local classifiers on the leaves for final classification. Partitioner trees are compared to five other ensemble methods in experiments on four different datasets. The results show that on large datasets partitioner trees have a similar classification accuracy as AdaBoost , and are superior to those of other meta classifier based ensemble methods. In addition, partitioner trees outperform most other ensemble methods in regard to training time and allow for adaptively tuning parameters.

Keywords: multiple classifier systems, decision tree, error-based training data variation, referee, voting with meta classifiers, large datasets.

1 Introduction

Classifier ensembles are based on the idea of combining error diverse classifiers. Partridge and Yates [15] provide definitions of error diversity and discuss approaches to create error diverse neural networks [9] . They conclude that the most promising approaches are to vary the training data and the network architecture of a neural network, while varying the initial weights and the number of nodes is least useful.

Georg Krempl · Vera Hofer
Department of Statistics and Operations Research, University of Graz, Austria
e-mail: {georg.krempl, vera.hofer}@uni-graz.at

O. Okun & G. Valentini (Eds.): Appli. of Supervised & Unsuper. Ensemble Meth., SCI 245, pp. 93–112.
springerlink.com © Springer-Verlag Berlin Heidelberg 2009

The approach of varying training data randomly is used by *Bootstrap aggregation*, or *Bagging*, a method proposed by Breiman [2]. Different subsets of the training data are created by randomly selecting instances from the original data. Since this is done with replacement, an instance present in the original training data might be present once or more often in a new training set, or might not be present at all. Then each classifier is trained on a different subset. Finally, new instances are classified by combining the prediction of all classifiers by majority voting.

Instead of leaving the variation of training data to chance, *Boosting* techniques use systematic variation. A very famous and powerful representative of this technique initially proposed by Schapire [12] is *AdaBoost* presented by Freund and Schapire [6]. Here each classifier focuses on the instances that his predecessors misclassified. This is done by assigning misclassified instances more weight within the dataset. New instances are then classified by combining the predictions of *all* classifiers by weighted majority voting. Due to weighting the data during the training process all but the first classifier are trained on data that systematically differ in their composition from the new data on which the classifiers are used later.

A more complex way of combining classifiers is to use meta classifiers . Chen and Stolfo [4] proposed such techniques and suggested *arbiters* and *combiners*. Initially, several error-diverse base classifiers are trained. In their first approach instances on which predictions are ambiguous are used to train a so-called *arbiter*. This arbiter then decides which base classifier to use for the final decision. In their second approach new features are derived from the base classifiers' predictions. On this derived features a *combiner* is trained to finally predict the correct class. Thus, the final decision can differ from all of the base classifiers' decisions. Another approach that uses such information about previous misclassification as input for a succeeding classifier is *Stacked Generalisation*, or *Stacking*, proposed by Wolpert [14]. In this approach the training data is first split into two disjoint subsets, then several base learners are trained on one subset and tested on the other. Their predictions are used as inputs for a second level of classifiers which are trained to predict the correct value of the response variable.

Combinations of the idea of *Boosting* and using meta classifiers to assess each classifier's expertise on a new instance are also discussed in literature. On the one hand, uncertain instances can be delegated to other classifiers based on the base classifier's own confidence in its predictions. Such a method of *Delegating* is suggested by Ferri et al. [5], where succeeding classifiers are solely trained on instances on which its predecessors where uncertain.

On the other hand, an external classifier can be trained to assess the quality of a classifier's prediction. Such an approach is *Arbitrating* between classifiers proposed independently by Ortega and by Koppel and Argamon [8]. They propose training a *referee* for each base classifier which estimates its reliability on a new instance. The estimates of all referees are used to arbitrate between the base classifiers and to select one classifier whose prediction is finally used.

Another approach that uses an external classifier is *grading*, proposed by Seewald and Fürnkranz [13]. Instead of selecting only the classifier with the highest reliability they combine the prediction of all trustworth classifiers. This combination of the predictions of a subset of base classifiers is done either by voting or by weighting these predictions by their confidence estimates.

The subsequent sections present a new approach using a referee as a *partitioner* to split the data into two disjoint subsets depending on the referee's prediction of a base classifier's reliability. Thus, the subsets are assumed to be more diverse against each other in respect to the base classifier than the original dataset in the sense that one subset contains the instances which are classified correctly by the base classifier , whereas the other subset comprises the instances for which the base classifier fails. Instead of keeping the initial base classifier for instances where it succeeded, the *preliminary* base classifier is discarded. Thus, on *both* subsets a new specialised local classifier must be trained. New instances are then assigned by the partitioner to one of the two new *local classifiers*. By iterating this approach a *partitioner tree* with *partitioners* on the inner nodes and *local classifiers* on the leaves is created.

The rest of this chapter is structured as follows. The next section provides a discussion of technical details and questions imposed by the construction and application of such partitioner trees. The section begins with the presentation of an algorithm for constructing and applying partitioner trees and continues with a discussion of its background. Further considerations deal with details on parameter settings and finally an analysis of the computational complexity is provided. In the third section, questions on the performance of the algorithm are addressed by an experimental comparison with other ensemble methods. Finally, the results are discussed and a conclusion with outlook to future research is given.

2 Partitioner Trees

2.1 Algorithm

Partitioner trees construct a decision tree by means of meta classifiers that predict the performance of a base classifier , called preliminary classifier. According to this idea, the data is split into two disjoint subsets that differ in regard to their difficulty of classification using the preliminary classifier. Thus, the classification results of a preliminary classifier are used to train a partitioner. If the partitioner predicts that the preliminary classifier is correct, an instance is assigned to the left node, otherwise to the right one. Hence, any partitioner collects the instances that are expected to be easily classified by the preliminary classifier on the left hand side and the more difficult instances on the right hand side. Once a split has been designed, the preliminary classifier is no longer necessary and therefore dismissed. Then the process is restarted on each node by creating new specialised preliminary classifiers on the assigned training instances. Thus, the training data is systematically varied by error-based splitting , i.e. by splitting the data depending on the classification error of the preliminary classifier.

Splitting by the partitioner results in a usually unbalanced binary tree. On the inner nodes only meta classifiers , the partitioners, exist, whereas on the leaves the local base classifiers are applied to finally predict the class membership of an instance. Going deeper into the tree the degree of specialisation increases. Since more difficult cases are separated from others, it is possible to apply appropriate local base classifiers on the leaves.

A tree of classifiers is constructed by iterating the process of

1. creating a preliminary classifier,
2. testing and dismissing this preliminary classifier,
3. using information on the accuracies of the preliminary classifier to train a meta classifier (a so-called partitioner),
4. using this partitioner to split the data into two disjoint subsets,
5. repeating this process by training a preliminary classifier on each subset,
6. deciding for each branch whether to continue with partitioning and branching or to stop by using the preliminary classifier as local classifier.

New instances are then predicted by

1. using the (first) partitioner to decide which branch to follow,
2. using the next partitioner on this branch to descend further in the tree and iterating this process until
3. a leaf is reached and its local classifier is used to predict the instance.

Remaining questions concern the construction phase on the one hand and the classification phase, on the other hand: each splitting step involves finding two classifiers, whereby only the partitioner is important for the assignment of an instance to the final local classifier. The question is on which dataset the two classifiers should be trained. A further question is how to apply the partitioner to a specific instance. The construction phase and the classification phase are addressed below.

2.1.1 Construction Phase

The data available for training are (x_i, y_i), where $x_i = (x_{i1}, \ldots, x_{ip}) \in X$, and $y_i \in \{-1, 1\}$ indicating class membership, and $i = 1, \ldots, n$. Presently, only the binary classification problem is considered. X is an appropriate feature space. The algorithm does not require the data to be quantitative.

Initially, a preliminary classifier $G(x)$ is designed on the training data (x_i, y_i), and also applied to (x_i, y_i) to predict the class membership of the data by resubstitution

$$\hat{y}_i = G(x_i).$$

From this prediction, a new dichotom response variable, ρ, is derived

$$\rho_i = \begin{cases} 1 & \ldots \quad \hat{y}_i = y_i \\ -1 & \ldots \quad \text{otherwise} \end{cases}$$

Hence, ρ measures whether the prediction of G was correct or not. This is shown in the left column of Fig. 1.

Now a new classifier, $M(x)$, referred to as partitioner, is trained on the entire dataset (x_i, y_i) to predict the value of ρ. The predicted values $\hat{\rho}$ are then used to split the training data into two disjoint subsets C and D such that

$$C = \{(x_i^C, y_i^C)\} = \{(x_i, y_i) \mid \hat{\rho} = 1\} \qquad D = \{(x_i^D, y_i^D)\} = \{(x_i, y_i) \mid \hat{\rho} = -1\}$$

The first subset contains all instances where the partitioner predicts that a base classifier (preliminary classifier) is correct. This process is shown in the right column of Fig. 1.

If sufficient instances are in each of the two subsets, the process can be iterated. In this case, a new preliminary learner is trained and applied on each of the subsets,

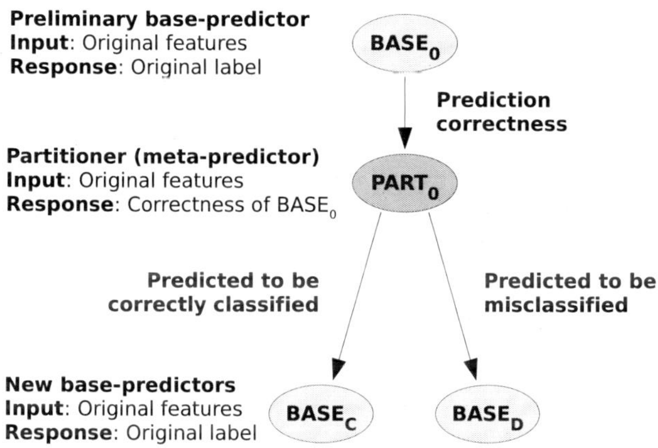

Fig. 1 Creation of the first partitioner

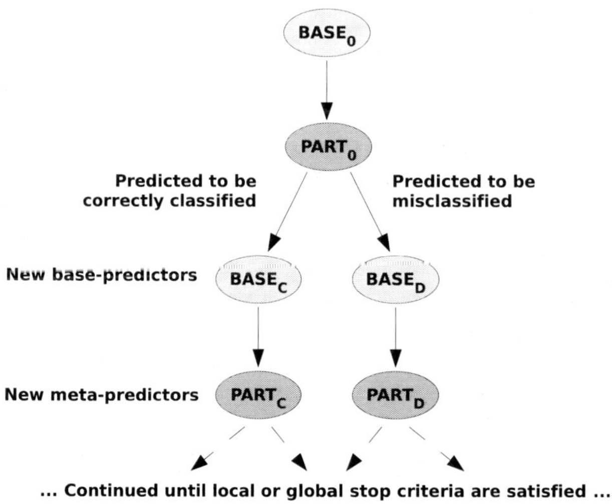

Fig. 2 Partitioner tree construction

C and D, to obtain new partitioners $M_C(x)$ and $M_D(x)$. If one subset becomes too small, there are two possible ways to proceed. First, the partitioner can be modified to create subsets more commensurated in size. Second, the process can be finalised by learning a local classifier $L(x)$ on the complete training set. In the latter case both subsets C and D are merged into a single training set. However, even the preliminary learner might be used as a local classifier to reduce the training time. Currently, the problem of sample size in a node is addressed by training a local classifier on the data in this specific node.

The iteration on the second level of the partitioner tree is depicted in Fig. 2. In the case of a dichotom response variable, the resulting tree will be a binary tree as shown in Fig. 3. This tree will typically not be balanced, since the right side with more difficult instances will require more layers of classifiers than the left side with easier instances.

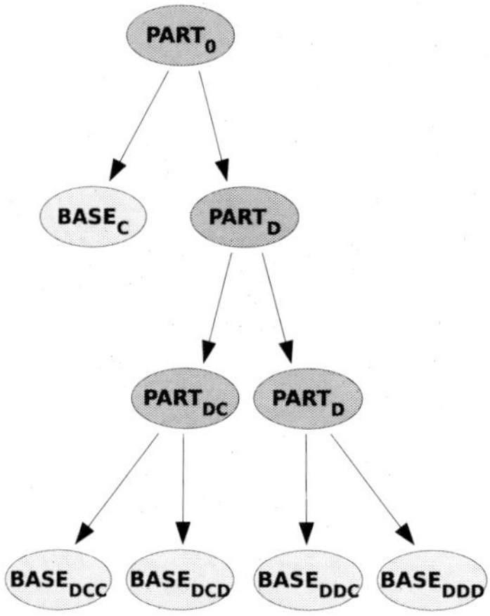

Fig. 3 Finished partitioner tree of tree depth 3

2.1.2 Application to New Data

New instances x_i^N are first assigned to a local classifier on a leaf of the partitioner tree which is then used to predict their class membership. The assignment starts at the first partitioner which is connected to two other nodes in the tree. Differently to the construction phase, not the value of $\hat{\rho}_i = M(x_i^N)$, but the probability estimation $\hat{p}_i$ of the first partitioner $p_i = P(\rho_i = 1)$ is used to select one of the two succeeding nodes. This succeeding node is either an inner node with a partitioner. In this case the process is repeated by using this local partitioner to estimate a new value of $\hat{p}_i$ and the algorithm descends to the corresponding node. Or the node is a leaf with a local classifier. Then the local classifier $L(x)$ is used to predict the final class membership of the new instance. If $\hat{p}_i = 0.5$ for a split into the left node or the right node, both nodes are selected and the final results is derived from weighing the results of both branches. More details on this problem are given below in Sect. 2.2.1.

2.2 Discussion

The algorithm described uses three types of classifiers: the *preliminary classifiers* $G(x)$ which are only used to train the *partitioners* $M(x)$ that predict the performance of the preliminary classifiers, and the *local base classifiers* $L(x)$ on the leaves used for final classification . Whereas the local base classifiers should exhibit high accuracy, the partitioners are assumed to make consistent decisions. However, the size of the tree will be smaller for more accurate preliminary classifiers. Furthermore, the partitioner tree does not require highly accurate preliminary classifiers, since they are dismissed later. This means that misclassifications by preliminary base classifiers at the beginning can be compensated by later classifiers. Therefore it is possible to start with fast and simple classifiers and increase the degree of sophistication with the tree depth. The additional computational costs of more sophisticated classifiers in deeper layers of the tree are compensated by the smaller number of instances that will arrive in those layers. Thus the speed of the ensemble should not be significantly affected.

The dismission of the preliminary classifier might need further justification, since the following simpler approach would also be possible: What happens if the preliminary classifier is kept for later classifications and, like before, the meta classifier is trained to predict its misclassifications? Then the meta classifier can be applied on new data, maybe even before the preliminary classifier, to predict misclassifications of the preliminary classifier and to filter those instances out, which are likely to be misclassified. In contrast to a partitioner tree , the instances are filtered out *after* the classifier used for final prediction has been trained on them. This raises an important question: are the remaining instances, which are *predicted to be correctly classified*, *representative* for the instances, which the preliminary classifier *classifies really correctly*? The answer to this question would only be yes, if the meta classifier would not make any wrong predictions. In such a case, a perfect ensemble could be created by using the preliminary classifiers' prediction or its inverse, depending on the judgement of the meta classifier . In practice, however, this is very unlikely.

Partitioner trees relax this problem of accurately predicting misclassifications to the easier problem of *splitting* data *systematically* and *consistently* into two subsets. Dismissing the preliminary classifier means that misclassifications of the partitioner are irrelevant as long as they occur consistently during training and on new data: During training and application the same partitioner assigns the instances to a local classifier. Thus, the instances assigned during training are expected to be representative for the instances assigned to the local classifier when the ensemble is applied to new data. This higher representativeness compared to the simpler approach justifies the additional complexity, since it might yield more accurate classifiers and thus a lower *test* error.

2.2.1 Parameters

The algorithm requires two kinds of parameters to be set: First, the types of classifiers used as partitioners and local classifiers. Second, the stop criteria for branching.

Choice of Classifiers Types

Base classifiers are used for two different tasks in the algorithm, for partitioning (actually predicting misclassifications) on the one hand and for classifying the data on the other hand. For the first task, *consistency* in predictions is crucial, while accuracy is not. For the second task, which concerns preliminary classifiers and final local classifiers, *accuracy* is the primary issue. Although it would be possible to use a different types of classifier for each task, this paper focuses on partitioner trees using the same type of classifier for both tasks.

While all kinds of classification techniques including ensembles of classifiers themselves can be considered to be used as base classifiers, the choice has to be made with regard to three aspects: First, the time needed to train and to apply the classifier. Second, the capability of a classifier to estimate class probabilities and third the levels of the response variable. Further necessary considerations are the classifiers' complexity and its dependence on the full training data. In [7] the choice of this parameter was discussed and it was shown, that among others decision trees are a good choice as classification method for local classifiers and partitioners.

Tree Depth and Stop Criteria

The tree depth d, defined as the number of levels of partitioners in the tree, can either be fixed in advance or determined indirectly. In the first case, the tree is branched until the defined depth is reached. In the second case, the definition of additional stop criteria is required. They can be based on a *minimal subset size*, on a *maximum time limit* or on the *accuracy* of the predictors in a branch. While the first and second are obvious, two approaches exist for using the accuracy: either a maximum accuracy is reached, for instance, when a classifier is perfect (this can happen if in one subset only instances of one class are left or if the instances there are easily separable), or the accumulated performance of both branches drops below the performance of its preceding branch.

These four parameters have to be chosen with regard to the size of the training set, the number of features in the data, the base classifiers used and the available training time. Since the training data is split into smaller subsets, in general, 2^d -times more training data is needed than for a single classifier[1]. Using other stop criteria than a fixed tree depth leads to a typically unbalanced, but better adopted tree, since the tree focuses on the more difficult instances. Using a decrease in the prediction accuracy as stop criterion suffices, as will be shown in Sect. 3.2. However, a reasonable minimum subset size might be defined, depending on the dataset size. Nevertheless training a partitioner tree can be done without explicit parameter tuning by simply requiring a decreasing error in each branch.

[1] This factor is a lower bound, provided that the data is splitted always into two equally sized subsets at each node. In practice, even more training data might be required. The imbalance of the binary tree increases this effect, since the deeper right side will typically get less than 50 % of all instances assigned. This is due to the fact that the preliminary classifiers will achieve an accuracy of at least 50 %.

Cut Threshold

In Sect. 2.1.2 a partitioner assigns a new instance based on its estimated conditional probability $\hat{p}_i$ to belong to the left side. If $\hat{p}_i \geq 0.5$, an instance is assigned to the successing node on the left side. If $\hat{p}_i \leq 0.5$, an instance is assigned to the successing node on the right side. As discussed above, this can lead to cases, where an instance is assigned to several leaves, whose predictions are combined using the instances' estimated probabilities of belonging to each of the leaves. It should be noted that this only occurs if $\hat{p}_i$ is exactly equal to 0.5 in at least one partitioner. The justification in such cases is obvious: Since both succeeding branches have exactly the same estimated conditional probability, both should be treated equally. However, it might be useful to consider both succeeding branches also in cases, where $\hat{p}_i$ is not exactly but close to 0.5, thus in cases, where both succeeding nodes are similarly likely. Nevertheless the more likely branch should be weighted higher. Since its higher estimated conditional probability will increase the relative weight of its succeeding branch, this is guaranteed.

The question which arises is, up to which point a succeeding branch should be considered. Should both branches be also considered in extreme cases, where the estimated conditional probability of one succeeding branch is nearly zero and of the other is nearly one? One option is to rely on the self-regulating mechanism of the weights. If one estimated conditional probability is very low in a branch, the estimated probability of the resulting leaf and thus its weight will be automatically low. In extreme cases, branches might get zero weight, if in the branch the conditional probability is zero.

Another option is to introduce a threshold for the probabilities to be considered. If for a new instance the estimated conditional probability of a branch is below this threshold, this succeeding branch is not considered and treated as it was cut out of the model. Such a *cut threshold* thus guarantees that only branches with a certain minimum likeliness are considered.

Technically this *cut threshold* can be applied either to each of the conditional probabilities estimated by the partitioners in the tree or to the resulting estimated probabilities of the leaves. In this case, leaves whose estimated probability is below the threshold, are not considered. While both approaches are possible, only the first one is considered in this paper. Thus before assigning a new instance to one branch, the conditional probability $\hat{p}_i$ assigned by partitioner M_i is adjusted.

Three cases can be distinguished. First, $\hat{p}_i$ is nearly 0.5. In this case, the instance will be assigned to both succeeding branches, but each branch will be weighted with its estimated probability. Second, $\hat{p}_i$ is below the cut threshold τ, in which case the left succeeding branch is ignored and the instance is assigned to the right branch. Thus the succeeding branch on the right side gets all the weight. Finally, the opposite is done, when $\hat{p}_i$ is greater than $1-\tau$. In this case the succeeding branch on the right side is very unlikely, thus ignored. This adjustment of the conditional probability based on the cut threshold can be written as

$$\hat{p}_i = \begin{cases} \hat{p}_i & \dots \quad \tau \leq \hat{p}_i \leq 1-\tau \\ 0 & \dots \quad \hat{p}_i < \tau \\ 1 & \dots \quad \hat{p}_i > 1-\tau \end{cases}$$

The cut threshold τ can thus take values between 0 and 0.5. For $\tau = 0$, the weighted predictions of all leaves are combined for the final classification of a new instance. For $\tau = 0.5$, only the best branch is considered, except if two branches have exactly the same predicted probability, in which case both are considered. Figure 4 illustrates the effect of τ on the voting weights in the tree.

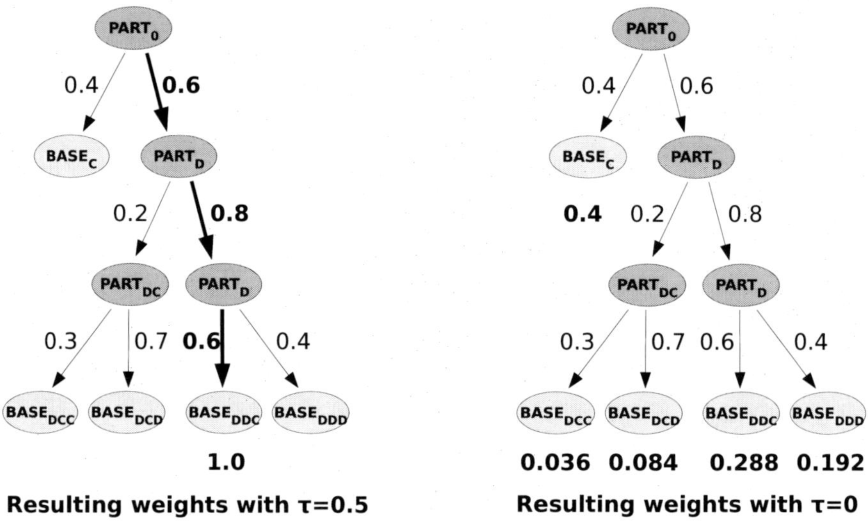

Fig. 4 Effect of the cut threshold τ on the voting weights

2.2.2 Computational Complexity

Training

In the following, d denotes the tree depth, i.e. the maximum number of arcs between the topmost node and a node at the bottom of the tree; n denotes the number of training instances available, i.e. the size of the training set; i denotes the index of a level, where the topmost level is denoted as level 0. Thus the index of the lowermost level is identical to the tree depth d.

Assuming a perfectly balanced binary tree, each level holds 2^i nodes. This means that a (balanced) partitioner tree of tree depth d holds $2^{(d+1)} - 1$ nodes, where on all but the nodes on the last layer only partitioners occur. Such a tree would hold $2^d - 1$ partitioners and 2^d local classifiers. For training such a tree, $2^{(d+1)} - 1$ (preliminary and local) base classifiers and $2^d - 1$ partitioners must be trained, altogether $3 * 2^d - 2$ classifiers. Considering only the number of classifiers would lead to a time and space complexity of $O(2^d)$ for training such a partitioner tree . This is, however, inadequate, as discussed below.

The training time of each single classifier for a given parameter setting mainly depends on the number of its training instances. This number is for all the classifiers on one level limited by the total number of training instances. For a perfectly balanced partitioner tree , on each level the total number of training instances used by classifiers of this level is equal to n. For unbalanced partitioner trees , which are more likely to occur, this number will even decrease with tree depth, since more and more classifiers become leaves and their training instances are not used in deeper levels. Mathematically, the number n_i of training instances used in a level i can be written as $n_i = \sum_{k \in L_i} n_k$, where L_i is the set of classifiers on level i and n_k is the number of instances used by classifier k for training.

If the computational cost of training a classifier C_k would linearly depend on n_k, the computational cost of training all classifiers of level i would be bounded by $O(n_i)$. Thus the cost of training the ensemble L_i of classifiers on level i would be equal to the cost of training a single classifier on all n_i instances. However, if the *computational complexity* of training a classifier *increases*, the *relative cost* of training the ensemble compared to a single classifier *decreases*. This is due to the fact that the training data was split into 2^i subsets, as discussed in Sect. 2.2.1. This means that the computational time complexity of training the ensemble has an upper bound of $Q(nd)$, where $Q(\cdot)$ denotes the computational time complexity of training a single classifier.

It should be noted that this asymptotic bound does not consider the positive effect discussed above, namely the decrease of n_i for deeper levels in unbalanced partitioner trees . Furthermore it should be remarked, though, that this training process can easily be parallelised, since all classifiers of one level can be trained parallely. Thus, the training time complexity could be even further reduced on a multiprocessor grid, unlimited parallel processing capacities presumed. To summarise it can be concluded, that the training time of a partitioner tree should be lower than the training time of other serially constructed ensembles, which do not split the training data, for example AdaBoost[2].

Prediction

Assuming a cut threshold of $\tau = 0.5$, the prediction of a new instance will involve at most $d+1$ classifiers. Thus prediction time complexity is bounded by $O(d)$, which is equivalent to the asymptotic time complexity of an AdaBoost ensemble with d classifiers. Computational cost will increase with a decrease of τ and reach $O(2^d)$ for $\tau = 0$.

However one can conclude, that for $\tau = 0.5$, the computational time complexity of training and applying a partitioner tree with tree depth d is not bigger than the computational time complexity of training and applying an AdaBoost ensemble.

[2] AdaBoost with d classifiers will also have an asymptotic training time complexity of $Q(nd)$, but since the training data is not split into smaller subsets, it will not profit from the effects described above.

3 Experiments

The experiments with this new meta-algorithm should answer three main questions. First, how should the parameters discussed in the previous section be chosen? Second, does this new way of employing a meta classifier yield any benefit compared to other meta classifier approaches? This leads to the third question, namely if there are datasets on which this algorithm performs better than other approaches. If so, is it possible to determine the characteristics of those datasets?

A parameter which needs to be chosen is the *cut threshold*. To study the effect of this cut threshold on the performance, different parameter settings (50%, 40%, 25%, 10% and 0%) are compared. The choice of the other parameters was discussed in the conference paper presented at SUEMA 2008 ([7]). It was shown that decision trees serve as a suitable *classification method* for both local base classifiers and partitioners. Since this method yields good results without explicit parameter fine-tuning, it is used to compare different ensemble techniques. As discussed in this conference paper, an increase in the misclassification rate is used as *stop criterion*. This assures adaptive behaviour with regard to the training data but might cause overfitting. Approaches to avoid this overfitting are subject of current research.

3.1 Experimental Setup

To answer the questions posed above, an experimental comparison is made between the partitioner tree approach and other ensemble techniques. All experiments are done using the R environment for statistical computing [11]. All ensemble methods are coded directly in R and tested on identical hardware to guarantee an unbiased training time comparison. For this paper, the following ensemble methods are compared:

- Boostrap aggregation (Bagging , [2]), with different numbers of base classifiers (5, 8, 11, 20, 50, 100).
- Adaptive boosting (AdaBoost , [6]), with different numbers of base classifiers (5, 8, 11, 20, 50, 100).
- A simplified version of the arbitrating approach (Simplified Arbitrating, explained below), with different numbers of base classifiers (2, 5, 8, 11, 20).
- Grading (Grading, [13]), with different numbers of base classifiers (2, 5, 8, 11, 20).
- Delegating (Delegating, [5]), with different delegating proportions (40 %, 50% and 60 %).

Arbitrating was simplified by using the class probabilities predicted by the base classifiers as features for the meta classifiers . Unlike the method proposed in [8], additional information about the inner states of the base classifiers cannot be used by the meta classifiers . Thus, this simplified version might have a lower performance. However this simplification allows to compare also the effect of employing meta classifiers at different positions. This comparison would be biased if additional information was provided to the meta classifiers of only one ensemble method.

The delegating proportion was chosen as suggested by Ferri, Flach and Hernandez-Orallo to values near 50 %. All ensembles used decision trees as classifiers, which were provided by the R package *rpart* [3].

3.1.1 Datasets and Methodology

For a comparison with other ensemble methods, experiments on various datasets are required. These datasets should differ in their characteristics, i.e. in the number of instances and the type of attributes. Therefore the performance on four different datasets from the University of California at Irvine (UCI) Machine Learning Repository ([1]) is compared. These datasets are the adult dataset, the spambase dataset, the credit approval dataset and the sonar (mines versus rocks) dataset. Their characteristics are listed in Table 1. All datasets are binary classification classification!binary problems, even though the partitioner tree approach could with minor changes also be used with multiclass classification problems, but this would be beyond the scope of this work.

Table 1 Datasets characteristics

Dataset	Attribute Types	Instances	Attributes	Balance[a]
Adult	Numerical, Categorical	65122	14	24.08 %
Spambase	Numerical	4601	57	39.40 %
Credit Approval	Numerical, Categorical	690	15	44.49 %
Sonar	Numerical	208	60	46.63 %

[a] Percentage of positives in the dataset.

For the experiments a 5-fold-cross-validation was used and the performance on the test sets of each of the five runs was used for evaluation. As prediction quality performance measures, the area under the curve (AUC, [10]) and the overall misclassification rate (OMR) are calculated using the predicted class membership probabilities over all five test sets. In addition, the time needed for training the ensemble is reported. For each ensemble method the results of the best parameter setting with regards to the area under the curve are used for the comparison with the other methods.

To study the effect of the number of training instances all methods were also evaluated on reduced variants of the two biggest datasets (adult and spambase dataset). For these reduced variants only the first 50 %, 25 % or 10 % of the instances in the training set of each run were used for training. However, the test sets for each run were the same as for the unreduced variants.

3.2 *Results*

The experimental results indicate that the larger the number of training instances available, the better is the performance of the partitioner tree compared to all other

Table 2 Comparison of AUC-best methods

Area Under the Curve

	Adult				Spambase				Credit	Sonar
Method	100%	50%	25%	10%	100%	50%	25%	10%	100%	100%
Single	0.843	0.850	0.844	0.855	0.893	0.904	0.903	0.871	0.897	0.758
Part.Tree	0.896	0.895	0.895	0.859	0.969	0.948	0.931	0.881	0.889	0.767
Bagging	0.862	0.864	0.8655	0.871	0.941	0.949	0.949	0.952	0.927	0.859
Adaboost	0.902	0.907	0.905	0.903	0.987	0.983	0.977	0.969	0.915	0.916
Simp.Arbit.	0.739	0.734	0.741	0.750	0.899	0.879	0.874	0.863	0.844	0.710
Grading	0.858	0.862	0.864	0.867	0.942	0.946	0.944	0.947	0.919	0.843
Delegating	0.850	0.854	0.859	0.856	0.898	0.897	0.907	0.869	0.892	0.778

Overall Misclassification Rate

	Adults				Spambase				Credit	Sonar
Method	100%	50%	25%	10%	100%	50%	25%	10%	100%	100%
Single	0.217	0.211	0.216	0.160	0.103	0.111	0.107	0.138	0.154	0.293
Part.Tree	0.141	0.180	0.184	0.151	0.080	0.106	0.115	0.146	0.180	0.298
Bagging	0.212	0.207	0.196	0.156	0.094	0.094	0.096	0.104	0.139	0.255
Adaboost	0.184	0.183	0.182	0.171	0.046	0.050	0.064	0.075	0.144	0.168
Simp.Arbit.	0.154	0.153	0.156	0.151	0.095	0.109	0.119	0.140	0.167	0.284
Grading	0.216	0.203	0.207	0.156	0.095	0.095	0.100	0.107	0.141	0.269
Delegating	0.147	0.146	0.147	0.150	0.090	0.095	0.100	0.116	0.148	0.245

Training time (in seconds)[e]

	Adults				Spambase				Credit	Sonar
Method	100%	50%	25%	10%	100%	50%	25%	10%	100%	100%
Single	122.8	49.21	21.01	5.46	14.16	6.44	2.99	1.03	0.54	0.67
Part.Tree	913.6	442.5	90.40	21.75	65.82	38.50	15.95	6.36	4.96	4.37
Bagging	6150	4902	1830	255.2	1442	649.3	294.3	105.5	55.85	76.78
Adaboost	3701	1527	1096	120.9	1082	877.3	362.2	67.92	89.86	111.8
Simp.Arbit.	747.9	728.6	697.7	15.68	173.7	30.69	34.32	18.40	7.44	4.15
Grading	3795	2255	671.8	150.8	722.4	313.3	136.3	47.48	30.41	36.80

Parameter setting of AUC-best method

	Adults				Spambase				Credit	Sonar
Method	100%	50%	25%	10%	100%	50%	25%	10%	100%	100%
Part.Tree[a]	(0.50)	(0.00)	(0.00)	(0.40)	(0.00)	(0.10)	(0.00)	(0.00)	(0.10)	(0.00)
Part.Tree[d]	66.00	62.80	12.20	3.60	6.20	7.80	4.80	4.00	4.20	3.00
Bagging[b]	(50)	(100)	(100)	(50)	(100)	(100)	(100)	(100)	(100)	(100)
Adaboost[b]	(20)	(11)	(50)	(20)	(50)	(100)	(100)	(50)	(100)	(100)
Simp.Arbit.[b]	(2)	(2)	(11)	(2)	(5)	(2)	(5)	(8)	(5)	(2)
Grading[b]	(20)	(20)	(20)	(20)	(20)	(20)	(20)	(20)	(20)	(20)
Delegating[c]	(0.40)	(0.40)	(0.50)	(0.50)	(0.60)	(0.60)	(0.60)	(0.60)	(0.40)	(0.50)

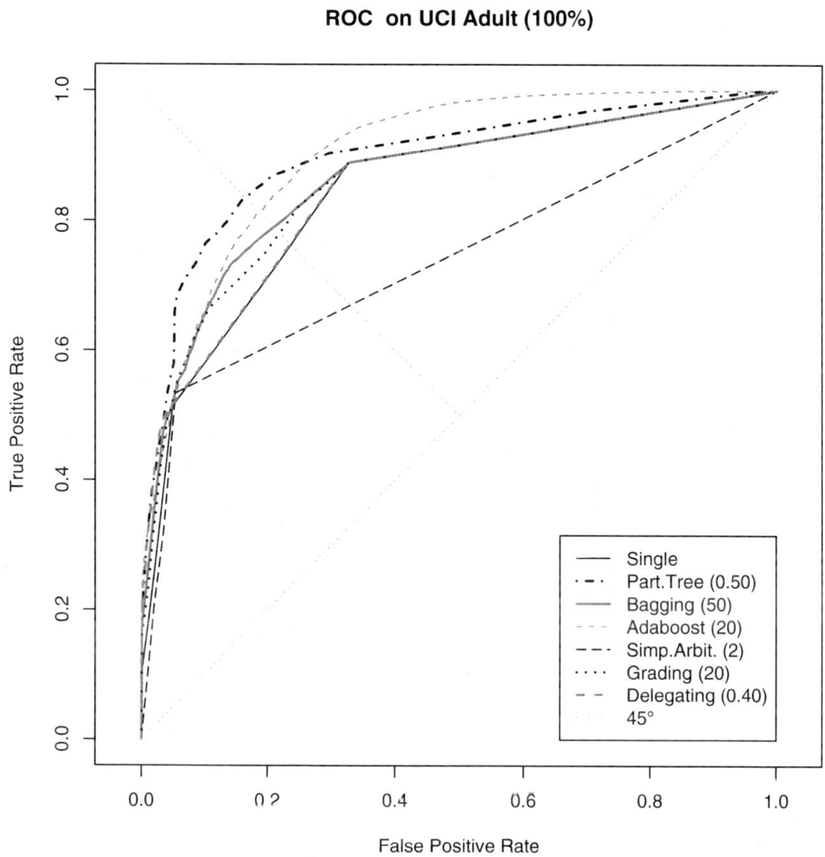

Fig. 5 Receiver Operating Characteristics on the UCI Adult dataset

methods. This is true for both the quality of prediction, measured in AUC or (Overall Misclassification Rate) OMR, and for the training time. While the partitioner tree method is outperformed on small datasets, the AdaBoost and the partitioner tree approach show the best prediction quality among all compared ensemble methods on large datasets. However the partitioner tree approach needs far less training time than AdaBoost . On the largest dataset (Adult), the partitioner tree reaches an AUC of 0.8957, which is 0.0059 less than the AUC of 0.9016 achieved by AdaBoost . In terms of OMR, the partitioner tree achieves the lowest misclassification rate of 0.1406 and thus outperforms AdaBoost (OMR of 0.1836) by 0.043. In terms of the training time the partitioner tree needed 913.62 seconds for training, thus only approximately one fourth of the training time of AdaBoost (3701.28 seconds). On the spambase dataset, AdaBoost was better in terms of the prediction accuracy (AUC of 0.9870 compared to 0.9687, OMR of 0.0459 compared to 0.0798), but again worse in terms of training time (1081.81 seconds compared to 65.82 seconds). For those

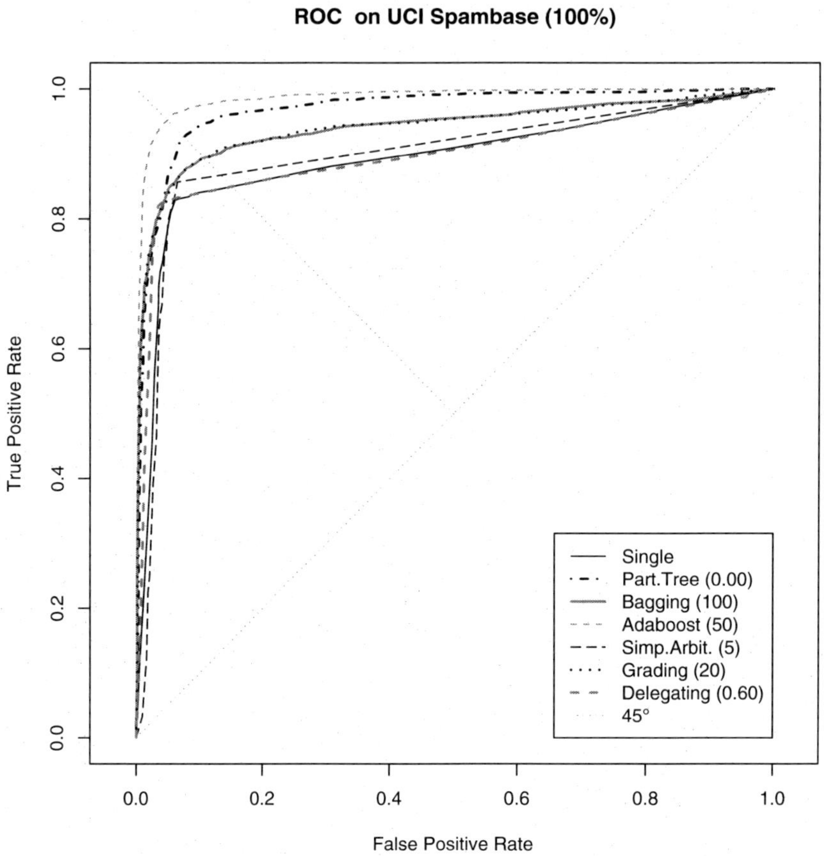

Fig. 6 Receiver Operating Characteristics on the UCI Spambase dataset

two datasets, all other ensemble methods yielded worse classification accuracy than the partitioner trees . Figure 5 shows the receiver operating characteristics (ROC) on the adult dataset, Fig. 6 shows the same for the spambase dataset. It should be remarked that on the large datasets AdaBoost did yield its best results with less than the possible 100 classifiers, thus raising the number of classifiers further would not have increased its performance. Grading might have profitted from additional classifiers, but was due its need of a meta classifier per base classifier already as slow as AdaBoost with 50 instances. Therefore no additional experiments with a higher number of base classifiers were done.

On the datasets with a reduced number of training instances the performance of the partitioner tree dropped very fast compared to the other ensemble methods. On the two small datasets (Credit, Sonar) the partitioner tree did not yield a satisfying accuracy compared to the other methods. The detailed experimental results are shown in Table 2.

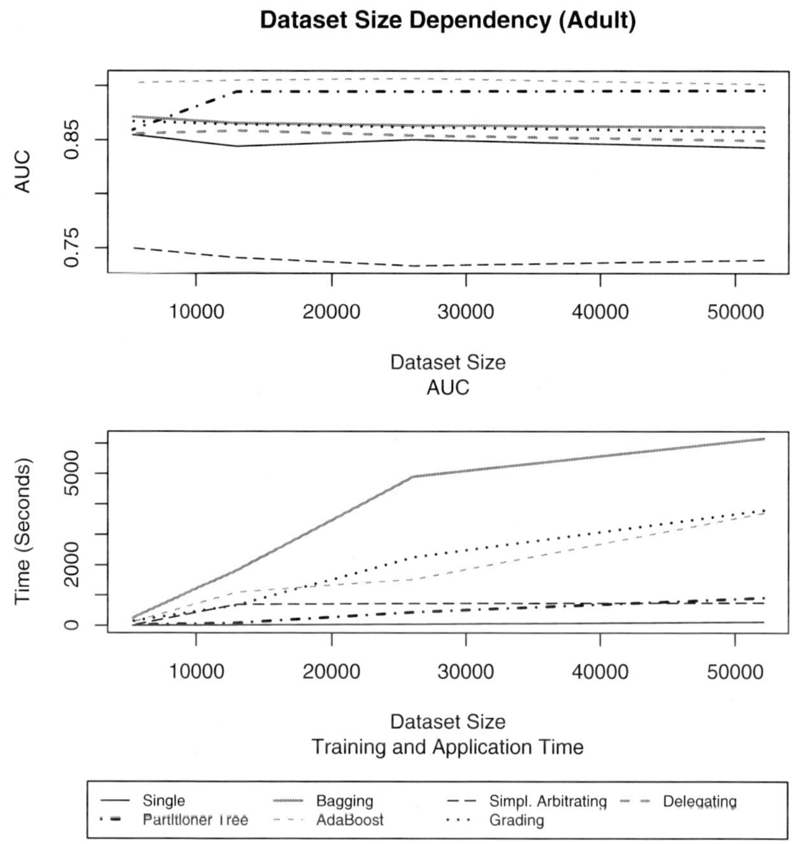

Fig. 7 Dependency on dataset size (Adult dataset)

In Table 2, the tuned parameters were: when marked with [a] - the cut threshold τ; when marked with [b] - the number of base classifiers; when marked with [c] - the delegating proportion; when marked with [d] - the observed average number of base classifiers in the model. When marked with [e], it meant that due to a programming error, the training time was not calculated for delegating.

3.2.1 Comparison to Other Meta Classifier Ensembles

The partitioner tree approach achieved better prediction accuracy than the other meta classifier-based ensemble methods on the two large datasets. On the adult and spambase dataset, the other best meta classifier ensemble was Grading with an AUC by 0.0379 lower on the adult and by 0.028 lower on the spambase dataset. The other best meta classifier-ensemble in terms of the OMR was Delegating with a misclassification rate worse by 0.68 % on the adult and worse by 1.06 % on the spambase

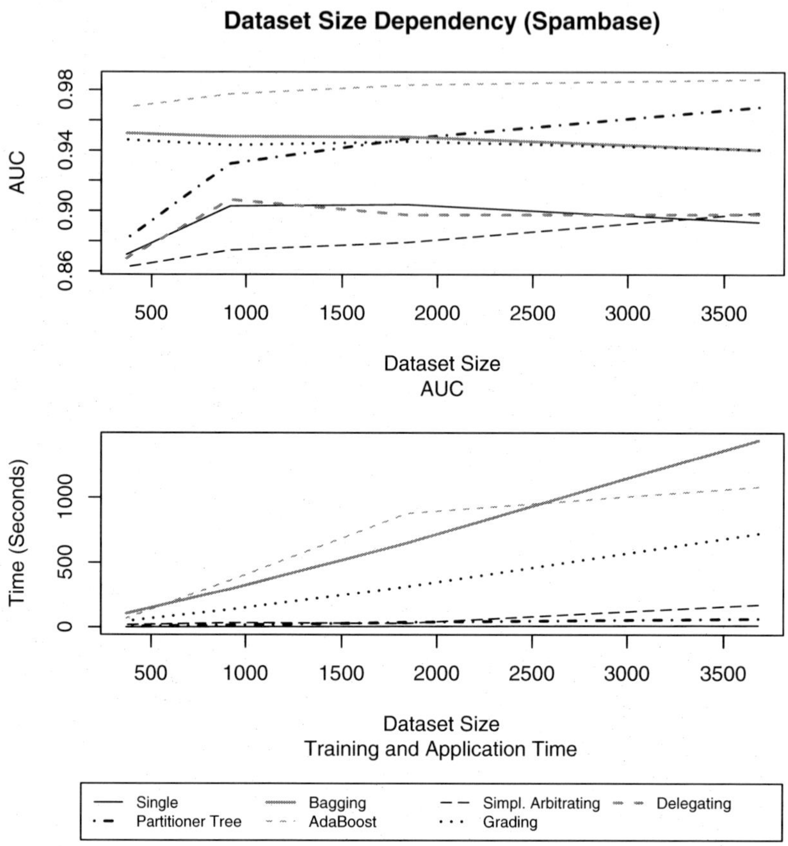

Fig. 8 Dependency on dataset size (Spambase dataset)

dataset. On the smaller datasets, these methods performed better than the partitioner tree approach but also worse than Bagging.

3.2.2 Dataset Size Dependency

The strong dependency on the dataset size is in accordance to the theoretical considerations in Sect. 2.2: since the training data is partitioned into two disjoint subsets in each inner node, the number of training instances available for a predictor decreases very fast. Therefore the algorithm requires a bigger number of training instances than other ensemble techniques if overfitting should be avoided. However, this behaviour is also the cause of a very positive effect: the very fast training speed. This correlation between the number of training instances available and the quality of prediction and the training time is shown in Fig. 8 for the spambase dataset and in Fig. 7 for the adult dataset.

This dependency also allows to determine large datasets as the domain for this partitioner tree . As a rule of thumb, the number of training instances should be at least greater than five thousand instances, depending on the dataset characteristics. Below this number of instances this ensemble method might suffer from overfitting and a lack of training instances in deeper nodes. To overcome this limitations, two strategies might be of interest for further research.

First, overfitting might be avoided by using a validation set several ways for obtaining such a subset during training from the training data and using it during the construction of the partitioner tree are possible and are currently under research.

Another strategy which might lead to a better performance on smaller datasets is the relaxation of the partitioning by the partitioners. Instead of strictly disjoint subsets smoother partitionings could be used, in a similar way as the cut threshold proposed for the application phase. Such a soft partitioning could be achieved by weighting instances differently in each subset instead of assigning them to only one subset. However, such an approach would also forfeit the fast training speed of the ensemble. Since the training time is often negligible on small datasets this could be still an interesting extension.

3.3 Conclusion

The new binary tree based ensemble method presented in this chapter has shown that it is a good ensemble method for classification when sufficiently training instances are available. Basic considerations anticipate the empirically proven good performance with regards to classification accuracy and speed on large datasets, where the method achieves a similar classification accuracy as AdaBoost but is at least four times faster in training. Compared to other meta classifier based ensemble methods, this method achieves a far better accuracy on large datasets in the experiments. Furthermore this method can be trained in an adaptive way without explicit parameter tuning, which leads to an additional training time advantage over other ensemble methods.

However the accuracy on smaller datasets was not competitive and methods to overcome this weakness will be subject of future research. One promising strategy is to prevent overfitting by specially adjusted methods. Another approach to further improve accuracy might be the use of more sophisticated methods on deeper levels in the tree. Finally it will be investigated whether the effects of dataset size observed with this algorithm can be generalised.

Although in this chapter only binary classification classification!binary problems were discussed, this algorithm could also be applied to multiclass classification or regression problems, which is subject of current research.

Acknowledgements. First, we would like to thank the referees and Oleg Okun and Giorgio Valentini for their helpful comments. We would also like to thank Roman Belavkin from Middlesex University for the interesting discussions on possible future extensions of this method. Finally, we would like to thank the maintainers of the University of California Machine Learning Repository [1] and the providers of the datasets used in this paper.

References

1. Asuncion, A., Newman, D.J.: UCI machine learning repository, http://www.ics.uci.edu/~mlearn/MLRepository.html
2. Breiman, L.: Bagging predictors. Tech. Rep. 421, Dept. Stat., Univ. California, Berkeley, CA (1994)
3. Breiman, L., Friedman, J.H., Olshen, R., Stone, C.J.: Classification and Regression Trees. Chapman & Hall, Boca Raton (1984)
4. Chan, P., Stolfo, S.J.: Learning arbiter and combiner trees from partitioned data for scaling machine learning. In: Fayyad, U.M., Uthurusamy, R. (eds.) Proc. 1st Int. Conf. Knowl. Discovery and Data Mining, Montreal, QB, pp. 39–44. AAAI Press, Menlo Park (1995)
5. Ferri, C., Flach, P.A., Hernandez-Orallo, J.: Delegating classifiers. In: Brodley, C.E. (ed.) Proc. 21st Int. Conf. Mach. Learn., Banff, AL. ACM, New York (2004)
6. Freund, Y., Schapire, R.E.: A decision-theoretic generalization of on-line learning and an application to boosting. In: Vitányi, P.M.B. (ed.) EuroCOLT 1995. LNCS, vol. 904, pp. 23–37. Springer, Heidelberg (1995)
7. Krempl, G., Hofer, V.: Partitioner trees: combining boosting and arbitrating. In: Okun, O., Valentini, G. (eds.) Proc. 2nd Workshop Supervised and Unsupervised Ensemble Methods and Their Appl., Patras, Greece (2007)
8. Ortega, J., Koppel, M., Argamon, S.: Arbitrating among competing classifiers using learned referees. Knowl. Inf. Syst. 3(4), 470–490 (2001)
9. Partridge, D., Yates, W.B.: Engineering multiversion neural-net systems. Neural Computation 8(4), 869–893 (1995)
10. Provost, F., Fawcett, T.: Analysis and visualization of classifier performance: comparison under imprecise class and cost distributions. In: Heckerman, D., Mannila, H., Pregibor, D. (eds.) Proc. 3rd Int. Conf. Knowl. Discovery and Data Mining, Newport Beach, CA, pp. 42–48. AAAI Press, Menlo Park (1997)
11. R Development Core Team: R: a Language and Environment for Statistical Computing (2006)
12. Schapire, R.: The strength of weak learnability. Mach. Learn. 5(2), 197–227 (1990)
13. Seewald, A.K., Fürnkranz, J.: An evaluation of grading classifiers. In: Hoffmann, F., Adams, N., Fisher, D., Guimarães, G., Hand, D.J. (eds.) IDA 2001. LNCS, vol. 2189, p. 115. Springer, Heidelberg (2001)
14. Wolpert, D.H.: Stacked generalization. Neural Networks 5(2), 241–259 (1992)
15. Yates, W.B., Partridge, D.: Use of methodological diversity to improve neural network generalisation. Neural Computing and Appl. 4(2), 114–128 (1996)

Disturbing Neighbors Diversity for Decision Forests

Jesús Maudes, Juan J. Rodríguez, and César García-Osorio

Abstract. Ensemble methods take their output from a set of base predictors. The ensemble accuracy depends on two factors: the base classifiers accuracy and their diversity (how different these base classifiers outputs are from each other). An approach for increasing the diversity of the base classifiers is presented in this paper. The method builds some new features to be added to the training dataset of the base classifier. Those new features are computed using a Nearest Neighbor (NN) classifier built from a few randomly selected instances. The NN classifier returns: (i) an indicator pointing the nearest neighbor and, (ii) the class this NN predicts for the instance. We tested this idea using decision trees as base classifiers . An experimental validation on 62 UCI datasets is provided for traditional ensemble methods, showing that ensemble accuracy and base classifiers diversity are usually improved.

Keywords: diversity, ensemble of decision trees, Kappa-Error movement diagrams.

1 Introduction

Ensembles are classifiers that combine predictions from other classifiers. The combined classifiers in an ensemble are called base classifiers. Some ensemble combination schemes have became popular and they have proved to be successful. Many of them use a set of base classifiers which are computed using the same algorithm.

Ensembles overall accuracy requires base classifiers not to predict wrongly the class of the same instances. They need to be diverse in order to complement each other. So, how can a set of base classifiers generated from the same algorithm provide those different outputs from the same inputs? Diversity has been achieved on ensembles using different strategies, most of them are based on modifying the training dataset of the base classifiers.

Jesús Maudes · Juan J. Rodríguez · César García-Osorio
Department of Ingenieria Civil, University of Burgos, Escuela Politécnica Superior
C/Francisco de Vitoria s/n 09006
e-mail: {jmaudes,jjrodriguez,cgosorio}@ubu.es

O. Okun & G. Valentini (Eds.): Appli. of Supervised & Unsuper. Ensemble Meth., SCI 245, pp. 113–133.
springerlink.com © Springer-Verlag Berlin Heidelberg 2009

In Bagging [4] diversity comes from randomly picking different instances for training each base classifier . The Random Subspaces method [12] chooses different subsets of attributes for training each base classifier. Random Forest [5] is a variant of Bagging using Decision Trees as base classifiers. In this type of random trees, the selection of the attribute for a decision node is done using only a random subset of the attributes.

Boosting [10] iteratively trains the base classifiers by modifying the weights of instances to train the current classifier. These new weights are computed from the training error of the previous base classifier, so each new base classifier becomes more specialized in instances that have been misclassified before.

Some of these methods are constrained to use a specific base classifier that provides the desired diversity (e.g. Random Trees), and some others get diversity in a way that cannot be used into other ensemble methods (e.g., Boosting re-weighting). One important advantage of our approach is that it can be applied to any base method within any ensemble technique. For example, in this work the experimental validation is focused mainly on ensembles of Decision Trees.

On the other hand, resampling in Bagging and random feature selection in Random Subspaces could be considered as ways of getting diversity that can be adopted directly in other combination schema. Diversity in these methods is acquired by adding some randomness to the training process of the base classifiers (i.e. random resampling, random features selection). In this work, we extend this collection of methods with another approach to supply diversity. We think our method belongs to such group because:

- Our method does not need to take into account the ensemble method in which it is going to be used (like it happens with resampling and random feature selection). Moreover, we can apply our method to these ensemble algorithms mentioned before, which have their own way of getting diversity, making their base classifiers even more diverse and improving the overall ensemble method performance.
- It brings a random component that makes the base classifiers to be built in a different way each time.

In order to inject randomness, we use the prediction from a not very accurate classifier. This classifier is built each time with a very small subset of instances of the whole training set picked randomly. We have used a *Nearest Neighbor* classifier for this purpose. The NN output is used to build new features that *disturb* or alter the predictions that the base classifier would do using the raw dataset alone. That is why we call our method *Disturbing Neighbors*.

The chapter is organized as follows. Section 2 describes the Disturbing Neighbor method. Section 3 compares our method with the state of the art ensembles of Decision Trees and contains the experimental validation. Section 4 analyzes what components of our method are essential and what algorithm steps could be discarded without getting significantly worse results. Section 5 gives the conclusions.

Algorithm 1. *Disturbing Neighbor* Base Classifier Training. 1-Nearest Neighbor function has been specified separately in order to emphasize that (i) it only returns the nearest neighbor index (not the predicted class), and (ii) distance is computed by taking into account only a random subset from the original features

Function *DN*-BaseClassifierTrainer

input : D: Training Dataset with l features and n instances,
m: Small integer,
BC_T: Training algorithm of a base classifier BC
output: A classifier trained using a *DN* variant of *BC* base method that can be used as a base classifier in an ensemble method

variables:
RndDimensions: Array $[1..l]$ of Boolean
RndNeighbors: Array $[1..m]$ of instances from D
D' : Augmented Dataset initially empty

begin
Set randomly more than $l/2$ elements of *RndDimensions* to *True* and the rest to *False*
Fill randomly *RndNeighbors* with m instances belonging to D **forall** $x \in D$ **do**
$x' \leftarrow x$ $i \leftarrow$ `NearestNeighbor` (x, *RndNeighbors*, *RndDimensions*) Append m new boolean attributes into x', taking all false values except the one in i position
$p \leftarrow$ class of *RndNeighbors*$[i]$ Append p as a new feature of x' Insert x' into D' dataset
end
Train a *BC* classifier using D' and BC_T Return *BC*;
end

Function NearestNeighbor
input : x:training dataset instance,
Neighbors: Array $[1..m]$ of instances,
BooleanMask: Array $[1..l]$ of Boolean
output: i:integer indicating the nearest neighbor
begin
Get the Nearest Neighbor to x in *Neighbors* by computing the Euclidean Distance where only dimensions set to *True* in *BooleanMask* are used Return the index in *Neighbors* of the 1-NearestNeighbor
end

2 Method

The method presented here can generate different base classifiers by adding new features to the training data. This new features are different each time, making the base classifiers diverse. The algorithm is shown in Fig. 1.

Given a training dataset D, the method:

1. takes m randomly selected instances from D to build a 1-NN classifier, where m is a small integer provided as parameter,

2. discards, randomly too, less than 50% of attributes of D; Euclidean distances for the 1-NN classifier are calculated without using the discarded attributes. This is like calculating the distances in a projected space.

Then, for each training instance x in D, we add the following $m+1$ new features:

1. The class predicted by the 1-NN classifier is added as a new feature.
2. Additional m boolean features are added, one for each of the m selected instances. These features are all *false* but the one corresponding with the nearest neighbor to instance x.

Table 1 Some instances from the Iris dataset augmented by new dimensions computed using Disturbing Neighbors (*DN*). Features prefixed by *Nearest* have been added by the *DN* method. *Nearest Class* attribute represents the Nearest Neighbor prediction, and *Nearest i* is *T* (true) when the *i*-th neighbor is the nearest one

Nearest Class	*Nearest 1*	*Nearest 2*	*Nearest 3*	*Nearest 4*	*Nearest 5*	*Nearest 6*	*Nearest 7*	*Nearest 8*	*Nearest 9*	*Nearest 10*	*sepal length*	*sepal width*	*petal length*	*petal width*	*class*
setosa	F	F	F	F	F	F	F	**T**	F	F	5.1	3.5	1.4	0.2	setosa
setosa	F	F	F	F	F	F	F	**T**	F	F	4.9	3.0	1.4	0.2	setosa
setosa	F	F	F	F	F	F	F	**T**	F	F	4.7	3.2	1.3	0.2	setosa
versicolor	F	F	F	F	F	F	**T**	F	F	F	7.0	3.2	4.7	1.4	versicolor
versicolor	F	F	**T**	F	F	F	F	F	F	F	5.5	2.3	4.0	1.3	versicolor
virginica	**T**	F	F	F	F	F	F	F	F	F	5.9	3.2	4.8	1.8	versicolor
virginica	F	F	F	**T**	F	F	F	F	F	F	7.4	2.8	6.1	1.9	virginica
virginica	F	F	F	F	F	F	F	F	**T**	F	6.1	2.6	5.6	1.4	virginica
versicolor	F	F	F	F	F	F	**T**	F	F	F	4.9	2.5	4.5	1.7	virginica

Table 1 shows some instances from the Iris dataset with these new added features. In this table m has been set to ten. The attribute *Nearest Class* is the 1-NN prediction, and the boolean attributes *Nearest n* represent which one of the ten randomly selected instances is the nearest neighbor. The result is an augmented dataset that can be used for training any base method in an ensemble, regardless the ensemble and the base classifier.

Figure 1 shows the effect of using the boolean attributes on the 1-NN classification of the conus-torus artificial data set [13]. In this dataset it is difficult to compute a boundary separating each region. We can use this figure to try to analyze how the diversity is provided:

- Figure 1 (bottom) shows how the original space in the upper part of Fig. 1 is divided into Voronoi regions corresponding to each neighbor. These regions are different each time a *DN* classifier is built. The added boolean attributes determine the region each instance belongs to. The division of the space in regions

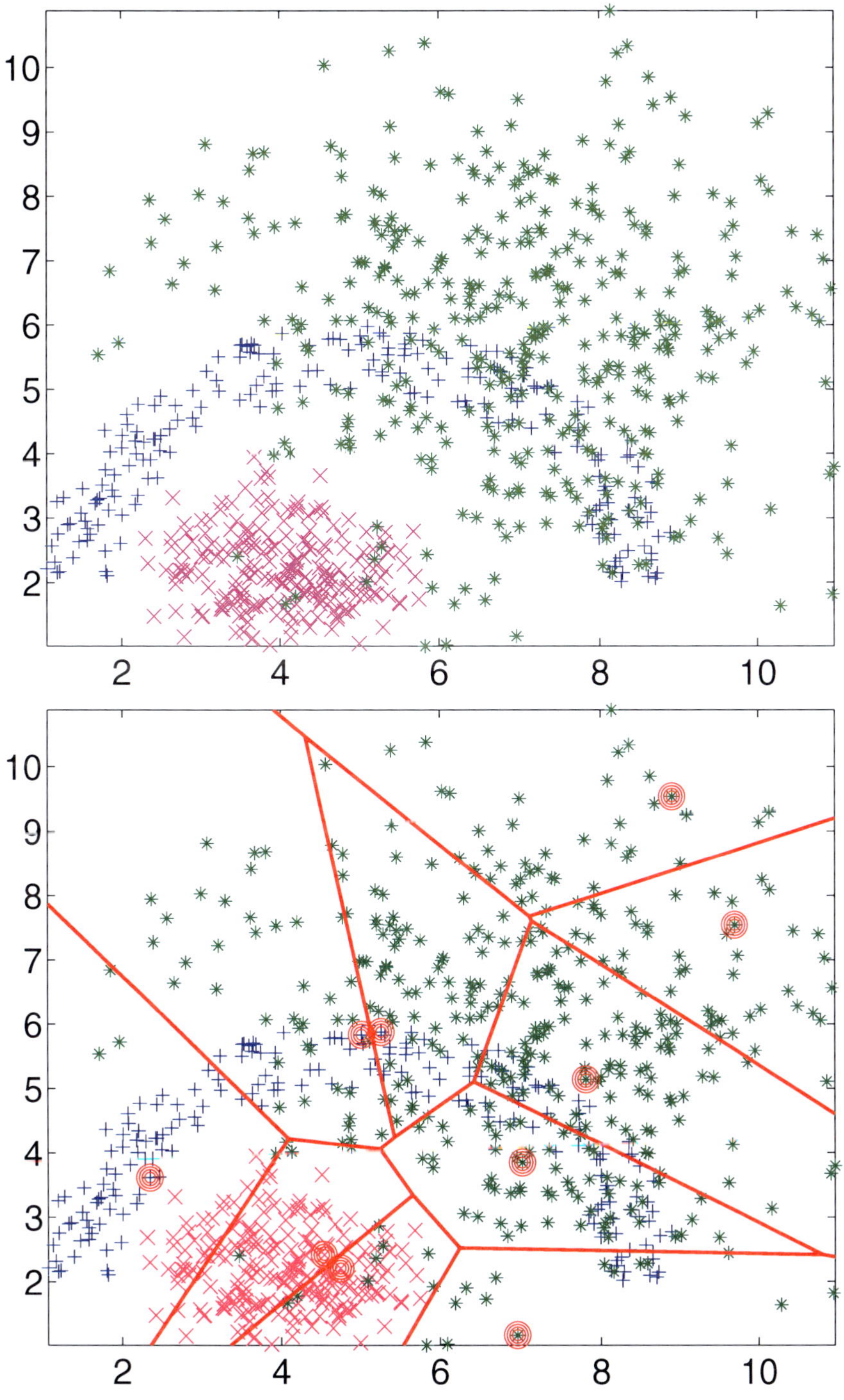

Fig. 1 Conus-torus dataset (top) and its Voronoi regions derived from 10 Neighbors (bottom)

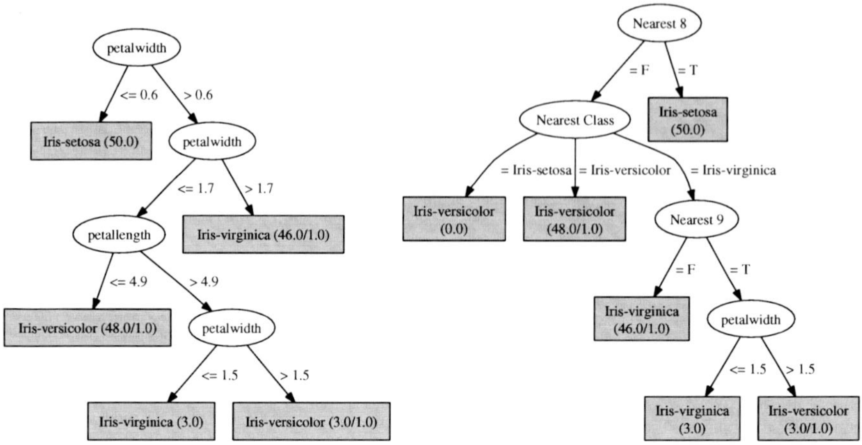

Fig. 2 C4.5 tree (left) and *DN*-C4.5 tree (right) for *Iris* dataset. Note that some splitting nodes from the *DN*-C4.5 use the *Nearest Neighbor* and the *Nearest Class* (Nearest Neighbor prediction)

provides extra expressiveness to the base method. For example, if *DN* is applied to a Decision Tree , the splitting nodes not only can split data by an attribute value, but they can change a membership of patterns as shown in Fig. 2. If a region has a lot of instances belonging to a class, it could be used to split a tree node. So, each time, different neighbors will produce different boolean values, and these values will produce a different base classifier, making the whole set of base classifiers diverse.

- Although it is not expected that the 1-NN class prediction would be very accurate, it could be accurate enough to make it one of the principal features to be taken into account by the base method. For example, if Decision Trees are used as base classifiers probably there will be one splitting node using 1-NN class prediction next to the tree root (see *Nearest Class* node in Fig. 2). Different neighbors will make different predictions, and these predictions will produce again diverse base classifiers .
- Finally, the random feature selection for computing Euclidean distances increases that diversity. Even if two classifiers would use almost the same m neighbors, both the 1-NN output and the Voronoi regions would be different if the distances are calculated using different subsets of attributes.

So the 1-NN alters or disturbs the base construction method on these three ways, and that is why we call it Disturbing Neighbors (*DN*). A base classifier trained with the augmented dataset probably does not perform better than when it is trained with the original dataset, but when *DN* is used in base classifiers within an ensemble, all the randomness in *DN* makes this set of base classifiers diverse, and the overall accuracy of the ensemble is improved, as shown in Sect. 3.

Finally, note that computationally parallelizable ensemble methods (i.e. Bagging, Random Subspaces or Random Forest), keep this algorithmic property when

Disturbing Neighbors are used. For our experiments we use $m = 10$, so computational cost does not grow significantly by using our variant of classifiers based on decision trees .

3 Results

Disturbing Neighbors were implemented in Java within WEKA environment [17]. We tested our method using WEKA ensemble implementations. Default WEKA parameters were used unless otherwise indicated. We compared our method with:

- Bagging [4].
- Random Forest [5].
- Boosting: We used AdaBoost [10] and MultiBoost [16]. In both Boosting versions we considered resampling and reweighting variants, which are respectively denoted as (S) or (W) in tables. In reweighting variant all the instances from the training dataset are used by each base classifier, but in each new round Boosting changes the weight distribution to focus on hard to classify instances. In resampling variant base classifiers are trained only with a sample of the training set according to such Boosting weight distribution.
- Random Subspaces [12]: We tested two configurations, picking 50% and 75% of the original problem dimensions.

The size of the ensemble was 50 in all the experiments. The base methods used for testing were:

- For Boosting , Bagging and Random Subspaces, the base method was J.48 Decision Trees (WEKA implementation of Quinlan C4.5 Decision Tree [15]). We tested the ensembles with plain J.48, and with J.48 disturbed by our method (*DN*-Decision Trees). The parameter m, the number of neighbors, was set to ten.
- For Random Forest obviously Decision Trees are used as base classifiers. The ensemble was also tested with the plain Decision Trees and with the disturbed variant (*DN*-Decision Trees).

We also included in the study an ensemble with fifty *DN*-Decision Trees as base classifiers to check if they perform well by their own without any sophisticated combination schema, just a simple average of the predictions generated by the individual base classifiers . We denote this method by *DN*-Ensemble from now on.

Finally, we wanted to know if k-NN accuracy was strong enough to be the main reason the disturbed classifiers could improve ensemble accuracy. So, we included IBk (WEKA implementation of k-NN [1]) to the test. We tested using fixed $k = 1$ and variable k configurations. This last configuration uses an optimized k value for each data set. The optimal k is obtained through cross validation. NN methods are very robust with respect to variations of data set, so they do not improve very much when combined with standard ensembles [9]. Thus, we have not considered ensembles of k-NN in our test. In particular, Bagging using 1-NN as base classifiers is equivalent

Table 2 Summary of the data sets used in the experiments. Field *id* will be used later in Figs. 5-9

id	Dataset	#N	#D	#E	#C
1	abalone	7	1	4177	28
2	anneal	6	32	898	6
3	audiology	0	69	226	24
4	autos	15	10	205	6
5	balance-scale	4	0	625	3
6	breast-w	9	0	699	2
7	breast-y	0	9	286	2
8	bupa	6	0	345	2
9	car	0	6	1728	4
10	credit-a	6	9	690	2
11	credit-g	7	13	1000	2
12	crx	6	9	690	2
13	dna	0	180	3186	3
14	ecoli	7	0	336	8
16	glass	9	0	214	6
16	heart-c	6	7	303	2
17	heart-h	6	7	294	2
18	heart-s	5	8	123	2
19	heart-statlog	13	0	270	2
20	heart-v	5	8	200	2
21	hepatitis	6	13	155	2
22	horse-colic	7	15	368	2
23	hypo	7	18	3163	2
24	ionosphere	34	0	351	2
25	iris	4	0	150	3
26	krk	6	0	28056	18
27	kr-vs-kp	0	36	3196	2
28	labor	8	8	57	2
29	led-24	0	24	5000	10
30	letter	16	0	20000	26
31	lrd	93	0	531	10
32	lymphography	3	15	148	4
33	mushroom	0	22	8124	2
34	nursery	0	8	12960	5
35	optdigits	64	0	5620	10
36	page	10	0	5473	5
37	pendigits	16	0	10992	10
38	phoneme	5	0	5404	2
39	pima	8	0	768	2
40	primary	0	17	339	22
41	promoters	0	57	106	2
42	ringnorm	20	0	300	2
43	sat	36	0	6435	6
44	segment	19	0	2310	7
45	shuttle	9	0	58000	7
46	sick	7	22	3772	2
47	sonar	60	0	208	2
48	soybean	0	35	683	19
49	soybean-small	0	35	47	4
50	splice	0	60	3190	3
51	threenorm	20	0	300	2
52	tic-tac-toe	0	9	958	2
53	twonorm	20	0	300	2
54	vehicle	18	0	846	4
55	vote1	0	15	435	2
56	voting	0	16	435	2
57	vowel-context	10	2	990	11
58	vowel-nocontext	10	0	990	11
59	waveform	40	0	5000	3
60	yeast	8	0	1484	10
61	zip	256	0	9298	10
62	zoo	1	15	101	7

#N: Numeric features, #D: Discrete features, #E: Examples, #C: Classes

to 1-NN [6]. Moreover, Bagging can slightly degrade the performance of stable algorithms (e.g., *k*-NN) [3].

For validation we used the 62 UCI datasets [2] in Table 2 and 5×2 stratified cross validation, which provides an acceptable number of repetitions [7]. Results are summarized in Tables 3-6.

Table 3 shows the methods using the average ranks from [8]. A number is assigned to each method and data set corresponding to its rank position in such dataset. If there are ties, average ranks are assigned. Then, for each method, the average position is calculated over all datasets (see first column of Table 3). The methods are

Table 3 Ensemble methods sorted by their average rank

Average Rank	Method
6.28	*DN*-MultiBoost (S)
6.73	*DN*-Random Forest
7.12	*DN*-MultiBoost (W)
8.01	*DN*-AdaBoost (S)
8.09	*DN*-AdaBoost (W)
8.27	MultiBoost (S)
8.65	Random Forest
8.84	*DN*-Subspaces (50%)
9.23	MultiBoost (W)
9.65	*DN*-Bagging
10.03	*DN*-Subspaces (75%)
10.11	AdaBoost (S)
10.23	AdaBoost (W)
11.54	k-Nearest Neighbor
11.90	Bagging
12.31	Subspaces (50%)
14.07	*DN*-Ensemble
14.31	Subspaces (75%)
14.61	1-Nearest Neighbor

then ordered using these values. We can see that all undisturbed ensemble methods were improved by their *DN* version.

Let us consider only methods in the experiment having a *DN* version and an undisturbed version. Table 4 shows rank positions of these methods according to Table 3. Table 4.a shows the *DN* versions, whereas Table 4.b shows the undisturbed versions. Relative order between methods is practically the same regardless of whether the *DN* version is used or not. The only exceptions are Bagging and Random SubSpaces(50%). So *DN* version improvements seem to be somehow independent of combination schemes. That is why *DN* can be thought of as an enhancement of an existing ensemble method. This hypothesis is also supported by the ranking of the *DN*-Ensemble in Table 3. The average ranking of this ensemble is even worse than k-NN, and it shows that simply using *DN*-base classifiers does not ensure the best possible ensemble.

The improvement of using *DN* versions is quantified in Table 5 that shows wins, ties and loses of disturbed ensemble versions against undisturbed versions. According to the sign test [8], for 62 data sets, one method is better than other with significance level of 5%, if the number of wins plus half the ties is greater or equal than 39. Hence, for all the methods in Table 5 the *DN* version is significantly better.

Table 3 shows that 1-NN and k-NN seem to poorly perform in the experiments. Thus we can think that *DN* versions are not improved by the k-NN algorithm strength, but rather by the diversity induced by the random neighbors selection.

Table 4 Ranking based on average ranks of disturbed and undisturbed variants of ensemble methods. **Table a** ranks disturbed variants, whereas **Table b** ranks undisturbed variants. Methods not preserving the relative order are marked in bold

Rank	Disturbed Method
1	*DN*-MultiBoost (S)
2	*DN*-Random Forest
3	*DN*-MultiBoost (W)
4	*DN*-AdaBoost (S)
5	*DN*-AdaBoost (W)
8	***DN*-Subspaces (50%)**
10	***DN*-Bagging**
11	*DN*-Subspaces (75%)

a.

Rank	Undisturbed Method
6	MultiBoost (S)
7	Random Forest
9	MultiBoost (W)
12	AdaBoost (S)
13	AdaBoost (W)
15	**Bagging**
16	**Subspaces (50%)**
18	Subspaces (75%)

b.

Table 5 Comparison of methods with and without *DN* based diversity

Method	Win-Tie-Loss
Bagging	50-1-11
Subspaces (50%)	50-3-9
Subspaces (75%)	54-3-5
AdaBoost (W)	47-0-15
AdaBoost (S)	49-1-12
MultiBoost (W)	47-1-14
MultiBoost (S)	48-0-14
Random Forest	40-3-19

Table 6 Comparison of ensemble methods with the *k*-Nearest Neighbor classifier

Method	Win-Tie-Loss
DN-Ensemble	26-2-34
DN-Bagging	41-0-21
DN-Subspaces (50%)	40-1-21
DN-Subspaces (75%)	36-2-24
DN-AdaBoost (W)	39-0-23
DN-AdaBoost (S)	39-1-22
DN-MultiBoost (W)	39-0-23
DN-MultiBoost (S)	40-1-21
DN-Random Forest	44-2-16

Table 6 also shows the same by comparing each *DN* method against *k*-NN. We can see that *k*-NN is significantly worse than all the *DN*-methods except *DN*-Ensemble and *DN*-Subspaces (75%).

We also tested diversity improvement of *DN*-Trees using the Kappa statistic [14]. Kappa measures how diverse two classifiers are, it can take values ranged from -1 to 1. A Kappa value equal to 1 means that both classifiers agree in every example, a value equal to 0 means that there is no agreement above that expected by chance, and negative Kappa values happen when there is disagreement between the classifiers. Then Kappa values are used to draw Kappa-Error Diagrams [14]. Figure 3 shows an example for *krk* dataset with Bagging and AdaBoost methods. For each pair of base classifiers we plot a point (x, y), where x is kappa measure for these two classifiers, and y is the average error of them. So ideally pairs of base classifiers would be close to left bottom corner, because it means they are accurate and diverse.

In Fig. 3 we see *DN*-clouds slightly displaced to the left of undisturbed ensembles clouds. It means that *DN*-methods are more diverse. Figure 4 shows *Kappa-Error Movement Diagram*. This diagram is based on the corresponding Kappa-Error diagram for each dataset. All diagrams are scaled using the maximum and minimum values of Kappa and Error. Each method considered (i.e., Bagging, Adaboost with resampling, Multiboost with resampling, Random Forest and Random Subspaces 50% attributes variant) is represented by an arrow pointing from the centre of the undisturbed version cloud to the centre of the *DN* version cloud. We can see a lot of arrows pointing left, which means a generalized improvement of diversity. The longer the arrow, the bigger the relative difference.

Finally, Figs. 5 to 9 show *Kappa-Error Relative Movement Diagrams* for each ensemble method. These diagrams are obtained gathering all arrows in Fig. 4 for

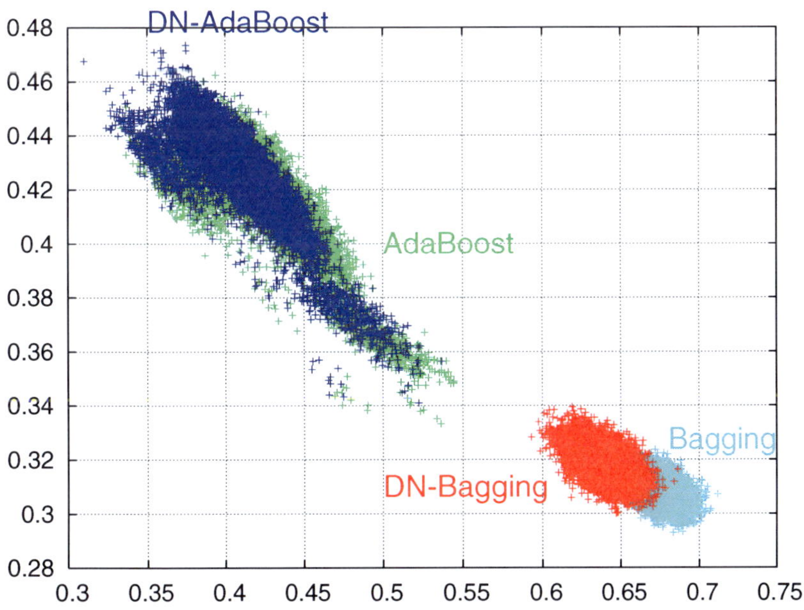

Fig. 3 Error vs. Kappa for AdaBoost and Bagging in *krk* dataset

Fig. 4 Kappa-Error Movement Diagram for the 62 datasets. Each arrow points from the center of the non-*DN* version cloud of a method to the center of the *DN* version cloud

an ensemble, and translating the starting point of every arrow to the origin of coordinates. Numbers represents the dataset *id* field in Table 2. This kind of diagram is a very convenient way of summing up the results shown in Fig. 4. The majority of arrows point to left, which is an indicator of diversity. Many arrows also point upward, showing that generally the increase of diversity is at the expense of the individual base classifier accuracy.

4 Lesion Study

There are three elements in the *DN*-base classifier construction: (i) Random Feature Selection is applied to compute 1-NN distances, (ii) m new boolean features are added indicating the Nearest Neighbor, and (iii) another extra feature is added

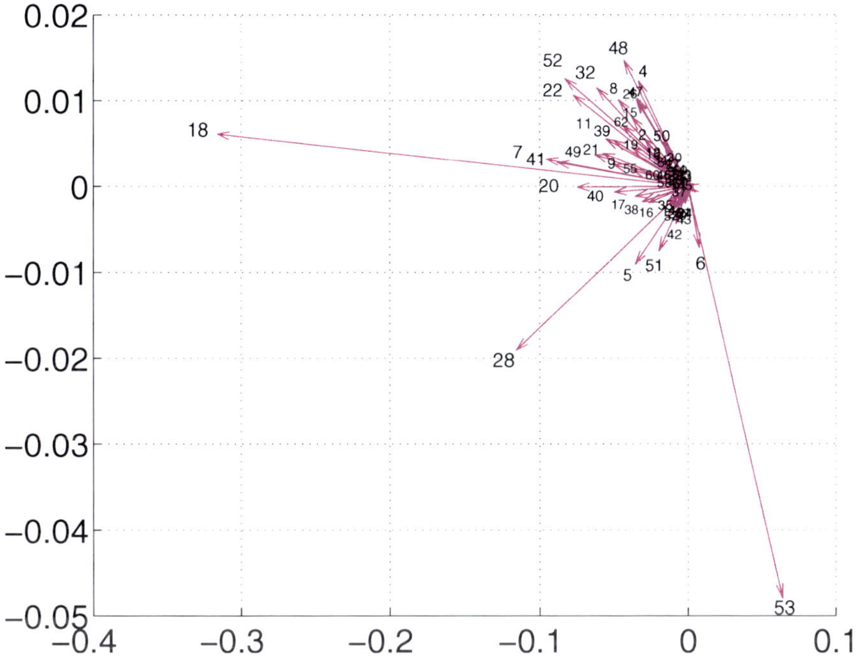

Fig. 5 Kappa-Error Relative Movement Diagram for Bagging

containing the Nearest Neighbor class prediction. It is very interesting to make an experimental study to point out which of these three elements are critical for *DN* accuracy, and which of them are not essential.

For this purpose, we create five new variants of *DN*:

1. DN_C is a variant that does not apply random feature selection, neither it takes into account which is the nearest neighbor. It only computes the nearest neighbor class prediction. DN_C variant can be shown as a type of Cascading [11] where the 1-NN classifier plays the role of $Level_1$ classifier, and the tree plays the role of $Level_2$ classifier. However, Cascading generally uses the whole dataset for training $Level_1$ classifier, whereas *DN* uses only a few, *m*, instances.
2. The DN_N variant does not apply random feature selection and does not takes the 1-NN prediction. It only uses the *m* binary attributes indicating which is the nearest neighbor.
3. The DN_{NC} variant does not apply random feature selection but considers both 1-NN prediction and *m* binary attributes.
4. The DN_{CR} variant is very similar to DN_C because the 1-NN classifier prediction is also taken into account. The difference is that DN_{CR} computes 1-NN distances using the random feature selection .

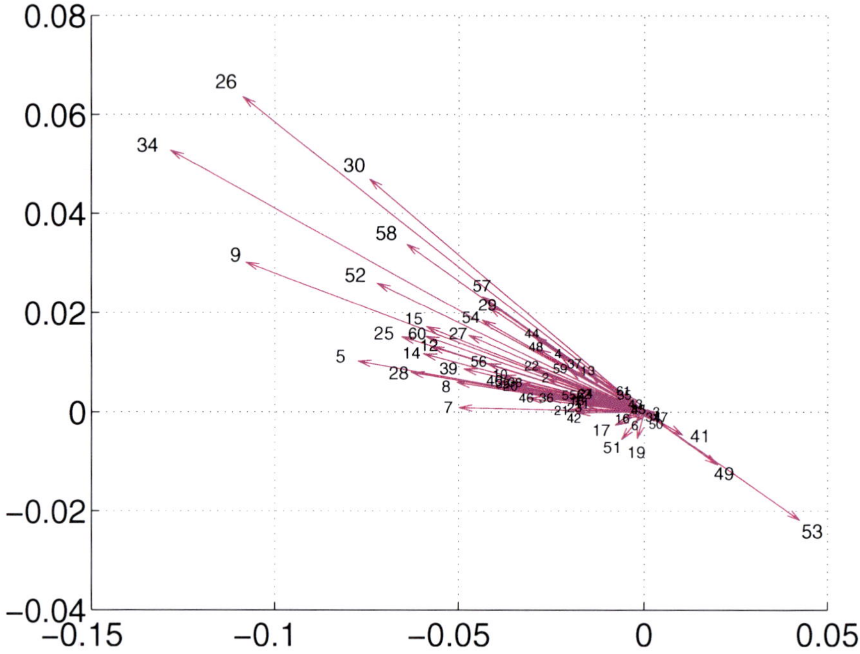

Fig. 6 Kappa-Error Relative Movement Diagram for Random Forest

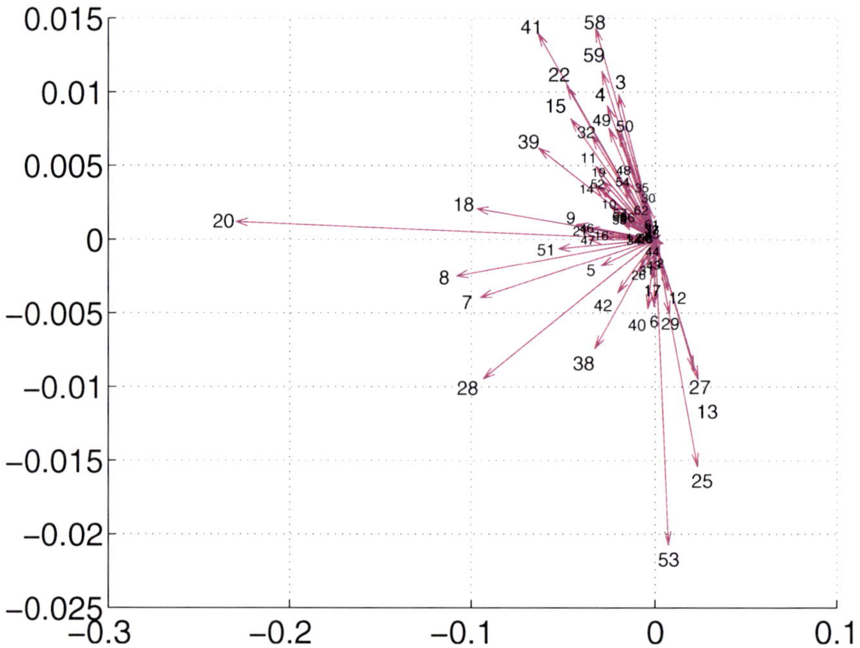

Fig. 7 Kappa-Error Relative Movement Diagram for Random Subspaces(50%)

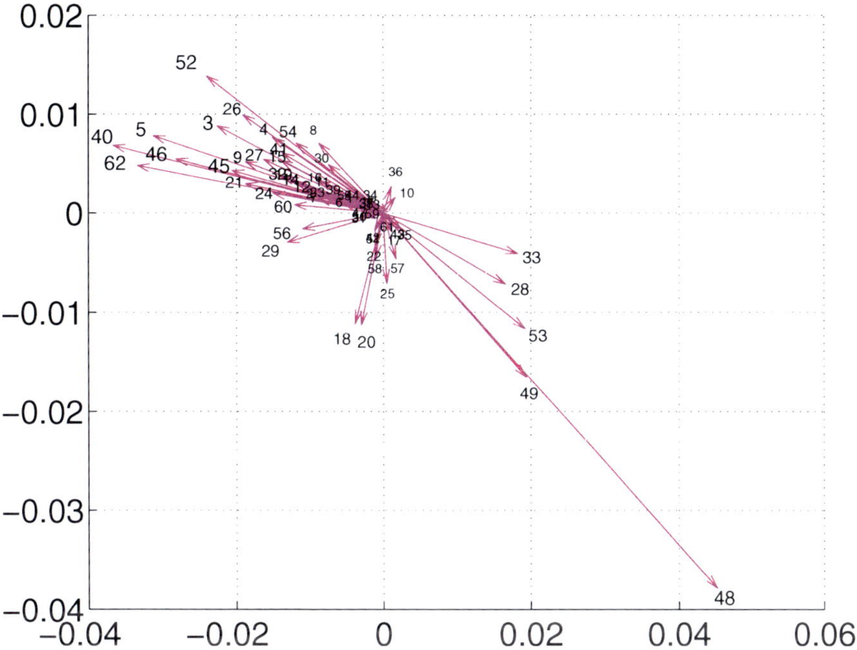

Fig. 8 Kappa-Error Relative Movement Diagram for AdaBoost(S)

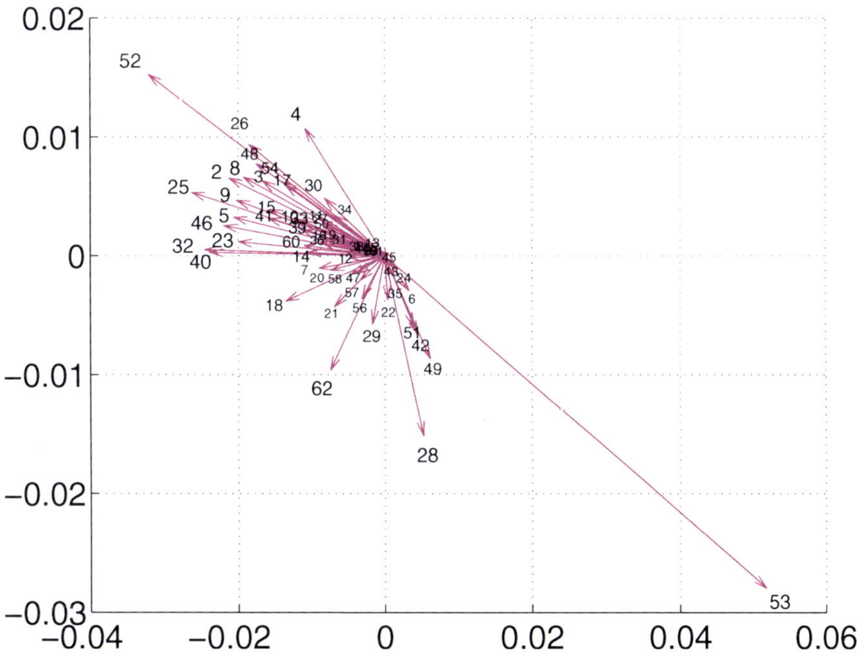

Fig. 9 Kappa-Error Relative Movement Diagrams for MultiBoost(S)

5. The DN_{NR} variant is very similar to DN_N because it also uses the m binary features. The difference is that DN_{NR} computes 1-NN distances using the random feature selection as well.

According to this nomenclature, the *DN* method itself could be rewritten as DN_{NCR} because it applies the three mentioned elements (**N**: the **N**earest neighbor from the m instances, **C**: the nearest neighbor **C**lass prediction, **R**: **R**andom feature selection).

The ensemble methods considered in Sect. 3 have been grouped into nine families:

1. Bagging.
2. Random Forest.
3. Random Subspaces using 50% of original features.
4. Random Subspaces using 75% of original features.
5. Reweighting AdaBoost.
6. Resampling AdaBoost.
7. Reweighting MultiBoost.
8. Resampling MultiBoost.
9. *DN*-Ensembles.

All families except *DN*-Ensembles consist of seven methods: the plain ensemble without any *DN* variant, the *DN* variant (DN_{NCR}), and the five variants described before. For *DN*-Ensembles only six methods are considered, because there is no plain ensemble.

Average ranks have been computed for each family using again the same 62 UCI datasets and 5×2 stratified cross validation. According to [8] using the two-tailed Nemenyi test, for 62 datasets a classifier performs better than other with significance level 5%, if the difference between their average ranks is greater than a critical value of 1.144 when seven methods are compared (i.e. all families except *DN*-Ensemble family), or 0.958 when six methods are compared (i.e. the *DN*-Ensemble family).

Tables 7-10 show the results for each family. Vertical lines with whiskers on the left side of each table group methods that are not significantly worse than the

Table 7 Average Rankings for Bagging, Random Forest and *DN*-Ensemble variants

Average Rank	Method
3.01	DN_{NC}-Bagging
3.16	*DN*-Bagging
3.88	DN_C-Bagging
3.91	DN_N-Bagging
3.94	DN_{NR}-Bagging
4.60	DN_{CR}-Bagging
5.50	Bagging

Average Rank	Method
3.40	*DN*-R Forest
3.60	DN_{NR}-R Forest
3.95	DN_{NC}-R Forest
4.10	DN_{CR}-R Forest
4.10	DN_C-R Forest
4.34	DN_N-R Forest
4.52	R Forest

Average Rank	Method
2.48	DN_{NC}-Ensemble
2.74	*DN*-Ensemble
2.97	DN_N-Ensemble
3.40	DN_{NR}−Ensemble
4.49	DN_C-Ensemble
4.91	DN_{CR}-Ensemble

Table 8 Average Rankings of Random Subspaces variants

Average Rank	Method
3.19	DN_{NC}-Subspaces (50%)
3.22	*DN*-Subspaces (50%)
3.44	DN_{N}-Subspaces (50%)
4.00	DN_{C}-Subspaces (50%)
4.02	DN_{NR}-Subspaces (50%)
4.32	DN_{CR}-Subspaces (50%)
5.81	Subspaces (50%)

Average Rank	Method
2.85	*DN*-Subspaces (75%)
2.98	DN_{NC}-Subspaces (75%)
3.59	DN_{N}-Subspaces (75%)
3.61	DN_{NR}-Subspaces (75%)
4.16	DN_{C}-Subspaces (75%)
4.64	DN_{CR}-Subspaces (75%)
6.17	Subspaces (75%)

Table 9 Average Rankings of AdaBoost variants

Average Rank	Method
3.23	*DN*-AdaBoost (W)
3.49	DN_{NR}-AdaBoost (W)
3.73	DN_{NC}-AdaBoost (W)
3.82	DN_{N}-AdaBoost (W)
4.01	DN_{C}-AdaBoost (W)
4.54	DN_{CR}-AdaBoost (W)
5.18	AdaBoost (W)

Average Rank	Method
3.09	DN_{N}-AdaBoost (S)
3.41	DN_{NR}-AdaBoost (S)
3.57	*DN*-AdaBoost (S)
3.83	DN_{NC}-AdaBoost (S)
4.16	DN_{C}-AdaBoost (S)
4.61	DN_{CR}-AdaBoost (S)
5.32	AdaBoost (S)

Table 10 Average Rankings of MultiBoost variants

Average Rank	Method
3.51	DN_{NC}-MultiBoost (W)
3.53	*DN*-MultiBoost (W)
3.62	DN_{NR}-MultiBoost (W)
3.69	DN_{N}-MultiBoost (W)
4.06	DN_{CR}-MultiBoost (W)
4.13	DN_{C}-MultiBoost (W)
5.45	MultiBoost (W)

Average Rank	Method
3.39	DN_{NC}-MultiBoost (S)
3.40	*DN*-MultiBoost (S)
3.77	DN_{N}-MultiBoost (S)
4.05	DN_{NR}-MultiBoost (S)
4.06	DN_{C}-MultiBoost (S)
4.22	DN_{CR}-MultiBoost (S)
5.12	MultiBoost (S)

first ranked method, whereas vertical lines with whiskers on the right side of each table group methods that are not significantly better than the last ranked method. Table 7 shows that all variants have no significant performance difference in the Random Forest family. But in the rest of families, *DN* variant is always in the top group whereas the plain method is always ranked at last position, except for *DN*-Ensembles, where there is no plain method.

Table 11 Average Rankings for all variants. Benefit is computed as the average rank difference when using a *DN* variant compared to using none for the same ensemble method

Average Rank	Benefit	Method	Average Rank	Benefit	Method
20.67	8.34	DN_{NC}-MultiBoost (S)	30.06	4.54	DN_C-AdaBoost (W)
21.95	7.06	*DN*-MultiBoost (S)	31.12	10.09	DN_C-Subspaces (50%)
22.64	6.37	DN_N-MultiBoost (S)	31.16	10.85	DN_N-Subspaces (50%)
22.84	6.69	*DN*-Random Forest	31.55	9.43	DN_{NC}-Bagging
22.86	6.15	DN_C-MultiBoost (S)	32.48		MultiBoost (W)
23.01	6.52	DN_{NR}-Random Forest	32.82	2.32	DN_{CR}-AdaBoost (W)
24.17	4.84	DN_{NR}-MultiBoost (S)	33.01	9.01	DN_{NR}-Subspaces (50%)
24.29	8.19	DN_{NR}-MultiBoost (W)	33.19	7.78	DN_C-Bagging
24.33	8.15	DN_N-MultiBoost (W)	33.31	7.66	*DN*-Bagging
24.40	8.08	DN_{NC}-MultiBoost (W)	33.58	8.44	DN_{CR}-Subspaces (50%)
24.61	4.40	DN_{CR}-MultiBoost (S)	33.75	14.55	DN_{NC}-Subspaces (75%)
24.83	4.70	DN_{NC}-Random Forest	34.26	14.04	*DN*-Subspaces (75%)
24.92	7.56	*DN*-MultiBoost (W)	34.48		AdaBoost (S)
25.43	7.06	DN_C-MultiBoost (W)	34.69	6.29	DN_N-Bagging
25.95	3.58	DN_N-Random Forest	35.15		AdaBoost (W)
26.90	7.59	DN_N-AdaBoost (S)	36.15	4.82	DN_{CR}-Bagging
26.99	7.49	DN_{NC}-AdaBoost (S)	36.15	4.82	DN_{NR}-Bagging
27.00	7.48	DN_{NR}-AdaBoost (S)	37.14	11.16	DN_N-Subspaces (75%)
27.02	5.47	DN_{CR}-MultiBoost (W)	37.60	10.69	DN_C-Subspaces (75%)
27.46	7.02	*DN*-MultiBoost (S)	37.93	10.37	DN_{NR}-Subspaces (75%)
27.52	2.01	DN_{CR}-Random Forest	39.11		k-Nearest Neighbor
27.72	1.81	DN_C-Random Forest	40.93	7.37	DN_{CR}-Subspaces (75%)
28.34	6.81	*DN*-AdaBoost (W)	40.98		Bagging
28.44	6.71	DN_N-AdaBoost (W)	42.02		Subspaces (50%)
29.01		MultiBoost (S)	45.98		DN_{NC}-Ensemble
29.19	5.29	DN_C-AdaBoost (S)	48.08		*DN*-Ensemble
29.19	12.82	DN_{NC}-Subspaces (50%)	48.30		Subspaces (75%)
29.23	5.91	DN_{NR}-AdaBoost(W)	48.38		DN_N-Ensemble
29.43	5.06	DN_{CR}-AdaBoost(S)	49.57		1-Nearest Neighbor
29.53		Random Forest	50.56		DN_{NR}-Ensemble
29.73	5.42	DN_{NC}-AdaBoost(W)	51.55		DN_C-Ensemble
30.32	11.69	*DN*-Subspaces (50%)	54.49		DN_{CR}-Ensemble

In all families, except for Random Forest, the variants that use the m binary features, indicating the Nearest Neighbor (from now on $Variants_N$), never fall into the group of methods that are not significantly better than the last ranked method. Moreover:

- $Variants_N$ always take at least the two top places in all families.
- $Variants_N$ are always in the group of methods that are not significantly worse than the first ranked one.

On the other hand, except for Random Forest, there is always a DN_{CR} or a DN_C method in the second from last place. In *DN*-Ensemble, Random Subspaces (75%) and the four Boosting families DN_{CR} and DN_C variants fill the bottom places along with the respective plain version if any. Many of these variants even do not show a significant improvement over their plain variant.

There are only three cases with significant difference between applying or not random feature selection to distances computation. In Bagging, and both AdaBoost variants the DN_C option is significantly better than the DN_{CR}. In the rest of families there is no significant difference between such options, and in all families there is no significant difference between DN_R and DN_{NR}, and between *DN*and DN_{NC} versions.

In four families the *DN* variant ranks better than the DN_{NC}, and in the other five happens the opposite. In all families except for Random Forest and both Boosting reweighting variants, DN_N variant ranks better than DN_{NR}, and DN_C variant ranks better than DN_{CR} only in two families. So, using or not a random subspace of features for 1-NN distances does not seem very important in the schemes considered. However, reducing input space dimensionality is an interesting option for big datasets, in order to achieve a slightly reduction in distance computation without significant accuracy loss.

We can conclude that using the m features pointing the Nearest Neighbor is the common characteristic to all successful variants in all families. Using the 1-NN prediction or the random feature selection do not seem to be essential.

Table 11 shows the average ranks for all methods and variants. The column benefit in the *DN* variant rows indicates the gain of the *DN* version regarding the plain ensemble. Although some relative positions in family ranks are swapped from the overall rank, swaps always happen between non-significantly different methods.

This overall rank shows that the order between the families of ensembles from Table 3 is preserved when the new *DN* variants are considered. So the improvement of using the *DN* method seems to be uniform among all the ensembles algorithms.

5 Conclusion

This paper presents the method for improving diversity of base classifiers in ensembles. Our method builds some new features to be added to the training set. These attributes are different for each base classifier, making base classifiers diverse.

The new features disturb the normal training of the base classifier and are obtained from the use of an 1-NN classifier. The instances (*Disturbing Neighbors*) used for the 1-NN classifier are a very small subset of the whole training dataset selected randomly for each base classifier and being the source of diversity for the ensemble.

An experimental validation has been provided showing that the idea presented in this paper improves accuracy in all considered ensembles of decision trees. Diagrams based on Kappa statistic have been introduced (Kappa-Error Movement

Diagram and Kappa-Error Relative Movement Diagram) showing that diversity is also improved.

A lesion study has been made in order to clarify what *DN* algorithm ingredients are essential. It seems that the 1-NN predictions and the use of a random subset of input space dimensions for computing the neighbors distances do not give any significant improvement, whereas using a set of extra binary features pointing which one is the nearest neighbor is the key of the success of the method. So 1-NN method does not seem to be replaceable by another method that just provides a class prediction. It is needed that such classifier also provides the way of dividing the input space into random regions, as the Nearest Neighbors do.

Acknowledgements. This work has been supported by the "Junta de Castilla y León" project BU007B08.

References

1. Aha, D., Kibler, D., Albert, M.K.: Instance-based learning algorithms. Mach. Learn. 6(1), 37–66 (1991)
2. Asuncion, A., Newman, D.J.: UCI machine learning repository., `http://www.ics.uci.edu/~mlearn/MLRepository.html`
3. Bauer, E., Kohavi, R.: An empirical comparison of voting classification algorithms: bagging, boosting, and variants. Mach. Learn. 36(1-2), 105–139 (1999)
4. Breiman, L.: Bagging predictors. Mach. Learn. 24(2), 123–140 (1996)
5. Breiman, L.: Random forests. Mach. Learn. 45(1), 5–32 (2001)
6. Bruno Caprile, B., Merler, S., Furlanello, C., Jurman, G.: Exact bagging with k-nearest neighbour classifiers. In: Roli, F., Kittler, J., Windeatt, T. (eds.) MCS 2004. LNCS, vol. 3077, pp. 72–81. Springer, Heidelberg (2004)
7. Dietterich, T.G.: Approximate statistical test for comparing supervised classification learning algorithms. Neural Comp. 10(7), 1895–1923 (1998)
8. Demšar, J.: Statistical comparisons of classifiers over multiple data sets. J. Mach. Learn. Res. 7, 1–30 (2006)
9. Domeniconi, C., Yan, B.: Nearest neighbor ensemble. In: Kittler, J., Petrou, M., Nixon, M.S. (eds.) Proc. 17th Int. Conf. Patt. Recogn., Cambridge, UK, pp. 228–231. IEEE Comp. Soc., Los Alamitos (2004)
10. Freund, Y., Schapire, R.E.: Experiments with a new boosting algorithm. In: Saitta, L. (ed.) Proc. 13th Int. Conf. Mach. Learn., Bari, Italy, pp. 148–156. Morgan Kaufmann, San Francisco (1996)
11. Gama, J., Brazdil, P.: Cascade generalization. Mach. Learn. 41(3), 315–343 (2000)
12. Ho, T.K.: The random subspace method for constructing decision forests. IEEE Trans. Patt. Analysis Mach. Intell. 20(8), 832–844 (1998)
13. Kuncheva, L., Whitaker, C.J.: Using diversity with three variants of boosting: aggressive, conservative, and inverse. In: Roli, F., Kittler, J. (eds.) MCS 2002. LNCS, vol. 2364, pp. 81–90. Springer, Heidelberg (2002)
14. Margineantu, D.D., Dietterich, T.G.: Pruning adaptive boosting. In: Fisher, D.H. (ed.) Proc. 14th Int. Conf. Mach. Learn., Nashville, TN, pp. 211–218. Morgan Kaufmann, San Francisco (1997)

15. Quinlan, J.R.: C4.5: Programs for Machine Learning. Morgan Kauffman, San Francisco (1993)
16. Webb, G.I.: MultiBoosting: a technique for combining boosting and wagging. Mach. Learn. 40(2), 159–196 (2000)
17. Witten, I.H., Frank, E.: Data Mining: Practical Machine Learning Tools and Techniques. Morgan Kaufmann, San Francisco (2005)

Improving Supervised Learning with Multiple Clusterings

Cédric Wemmert, Germain Forestier, and Sébastien Derivaux

Abstract. Classification task involves inducing a predictive model using a set of labeled samples. The accuracy of the model usually increases as more labeled samples are available. When one has only few samples, the obtained model tends to offer poor results. Even when labeled samples are difficult to get, a lot of unlabeled samples are generally available on which unsupervised learning can be done. In this chapter, a way to combine supervised and unsupervised learning in order to use both labeled and unlabeled samples is explored. The efficiency of the method is evaluated on various UCI datasets and on the classification of a very high resolution remote sensing image when the number of labeled samples is very low.

Keywords: semi-supervised learning, clustering, few labeled data.

1 Introduction

The number of labeled samples is a crucial issue for supervised classification. If too few examples are given to a classical algorithm, the induced predictive model will have poor performance. Sadly, in many real-world applications, labeled samples are difficult to obtain. This is often due to the cost of a human manual labeling. For example, we can cite all the problems where the user only gives few examples, and the system has to find similar objects in a database (content-base image retrieval, online web-page recommendation, ...). In these cases, only few labeled samples are available although many unlabeled data are present (all the other instances in the database). In web-page recommendation, a user labels interesting pages. It is not possible to ask for the user to produce other samples as he might not know them. The same problem appears with online shop services when a user buys a product and the system wants to automaticaly advise him on other related products. The system only knows which products the user bought and the product rating (collaborative filtering).

Cédric Wemmert · Germain Forestier · Sébastien Derivaux
LSIIT - University of Strasbourg
Pôle API, Bd Sébastien Brant - 67412 Illkirch, France
e-mail: {wemmert, forestier, derivaux}@lsiit.u-strasbg.fr

O. Okun & G. Valentini (Eds.): Appli. of Supervised & Unsuper. Ensemble Meth., SCI 245, pp. 135–149.
springerlink.com © Springer-Verlag Berlin Heidelberg 2009

Another challenging problem is to compute a high accuracy classification when the ratio between the number of available labeled data and the number of features is very small. If the number of features is high, the standard classifiers will need a lot of training samples to perform a relevant classification. This observation is known as the Hughes phenomenon [9]. In remote sensing, the hyperspectral sensors can produce data with a very large number of bands, up to 200 (which means 200 real values for each pixel). With such data, more details can be observed in the land cover, i.e., the number of classes of interest is increased. More features and more classes involve more samples, which are generally expensive and time consuming to acquire. The same observation can be highlighted with the *object-oriented* analysis of Very High Resolution images. With this kind of images, a first step of segmentation is commonly performed to build "geographical objects". These objects are then characterized by many different features (spectral, spatial or contextual). Thus, the dataset to classify is often composed of objects described by a lot of features, but with very few or no samples.

Despite the fact that labeled samples are rare and insufficient, compared to the dataspace dimension, unlabeled samples are generally available in great quantities. Some research work revealed that these samples can be used to improve supervised classification. The main idea is to partially classify the unlabeled samples using the labeled ones, and then to use them to induce a new model [2, 15, 17].

Our method is slightly different of the existing ones, as we use multiple unsupervised classifications to create new features describing the labeled samples. Then, a supervised classification is applied to this new data space. As unsupervised classification creates clusters that tend to maximize intra-cluster similarity and inter-cluster dissimilarity, no labeled sample is needed. The clustering may be seen as a way to resume the distribution of the samples. It is sometimes used to reduce data before classification step.

If two classes form a highly homogeneous cloud in the feature space, one can not expect a clustering to separate them. Even if our approach can weaken this condition, it must be kept in mind.

In this chapter, we first present some related works and justify our method in Sect. 2. Then, the algorithm is described in Sect. 3. In Sect. 4, we present many experiments made on UCI datasets. The results that we obtain are compared with classical supervised methods to quantify the improvement given by the unsupervised clustering . We also present some results obtained on the real-world data extracted from a VHSR (Very High Spatial Resolution) remote sensing image of the urban area of Strasbourg (France). Finally, we give conclusions and draw some perspectives about future work.

2 Related Works

Many previous works have shown that unlabeled data can help to improve the quality of classification when very few labeled samples are available [2, 4, 7, 11, 15].

The first way to exploit unlabeled objects is the co-training method proposed in [2]. The main idea is to use two complementary classifications to iteratively label the unlabeled data. This assumes that two independent and complementary feature sets exist on the data.

To extend this method and avoid the independence and redundancy of the feature sets, which is not realistic in real problems, Goldman and Zhou presented in [7] a co-training strategy which uses unlabeled data to improve the performance of a supervised classifier. Their method uses two different supervised learners which can select some unlabeled data to label the other learner in an iterative way. Experiments have shown that the method increases the accuracy of the ID3 algorithm. More recently, Raskutti et al. [16] present a co-training method that does not necessary need two complementary supervised learning algorithms. The idea is to produce an alternate view on the data by performing an unsupervised classification algorithm on the whole dataset (labeled and unlabeled). Then, the original view and the view built from the clustering are used to create two independent predictors for co-training. In [20], the authors present a co-training approach which assumes to have two views of the data. The method uses the correlation between the two views to produce extra positive and negative samples in an iterative process. Experiments, where only one labeled sample is available, show that the method outperforms other co-training approaches.

Unlike co-training, ASSEMBLE [1] can build semi-supervised ensembles of any size and does not require the domain to have multiple views. ASSEMBLE incorporates self-labeled examples in a Boosting framework. At each iteration of ASSEMBLE examples from the unlabeled set are labeled by the current ensemble and added to the training set. In [4], the empirical study of various semi-supervised learning techniques on a variety of datasets is presented. Different experiments are made to evaluate the influence of the size of the labeled and unlabeled sets, or the effect of noise in the samples. The paper concludes that the performance of the methods is heavily dependent of the field of application and the nature of the dataset. However, using labeled *and* unlabeled samples improves the accuracy in most of the cases.

A more recent approach was discussed in [3]. The idea is to simultaneously compute the clustering and the classification , instead of proceeding in two sequential steps. To achieve this goal, the authors define an objective function that evaluate not only the classification ability but also the clustering ability by mixing two terms: the misclassification rate (for the supervised part) and the clustering impurity (for the unsupervised aspect). A quite similar method is presented in [6] and applied to real marketing datasets.

In the field of remote sensing image classification, some previous work to take advantage of unlabeled data in the classification process was undertaken by Shahshahani and Landgrebe [18]. They proposed three methods to incorporate simultaneously labeled and unlabeled samples in parametric, nonparametric and semi-parametric classifiers. They also gave some results on a small extract of an AVIRIS remote sensing image to show the enhancement of the classification performance.

Later, in [8], the authors presented a new covariance matrix estimator. It produces classification with a higher accuracy that the standard covariance matrix estimation methods, when using a limited training data set. Some experiments on the classification of an agricultural zone of the Nevada, based on an AVIRIS image, showed the efficiency of the estimator on real problems.

More recently, Jia et al. [10] introduced a new method to deal with hyperspectral data. The idea is to cluster the training data and then to associate spectral clusters with information classes, thus building a cluster-space classification. Then, each pixel is classified according to its cluster membership and the membership of the cluster to information classes.

Morgan et al. [13] proposed another approach to solve the problem of the small amount of labeled data. It consists in using a feature reduction scheme that adaptively adjusts itself to the size of the samples dataset. Some experiments are presented on hyperspectral data obtained with the HyMap sensor (Hyperspectral Mapper). The results show that even if the feature space was reduced and the number of samples was very low, the classifications produced have a high accuracy.

The method presented in this paper is slightly different from the ones discussed above. All the co-training methods use the labeled and unlabeled samples together in the training step. If more labeled samples are available, the training step needs to be executed again, which is often costly. In our approach, the unsupervised classification can be seen as a pre-processing step, which is performed only once. Then, depending on the availability of labeled samples, the supervised classification can be computed. This training part is very quick as the number of samples is very small.

3 Description of the Method

3.1 Improving Supervised Classification with Clustering

To better understand how clustering can improve supervised classification, we present here an example of an artificial dataset built from the UCI iris dataset. We selected only two features (petalwidth and sepallength), but we retained the 150 instances and the three classes (iris-virginica, iris-versicolor and iris-setosa). We also selected 6 instances as labeled samples (2 labeled samples per class). The dataset to be classified is presented on Fig. 1.

We performed a simple C4.5 supervised classification of this dataset.Using the six labeled instances shown in Fig. 1, we obtained the following results (see also Table 1 for details on different accuracy rates and Table 2 for the confusion matrix of the result):

- Correctly classified instances: 51.3%
- Incorrectly classified instances: 48.6%
- Kappa statistic: 0.27
- Mean absolute error: 0.3244
- Root mean squared error: 0.5696
- Relative absolute error: 73.0%
- Root relative squared error: 120.8%

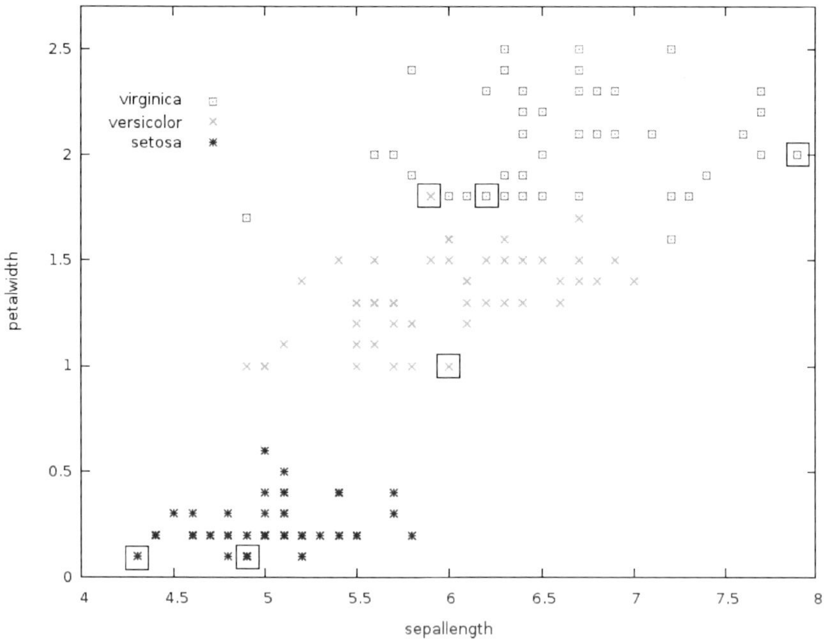

Fig. 1 The original dataset to be classified: the labeled samples used for classification are the 6 instances surrounded by a black squares

Table 1 Detailed accuracy by class for the first result

Class	TP Rate	FP Rate	Precision	Recall	F-Measure	ROC Area
Iris-setosa	0.12	0	1	0.12	0.214	0.56
Iris-versicolor	0.6	0.53	0.361	0.6	0.451	0.535
Iris-virginica	0.82	0.2	0.672	0.82	0.739	0.81
Weighted avg.	0.513	0.243	0.678	0.513	0.468	0.635

Table 2 Confusion matrix for the first result

	Iris-setosa	Iris-versicolor	Iris-virginica
Iris-setosa	6	44	0
Iris-versicolor	0	30	20
Iris-virginica	0	9	41

It is obvious that these bad results are due to the (arbitrary) bad choice of the labeled samples and the low ratio between the number of labeled samples and the number of instances to classify (4%). This is to illustrate the fact that the user does

Table 3 Detailed accuracy by class for the enhanced dataset

Class	TP Rate	FP Rate	Precision	Recall	F-Measure	ROC Area
Iris-setosa	1	0.07	0.877	1	0.935	0.965
Iris-versicolor	0.46	0.09	0.719	0.46	0.561	0.685
Iris-virginica	0.82	0.2	0.672	0.82	0.739	0.81
Weighted avg.	0.76	0.12	0.756	0.76	0.745	0.82

Table 4 Confusion matrix for the enhanced dataset

	Iris-setosa	Iris-versicolor	Iris-virginica
Iris-setosa	50	0	0
Iris-versicolor	7	23	20
Iris-virginica	0	9	41

not necessary know how the dataset is organized, and where the chosen samples are localed in the data space.

Now let us perform three clusterings of the complete dataset to enhance the data description. We used the classical k-means algorithm with 2, 3 and 4 clusters. Then, we ran again the C4.5 algorithm on the new enhanced dataset, now containing five features: the two original attributes and three nominal attributes corresponding to each clustering result, respectively: (sepallength, petalwidth, kmeans2, kmeans3, kmeans4). The results are the following:

- Correctly classified instances: 76%
- Incorrectly classified instances: 24%
- Kappa statistic: 0.64
- Mean absolute error: 0.16
- Root mean squared error: 0.4
- Relative absolute error: 36.0%
- Root relative squared error: 84.8%

The results are much better: all the accuracy rates and the statistical quality criteria are higher (Table 3), and as shown in Table 4, the problem with the iris-setosa class has been partially solved.

3.2 The Proposed Method

The main idea of the proposed method is to improve the classification by first producing a clustering on the dataset. The clustering, computed on all the labeled and unlabeled objects, regroups the similar objects together, maximizing the intracluster similarity and the intercluster dissimilarity. If the classes of the problem are well separated in the feature space, we should be able to associate to each cluster one of the classes, using the class of the labeled samples which belong to the cluster.

Unfortunately, in real problems, classes are generally not well separated. It is then possible to have samples from different classes in one cluster, or no sample in others. To address this issue, the proposed method uses a combination of multiple clusterings.

Let $\mathscr{X}$ denote a set of n data objects $x_j \in \mathscr{X}$. We consider a q-class classification problem with m labeled and l unlabeled objects where m is very low and $l >> m$.

Let $\mathscr{L}$ be the set of labeled objects of $\mathscr{X}$:

$$\mathscr{L} = ((x_1, y_1), \ldots, (x_m, y_m)) \ , \tag{1}$$

where $y_i \in \{1, \ldots, q\}$ are the target values of the samples.

Let $\mathscr{U}$ be the set of unlabeled objects of $\mathscr{X}$:

$$\mathscr{U} = (x_{m+1}, \ldots, x_{m+l}) \ . \tag{2}$$

A clustering is a partition of $\mathscr{X}$ into k clusters, and is represented as an n-dimensional cluster labeling vector

$$C = \left(C^j\right)_{j=1}^{n} \in \mathscr{C}^n \ , \tag{3}$$

where $\mathscr{C} = \{c_1, \ldots, c_k\}$. We consider here b clusterings of the dataset $\mathscr{X}$, represented as a $n \times b$ matrix of cluster labeling vectors. Let $\mathbf{C}$ denote this set of clusterings, $\mathbf{C} = \{C_1, \ldots, C_b\}$. The idea is to transform each labeled sample x_i, $\forall i < m$, a new feature vector

$$v(x_i) = \left(C_1^i, \ldots, C_b^i, y_i\right) \ , \tag{4}$$

where C_j^i is the cluster label assigned by the jth clustering method C_j to x_l. Then, a predictive model $\mathbf{P} : \mathscr{X} \to \{1, \ldots, q\}$ can be induced from this new dataset $V = \{v(x_i)\}_{i=1}^{m}$, using a classical supervised learning method. Finally, the label $\mathbf{P}(x_i)$ is assigned to each unlabeled object x_i of $\mathscr{U}$.

The complete process of classification is given in Algorithm 2.

Algorithm 2. Classification with few labeled data

1: apply b clustering algorithms to the dataset $\mathscr{X}$ and to form $\mathbf{C} = \{C_1, \ldots, C_b\}$
2: **for all** $x_i \in \mathscr{L}$ **do**
3: $\quad v(x_i) = \left(C_1^i, \ldots, C_b^i, y_i\right)$ where C_k^i is the cluster label assigned to x_i by the kth clustering method C_k and y_i is the class label of x_i
4: **end for**
5: apply a supervised learning method to produce a predictive model $\mathbf{P}$ from $V = \{v(x_i)\}_{i=1}^{m}$
6: **for all** $x_j \in \mathscr{U}$ **do**
7: $\quad$ assign $\mathbf{P}(C_1^j, \ldots, C_b^j)$ to x_j, where C_k^j is the cluster label given to x_j by the kth clustering method C_k
8: **end for**

---

4 Experiments

4.1 Artificial Benchmark Evaluation

The method described in the previous section has been evaluated on various datasets of the UCI repository [14]. Table 5 presents information about these datasets.

To apply the proposed method, we first had to choose how many clusterings will be run on the dataset (i.e., how many attributes each object in the new data space will have), and then the different clustering methods. We chose four different configurations to study the importance of the number of clusterings. The four configurations are referred as follows:

1. *simple*: one EM (Expectation-Maximization [5])
2. *low*: one EM and one KMeans [12]
3. *medium*: two EM and two KMeans
4. *high*: c EM and c KMeans (with c the number of classes of the datasets).

Table 5 Information about the different datasets

Dataset	Nb. classes	Nb. attributes	Nb. objects
iris	3	4	150
wine	3	13	178
ionosphere	2	34	351
diabetes	2	8	768
breast-w	2	9	699
anneal	5	38	898

Each method was run with a number of clusters equal to the number of classes actually present in the dataset, except for the *high* configuration, where the clustering method $k \in \{2,\ldots,c\}$ had k clusters. The four configurations were compared with other algorithms, taken in different families of learning algorithms: the standard tree inducer C4.5, Naive Bayes and 1-nearest-neighbor (1-NN) algorithm. Results are presented in Tables 6 and 7. The number of samples used is indicated at the beginning of each line. We chose to evaluate the method when $2, 4, 8$ and 16 samples per class were available.

For the four configurations of the proposed method, 50% of the remaining data were used for the unsupervised learning, and the other 50% for the evaluation of the method. We chose to use the Naive Bayes classifier as the supervised method in Algorithm 2.

As performance may greatly differ depending on the labeled set, the experiments were carried out 20 times and the results were averaged. At each run, the different sets are filled with randomly chosen samples.

The proposed method outperformed the supervised learning when the number of samples was very low (2 or 4 samples per class). For example, it can be noticed on the *breast-w* dataset that with two samples the *high* configuration reaches 95.73%

Table 6 Results of the proposed method with 4 configurations, according to the different datasets with 2, 4, 8 and 16 labeled samples available. Values correspond to the means and the standard deviations on 20 runs

Dataset		Simple	Low	Medium	High
iris	(2)	72.43(±16.7)	71.18(±21.5)	**84.10**(±**9.9**)	82.92(±12.0)
	(4)	83.62(±10.9)	87.75(±7.9)	84.57(±11.4)	**87.90**(±**5.8**)
	(8)	89.92(±5.2)	88.81(±5.2)	90.32(±5.5)	89.84(±3.9)
	(16)	88.43(±5.2)	89.02(±4.9)	89.51(±5.2)	91.47(±5.7)
wine	(2)	79.83(±22.0)	85.29(±14.5)	**93.78**(±**3.6**)	90.29(±9.6)
	(4)	90.90(±7.8)	92.47(±8.9)	**96.02**(±**2.2**)	94.52(±3.4)
	(8)	95.19(±2.6)	94.09(±4.0)	**96.36**(±**2.0**)	96.10(±2.1)
	(16)	94.69(±3.7)	95.46(±3.6)	**96.23**(±**2.4**)	94.77(±3.5)
breast-w	(2)	84.34(±18.8)	85.73(±18.1)	95.39(±1.1)	**95.73**(±**2.1**)
	(4)	90.09(±14.3)	92.30(±9.7)	94.51(±1.5)	**95.49**(±**1.7**)
	(8)	94.65(±1.7)	94.94(±1.7)	94.96(±1.0)	**95.60**(±**1.6**)
	(16)	94.93(±1.2)	94.70(±1.4)	95.25(±1.4)	**96.29**(±**1.4**)
diabetes	(2)	52.63(±9.5)	54.96(±9.7)	55.75(±5.6)	**59.78**(±**6.3**)
	(4)	52.38(±5.7)	53.53(±6.5)	58.05(±10.6)	57.89(±8.2)
	(8)	57.89(±8.6)	57.17(±8.0)	56.62(±8.1)	62.97(±6.9)
	(16)	58.53(±8.7)	58.89(±8.0)	61.86(±4.6)	62.65(±8.7)
ionosphere	(2)	57.53(±14.8)	59.08(±13.6)	**67.87**(±**13.1**)	62.39(±14.5)
	(4)	59.42(±12.0)	64.83(±12.0)	67.65(±10.6)	**69.30**(±**10.2**)
	(8)	67.47(±10.8)	68.07(±11.0)	70.36(±11.8)	70.39(±10.7)
	(16)	69.69(±7.8)	69.78(±9.5)	72.50(±4.3)	76.03(±7.1)
anneal	(2)	61.68(±9.1)	56.48(±14.5)	67.05(±8.4)	68.96(±7.0)
	(4)	67.60(±9.0)	62.51(±12.2)	73.57(±5.2)	75.65(±4.4)
	(8)	68.94(±7.9)	72.08(±5.6)	76.22(±4.0)	78.14(±4.1)
	(16)	71.76(±6.4)	68.78(±7.5)	76.35(±5.2)	79.95(±3.6)

instead of 83.52% for the best supervised approach (1-NN). One can notice that the best accuracy amongst the different configurations is reached with the *medium* and *high* ones.

This result enforces the intuitive feeling that adding more clusterings improves the result, as the objects are described with more details (i.e. have more attributes). Figures 2 (a),(b),(c) and (d) illustrate this result and show the increase of accuracy according to the number of available samples.

This figure also illustrates that when the number of samples increases, the supervised approaches give better results. For example, when 16 samples are available, the supervised methods outperformed the proposed approach on 5 of the 6 datasets. It confirms that supervised methods need several examples to produce efficient predictive models.

As stated in Introduction, if the data distribution is not correlated with the class information, using clustering is useless. This affirmation can be study on the *anneal* dataset, where the proposed semi-supervised approach obtains worse results than 1-nearest-neighbor, regardless of the number of available samples.

Table 7 Results for classical methods according to the different datasets with 2,4,8 and 16 labeled samples available. Values correspond to the means and the standard deviations on 20 runs

Dataset		C4.5	1-NN	NB
iris	(2)	55.97(±13.1)	79.58(±13.9)	65.00(±15.5)
	(4)	80.94(±11.9)	87.17(±12.0)	84.49(±11.7)
	(8)	90.16(±4.8)	**94.52**(±**3.7**)	92.06(±4.3)
	(16)	93.14(±4.2)	**95.98**(±**2.1**)	95.29(±2.9)
wine	(2)	50.35(±6.9)	88.14(±6.3)	55.47(±17.6)
	(4)	70.24(±12.4)	90.36(±6.5)	70.78(±11.6)
	(8)	84.61(±5.9)	93.12(±4.3)	92.53(±6.2)
	(16)	86.85(±3.8)	95.77(±2.2)	95.62(±3.2)
breast-w	(2)	63.38(±16.9)	83.52(±17.3)	83.03(±15.2)
	(4)	85.97(±9.1)	94.29(±3.0)	84.21(±16.8)
	(8)	89.34(±3.2)	94.72(±2.0)	94.46(±1.7)
	(16)	90.06(±2.6)	94.45(±2.7)	95.07(±1.7)
diabetes	(2)	52.45(±6.8)	55.56(±7.4)	51.83(±5.0)
	(4)	**63.03**(±**5.7**)	58.49(±5.6)	55.05(±6.0)
	(8)	63.99(±4.9)	63.34(±5.2)	**64.02**(±**5.86**)
	(16)	66.20(±5.0)	65.16(±3.7)	**68.82**(±**3.2**)
ionosphere	(2)	52.56(±10.2)	56.32(±9.6)	58.51(±9.4)
	(4)	63.58(±8.3)	67.18(±9.5)	63.02(±10.4)
	(8)	69.52(±8.7)	73.10(±9.2)	**82.05**(±**5.5**)
	(16)	81.56(±5.5)	78.88(±7.7)	**82.94**(±**4.2**)
anneal	(2)	54.55(±10.6)	**72.02**(±**8.6**)	45.09(±8.9)
	(4)	72.94(±10.0)	**81.89**(±**4.6**)	69.51(±7.9)
	(8)	83.80(±5.6)	**87.18**(±**3.3**)	85.16(±5.2)
	(16)	**92.27**(±**3.6**)	90.09(±2.1)	88.80(±2.8)

We also evaluated the influence of the size of the dataset available for the unsupervised learning. Figures 2 (e) and (f) show the evolution of the accuracy according to the size of the dataset used for the unsupervised learning (10%, 25% and 50% of the datasets).

Figure 2 (f), corresponding to the evaluation on the *wine* dataset, indicates that the increase of unlabeled samples available helps to produce better results. This is due to the ability of the clustering to better grasp the dataset when the density of objects increases.

4.2 Real Data Evaluation

As presented in Introduction, in the field of remote sensing classification , the problem of the low ratio between the number of features and the number of samples is really important when dealing with hyperspectral data or with very high resolution images. Indeed, in this last case, the classification is basically done in two steps: segmentation of the image is performed, producing a set of regions (i.e., groups of

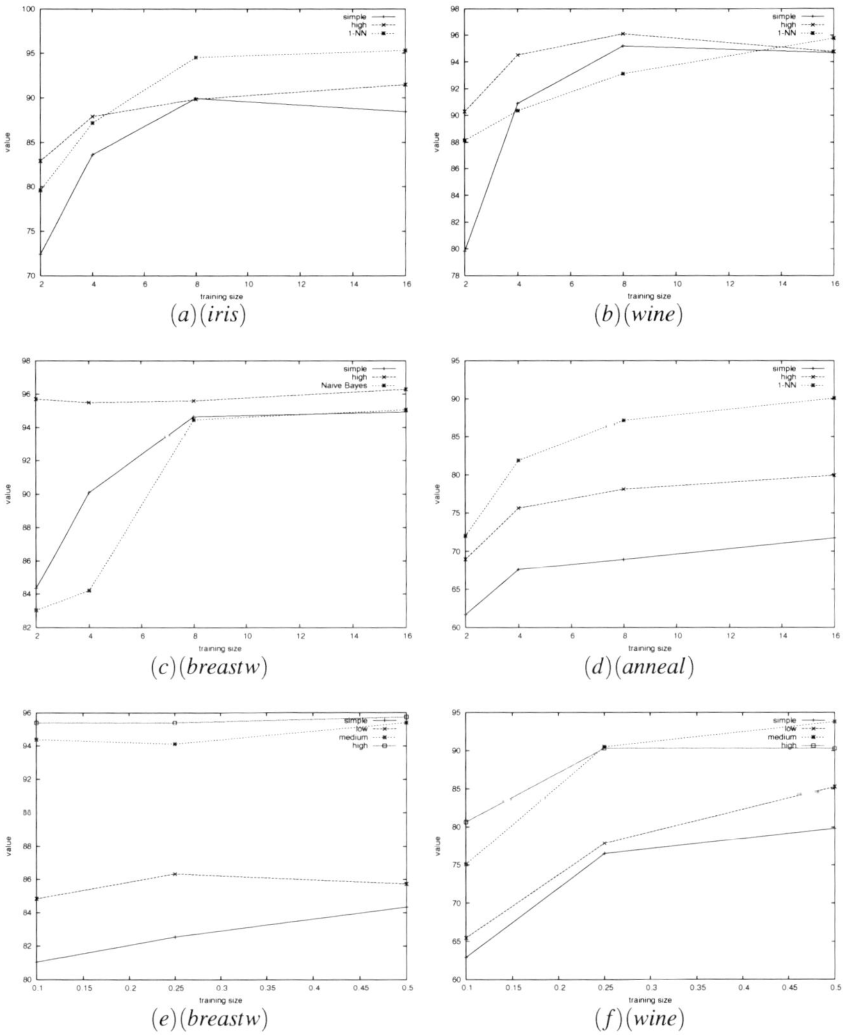

Fig. 2 (a),(b),(c) and (d) show the accuracy according to the number of available samples; (e) and (f) show the accuracy according to the size of the dataset used for the unsupervised learning

homogeneous pixels); then, these regions are characterized by many features (spectral attributes, geometrical features, spatial attributes, etc.). Thus, we have a new dataset to classify, composed of few regions (compared to the number of pixels in the image), but characterized by more features.

We present here an experiment of the classification of a very high remote sensing image showing an urban area of the city of Strabourg (France). The input data is a pan-sharpened Quickbird image with 4 spectral bands and a spatial resolution of 0.7

Fig. 3 Extract of the Quickbird image

meter, i.e., a pixel on the image represents a square of 0.7×0.7m on the ground. The image is given in Fig. 3.

After having computed the segmentation of this image by using the watershed segmentation algorithm [19], we characterized each region found by the following attributes:

- 8 attributes representing the mean and the standard deviation of each spectral channel of the pixels composing the region;
- 8 attributes representing the mean and the standard deviation of the value of each spectral channel over the sum of all channels values;
- 2 features representing the mean and the standard deviation of the mean of all spectral channels;
- 2 features calculated as the mean and the standard deviation of the NDVI (*Normalized Difference Vegetation Index*) of pixels composing the region;
- 1 attribute representing the area covered by the region;

Table 8 Results on the remote sensing dataset with different number of samples (1, 2, 4, 8, 16 and 32). Values correspond to the means and the standard deviations on 100 runs

Samples	Low	Medium	High	C4.5	1-NN	NB
1	65.6(±21.2)	90.7(±14.7)	89.2(±16.4)	35.2(±10.4)	67.1(±19.2)	57.4(±23.8)
2	88.7(±16.6)	97.0(±8.1)	97.4(±8.2)	48.9(±17.4)	83.0(±16.4)	61.6(±18.1)
4	96.7(±8.1)	98.7(±2.7)	99.0(±2.5)	72.2(±14.9)	97.4(±3.9)	83.1(±15.0)
8	98.6(±2.8)	98.8(±2.7)	99.0(±2.3)	87.0(±10.5)	98.9(±2.7)	95.7(±7.4)
16	98.4(±3.4)	98.4(±3.1)	99.0(±2.8)	91.0(±8.5)	99.5(±1.8)	98.8(±3.0)
32	98.4(±4.1)	98.1(±4.7)	99.5(±2.1)	95.6(±6.6)	99.9(±1.1)	99.2(±2.8)

- 1 attribute calculated as the elongation of the shape represented by the region;
- 1 attribute measuring the fitting of the shape represented by the region and its oriented bounding box.

As a result, the dataset generated is composed of 186 objects described by 23 features, divided in three classes. As in the previous experiment, we chose three different configurations to study the importance of the number of clusterings. The three configurations are referred to as *low* (two EM clusterers), *medium* (two EM and two KMeans) and *high* (6 EM and 6 KMeans). We also compare the result with three classical supervised classification methods (C4.5, 1-NearestNeighbor and

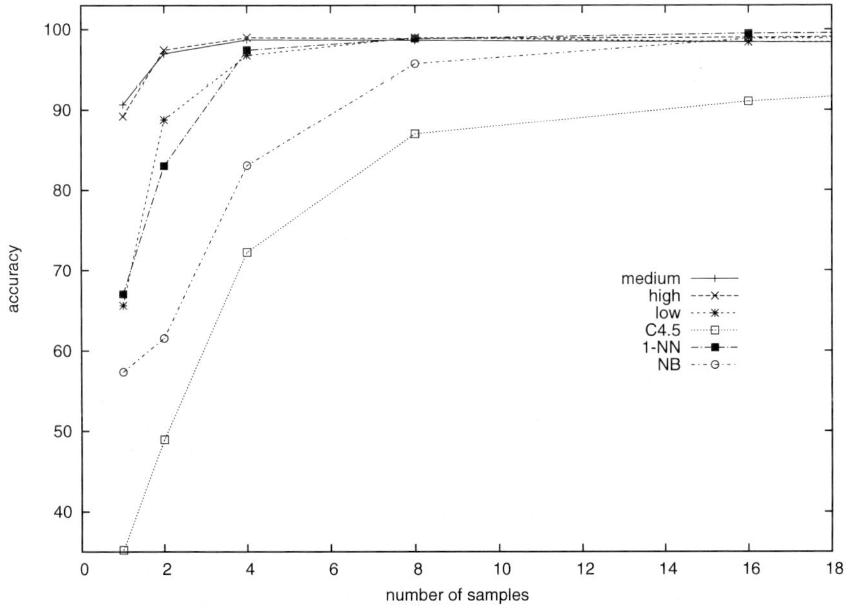

Fig. 4 Evaluation of the accuracy on the remote sensing dataset compared to the number of samples

Naive Bayes). Again, the experiment is performed 100 times and the results are averaged. At each run, the different sets are filled with randomly chosen samples. The results obtained with different number of samples are given in Table 8.

Again, these results confirm that more clusterings improve the result, as the *high* configuration is almost always the best when only few samples are available. We also notice that the result is only better than with the supervised 1-NN method when there are very few labeled objects (less than 16, which corresponds to less than 10% of the data). Figure 4 illustrates this and shows the increase of accuracy according to the number of available samples. It confirms that the minimal number of samples is needed for supervised methods in order to produce efficient results. But this lack of information can be balanced by the use of many clusterings on the unlabeled data.

5 Conclusion

In this chapter, it has been shown that many clustering results can be combined through supervised classification in order to achieve better accuracy. The method presented has shown good results when the number of labeled samples is very low and when many unlabeled samples are available.

Nevertheless some questions remain open. How many clustering algorithms should be used ? Is it better to enhance diversity amongst them ? How to detect in advance if a specific dataset can use this method ? These few questions give us some directions to consider in order to improve the present work.

References

1. Bennett, K.P., Demiriz, A., Maclin, R.: Exploiting unlabeled data in ensemble methods. In: Proc. 8th ACM SIGKDD Int. Conf. Knowledge Discovery and Data Mining, Edmonton, AB, pp. 289–296. ACM, New York (2002)
2. Blum, A., Mitchell, T.: Combining labeled and unlabeled data with co-training. In: Proc 11th Annual Conf. Comp. Learn. Theory, Madison, WI, pp. 92–100. ACM, New York (1998)
3. Cai, W., Chen, S., Zhang, D.: A simultaneous learning framework for clustering and classification. Pattern Recogn. (in press)
4. Chawla, N.V., Karakoulas, G.J.: Learning from labeled and unlabeled data: an empirical study across techniques and domains. J. Artif. Intell. Res. 23, 331–366 (2005)
5. Dempster, A.P., Laird, N.M., Rubin, D.B.: Maximum likelihood from incomplete data via the EM algorithm. J. the Royal Stat. Society Series B 39(1), 1–38 (1977)
6. Deodhar, M., Ghosh, J.: A framework for simultaneous co-clustering and learning from complex data. In: Berkhin, P., Caruana, R., Wu, X. (eds.) Proc. 13th ACM SIGKDD Int. Conf. Knowledge Discovery and Data Mining, San Jose, CA, pp. 250–259. ACM, New York (2007)
7. Goldman, S., Zhou, Y.: Enhancing supervised learning with unlabeled data. In: Langley, P. (ed.) Proc. 17th Int. Conf. Mach. Learn., Stanford, CA, pp. 327–334. Morgan Kaufmann, San Francisco (2000)
8. Hoffbeck, J.P., Landgrebe, D.A.: Covariance matrix estimation and classification with limited training data. IEEE Trans. Pattern Analysis Mach. Intell. 18(7), 763–767 (1996)

9. Hughes, G.: On the mean accuracy of statistical pattern recognizers. IEEE Trans. Inf. Theory 14(1), 5–63 (1968)
10. Jia, X., Richards, J.A.: Cluster-space representation for hyperspectral data classification. IEEE Trans. Geoscience and Remote Sensing 40(3), 593–598 (2002)
11. Joachims, T.: Transductive inference for text classification using support vector machines. In: Bratko, I., Dzerovski, S. (eds.) Proc. 16th Int. Conf. Mach. Learn., Bled, Slovenia, pp. 200–209. Morgan Kaufmann, San Francisco (1999)
12. MacQueen, J.: Some methods for classification and analysis of multivariate observations. In: Le Cam, L.M., Neyman, J. (eds.) Proc. 5th Berkeley Symp. Math. Stat. Probability, Berkeley, CA, pp. 281–297 (1967)
13. Morgan, J.T., Ham, J., Crawford, M.M., Henneguelle, A., Ghosh, J.: Adaptive feature spaces for land cover classification with limited ground truth data. Int. J Pattern Recogn. Artif. Intell. 18(5), 777–799 (2004)
14. Newman, D.J., Hettich, S., Blake, C.L., Merz, C.J.: UCI Repository of machine learning databases, University of California, Irvine, Dept. of Information and Computer Sciences, `http://www.ics.uci.edu/~mlearn/MLRepository.html`
15. Nigam, K., McCallum, A.K., Thrun, S., Mitchell, T.: Text classification from labeled and unlabeled documents using EM. Mach. Learn. 39(2–3), 103–134 (2000)
16. Raskutti, B., Ferrá, H.L., Kowalczyk, A.: Combining clustering and co-training to enhance text classification using unlabelled data. In: Proc. 8th ACM SIGKDD Int. Conf. Knowledge Discovery and Data Mining, Edmonton, AB, pp. 620–625. ACM, New York (2002)
17. Seeger, M.: Learning with labeled and unlabeled data. University of Edinburgh, Scotland (2002), `http://www.kyb.mpg.de/bs/people/seeger/papers/review.pdf`
18. Shahshahani, B.M., Landgrebe, D.: The effect of unlabeled samples in reducing the small sample size problem and mitigating the Hughes phenomenon. IEEE Trans. Geoscience and Remote Sensing 32(5), 1087–1095 (1994)
19. Vincent, L., Soille, P.: Watersheds in digital spaces: an efficient algorithm based on immersion simulations. IEEE Trans. Pattern Analysis Mach. Intell. 13(6), 583–598 (1991)
20. Zhou, Z.-H., Zhan, D.-C., Yang, Q.: Semi-supervised learning with very few labeled training examples. In: Proc. 22nd AAAI Conf. Artif. Intell., Vancouver, BC, pp. 675–680. AAAI Press, Menlo Park (2007)

The Neighbors Voting Algorithm and Its Applications

Gabriele Lombardi, Elena Casiraghi, and Paola Campadelli

Abstract. In the last ten years the tensor voting framework (TVF), proposed by Medioni at al., has proved its effectiveness in perceptual grouping of arbitrary dimensional data. In the computer vision and image processing fields, this algorithm has been applied to solve various problems like stereo-matching, 3D reconstruction, and image inpainting . In this paper we propose a new technique, inspired to the TVF, that allows to estimate the dimensionality and normal orientation of the manifolds underlying a given point set. These features are encoded in tensors that can be considered as weak classifiers, whose combination is then used as a strong classifier to solve different classification problems. To prove the effectiveness of the described algorithm, three problems are considered: clustering by dimensionality estimation, image classification by manifold learning, and image inpainting by texture learning.

Keywords: tensor voting, ensemble methods, clustering, classification, image inpainting.

1 Introduction

The Tensor Voting Framework (TVF) proposed in [5] by Medioni at al. and further developed over the past ten years [8, 11] is a computational framework that can address a wide range of computer vision problems in a unified way. The framework has been designed based on perceptual principles, formulated by Gestalt psychologists in order to infer salient structures from sparse and noisy data. It has been successfully applied to image processing problems, such as stereo matching [9, 12], image repairing [6], boundary inference [16], and motion segmentation [13].

The TVF is a general methodology suitable for problems of any dimensionality that can be formulated as a perceptual organization problem. In [10] Medioni

Gabriele Lombardi · Elena Casiraghi · Paola Campadelli
Dipartimento di Scienze dell'Informazione, Università degli Studi di Milano,
Via Comelico 39-41, Milano, Italy
e-mail: {lombardi,casiraghi,campadelli}@dsi.unimi.it

O. Okun & G. Valentini (Eds.): Appli. of Supervised & Unsuper. Ensemble Meth., SCI 245, pp. 151–173.
springerlink.com © Springer-Verlag Berlin Heidelberg 2009

at al. applied tensor voting to the problem of learning a target function from a set of points. To this aim, they proposed a new implementation which can deal efficiently with data in arbitrary dimensional spaces; their technique prevents voting fields pre-computation and storing, and prevents numeric integrations too. In [2] a new algorithm to efficiently compute tensorial votes is presented; it employs a new family of decay functions to reduce the additive noise effects on the inferred structure. In [15] Medioni and Tang described an augmentation of the TVF consisting in the estimation of the manifold curvature and its usage to the aim of improving the manifold inference precision.

Although TVF could be applied to data of arbitrary dimension, its high time and space complexity discourages its application in case of high dimensional spaces, in fact its time complexity $O(N^2D^3)$ is grows cubically with respect to the space dimensionality. To overcome this problem we have developed an approximation of TVF, named *Neighbors Voting* algorithm NV, which is iterative and consists in the generation of tensorial votes between neighboring points. NV is computationally more efficient, in fact, its time complexity is $O(N^2D^2)$, and the single vote generation operations can be easily implemented by using SIMD microprocessor instructions, resulting in a much faster computational tool.

Like TVF, the NV algorithm is iterative and consists in the generation of tensorial votes between neighboring points. In this paper, which is an extension of the previous work [3], where NV was at first introduced, we will show that each pair 'point-tensor' can be considered as a (weak) classifier for the classification problem; the combination of weak classifiers is then used as a strong classifier to infer the dimension of the underlying manifold. In this work, we demonstrate the efficiency and effectiveness of our approach by applying it to solve three different problems (clustering , classification , and image inpainting) on both synthetic and real data.

The paper is organized as follows. Section 2 briefly recalls the theory underlying the TVF. Section 3 presents the NV theory and describes the developed algorithm. Section 4 describes three applications based on the NV algorithm. Section 5 discusses the results achieved by the three applications. Section 6 reports about conclusions and future work.

2 The Tensor Voting Framework

The TVF is summarized in this section. Our notation uses capital letters for matrices, bold characters for matrices and vectors, all the vectors are column-vectors, calligraphic capital letters are sets. To maintain uncluttered notation, function parameters are not reported when their meaning is clear from the context.

The TVF is a mechanism that forces the interaction among input tokens (points) in order to infer salient perceptual structures.

Each token is associated with a local potential orientation of the manifold going through it, and the orientation information is propagated from each token to its neighbors via a voting operation. Votes are cast from one location to another, forcing orientation updates. This information diffusion process allows to characterize the tokens so that:

- tokens that receive no vote can be classified as outliers,
- tokens that receive votes from almost every direction identify a locally unoriented manifold,
- tokens that receive votes mainly from a well defined direction describe an oriented manifold.

More precisely, every token is a location where an orientation is defined. To manage orientations, symmetric non-negatively defined second order tensors[1] are used. The eigensystem of these structures, in $\Re^n$, consists of n orthonormal eigenvectors, and n non-negative associated eigenvalues. The eigenvectors describe the orientation of the underlying manifold, whilst the eigenvalues give a confidence for each direction; geometrically, each tensor represents a hyper-ellipsoid.

Each vote cast by emitters is itself a tensor.

Given the eigensystem $(\mathbf{X}, \Lambda)$, where $\mathbf{X}$ is an orthogonal matrix composed column wise by the orthonormal eigenvectors, and Λ is a diagonal matrix containing the corresponding eigenvalues, the associated tensor can be computed as: $\mathbf{T} = \mathbf{X}\Lambda\mathbf{X}^T = \sum_{i=1}^{n} \lambda_i \mathbf{e}_i \cdot \mathbf{e}_i^T$, where the λ_i are the eigenvalues in non-increasing order, and the $\mathbf{e}_i$ are the corresponding eigenvectors. $\mathbf{T}$ can be decomposed into base tensors as follows:

$$\mathbf{T} = \sum_{i=1}^{n-1} \left((\lambda_i - \lambda_{i+1}) \sum_{j=1}^{i} \mathbf{e}_j \cdot \mathbf{e}_j^T \right) + \lambda_n \sum_{j=1}^{n} \mathbf{e}_j \cdot \mathbf{e}_j^T \,, \tag{1}$$

with n non-zero eigenvalues and $1 \leq i \leq n$. The last term in Eq. 1 is an unoriented (ball) tensor, the other $n-1$ terms are oriented tensors. The $(\lambda_i - \lambda_{i+1})$ and the λ_n coefficients are called saliency values and are denoted by s_i.

Each decomposed tensor term is useful to identify a manifold dimensionality: as an example, in $\Re^2$ the ball tensor identifies filled regions, and the oriented tensor (stick) describes curves.

Tensor composition can be obtained summing tensors component by component, thus allowing to merge uncertainties and to reinforce coherent orientations.

In the TVF, each token casts tensorial votes to neighboring tokens, and it is updated by substituting its tensor with the sum of received votes.

The voting fields, emitted by tokens, play a central role in the data information propagation process, so that the vote generating function must be chosen carefully. The simplest voting field is the stick field in $\Re^2$: given a stick tensor $\mathbf{E}_{stick}$ (emitter), and a point $\mathbf{p}$ (receiver), the vote $\mathbf{V}_{stick}$ computed in $\mathbf{p}$, shown in Fig. 1 represents the most likely normal, in $\mathbf{p}$, to the curve that we want to infer, according to the curve's normal defined by the voter $\mathbf{E}_{stick}$.

[1] From now on, they will be simply referred as tensors.

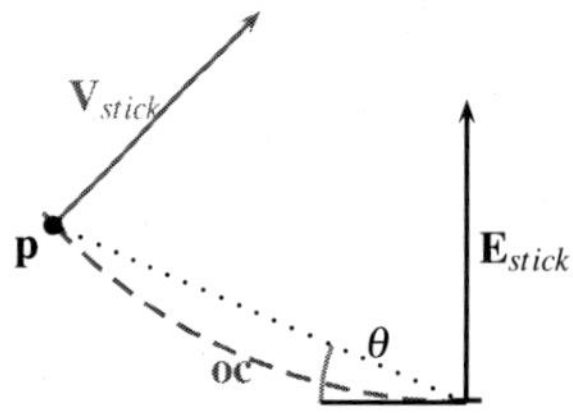

Fig. 1 Stick vote in $\Re^2$: in $\mathbf{p}$ the best orientation $\mathbf{V}_{stick}$ is the normal to the osculating circle **oc**

The vote $\mathbf{V}_{stick}$ can be computed as follows:

$$\mathbf{V}_{stick}(\mathbf{p}) = DF(\mathbf{p}) \begin{bmatrix} -\sin(2\theta) \\ \cos(2\theta) \end{bmatrix} \begin{bmatrix} -\sin(2\theta) & \cos(2\theta) \end{bmatrix} , \tag{2}$$

where the dependency of $\mathbf{V}_{stick}$ from $\mathbf{E}_{stick}$ is in the computation of the angle θ, and $DF(\cdot)$ is the decay function that controls the vote intensity with respect to the length s and curvature k of the arc $\mathbf{c}$. More precisely:

$$DF(\mathbf{p};\sigma) = \exp\left(-\frac{s(\mathbf{p})^2 + ck(\mathbf{p})^2}{\sigma^2} \right) . \tag{3}$$

The parameter σ controls the scale, whereas the constant c is used to maintain the isotropy. The arc length and the curvature are calculated as follows:

$$s(\mathbf{p}) = \frac{\theta \|\mathbf{p}\|}{\sin(\theta)},\ k(\mathbf{p}) = \frac{2\sin(\theta)}{\|\mathbf{p}\|} . \tag{4}$$

Moreover, the decay function is forced to zero when $\theta > \frac{\pi}{4}$.

Given the stick field in $\Re^2$, the ball field can be obtained integrating the stick one over all its possible rotations, that is:

$$\mathbf{V}_{ball}(\mathbf{p}) = \int_0^{\pi} \mathbf{R}_\alpha \mathbf{V}_{stick}(\mathbf{R}_\alpha \mathbf{p}) \mathbf{R}_\alpha^T d\alpha , \tag{5}$$

where $\mathbf{R}_\alpha$ represents a 2×2 rotation matrix.

The voting algorithm is conceptually a simple task: given a set of points $\{\mathbf{p}_i \in \Re^D\}$, a ball tensor (identity) is generated as initial emitter for each $\mathbf{p}_i$; then voting passes are iterated at least two times to improve the manifold orientation estimation encoded in the set of tensors $\mathbf{T}_i$ associated with the points. A voting pass consists of the following steps:

1. each receiver is initialized with a null tensor;
2. for each emitter $(\mathbf{p}_i, \mathbf{T}_i)$:
   a. determine the set of receivers $(\mathbf{p}_j, \mathbf{T}_j)$ distant less than 3σ from the voter[2];

[2] The radius depends on the voting field edge size. The value 3σ allows to capture more than the 99% of the decay volume.

b. for each receiver j compute the vote $\mathbf{V}_i^j$ cast to it by the emitter i as follows:
   i. decompose $\mathbf{T}_i$ as defined in Eq. (1), thus obtaining the base tensors $\mathbf{T}_{i,k}$, with $1 \leq k \leq D$, where D is space dimensionality;
   ii. for each base tensor $\mathbf{T}_{i,k}$ compute the tensorial vote $\mathbf{V}_{i,k}^j$ casted according to the associated voting field; as an example, in $\Re^2$ the computed votes are: the stick vote (see Eq. (2), and the ball one (see Eq. (5));
   iii. compute the emitted vote as $\mathbf{V}_i^j = \sum_k s_k \mathbf{V}_{i,k}^j$ where s_k are the saliency values of $\mathbf{T}_i$;
c. update the receiver's tensor by adding $\mathbf{V}_i^j$ to it.

After each voting pass execution, the output tensors accumulated in the receivers can be used as emitters during the next voting pass.

The TVF is a general technique that could be applied in arbitrary dimensional spaces; nevertheless, when the space dimensionality is high this method cannot be used in practice, due to its high time and space complexity. Indeed, due to the curse of dimensionality, its time and space complexity grows as $\Theta(N\log(N)D^3)$ and $O(N^2D^3)$, where N is the number of data points and D is the space dimensionality[3]. The technique proposed in this paper is much faster in high dimensional spaces; it has a lower time complexity ($\Theta(N\log(N)D^2)$), and has less memory requirements ($O(ND^2)$). The drawback is that the NV technique is an approximation of the TVF, thus generating less precise results.

3 The Neighbors Voting Algorithm

In this section we describe an algorithm that is a simplified version of the TVF; it obtains similar results with a lower time and space computational cost. To reduce the complexity we simplify the vote generation algorithm by avoiding the base tensors decomposition step, so that each emitter directly generates the vote on each receiver. Moreover, our voting scheme approximates tensor orientation, thus preventing the time-consuming tensor rotation operations. To compensate for the missing information we add a nonlinear filtering step that allows to generate local classification for the manifold dimensionality.

As in the TVF, the initialization of the NV algorithm associates ball tensors to each point $\mathbf{p}_i$ in the D dimensional space, so that "tokens" are defined with the pairs $(\mathbf{p}_i, \mathbf{T}_i)$. With this initialization no information is available about the manifold orientations, so that the first voting pass is different from the others; in [10] Medioni et. al. give the following equation to compute the ball tensorial vote cast from the emitter i to the receiver j:

$$\mathbf{T}_i^j = W_i^j \left(\mathbf{I} - \frac{(\mathbf{p}_j - \mathbf{p}_i) \cdot (\mathbf{p}_j - \mathbf{p}_i)^T}{\| \mathbf{p}_j - \mathbf{p}_i \|^2} \right) , \tag{6}$$

[3] The $N\log(N)$ part of the $\Theta(\cdot)$ mean-case complexity notation is due to the (realistic) hypothesis that there are approximately $\log(N)$ neighbors for each data point and the used data structure allows to find them in logarithmic time.

where $W_i^j = DF(\mathbf{p}_j - \mathbf{p}_i)$ is the ball decay function, that is the unnormalized Gaussian function. Summing over all the emitters i, we obtain the vote at the receiver j:

$$\mathbf{T}^j = \sum_{\forall i} \mathbf{T}_i^j = W^j \mathbf{I} - \sum_{\forall i} W_i^j \frac{(\mathbf{p}_j - \mathbf{p}_i) \cdot (\mathbf{p}_j - \mathbf{p}_i)^T}{\| \mathbf{p}_j - \mathbf{p}_i \|^2} = W^j \mathbf{I} - \mathbf{A}^j \tag{7}$$

where $\| \cdot \|$ is the Euclidean norm, $W^j = \sum_{\forall i} W_i^j$, and $\mathbf{A}^j = \sum_{\forall i} W_i^j \frac{(\mathbf{p}_j - \mathbf{p}_i) \cdot (\mathbf{p}_j - \mathbf{p}_i)^T}{\|\mathbf{p}_j - \mathbf{p}_i\|^2}$.

Based on the consideration that:

$$\frac{\mathbf{T}^j}{\| \mathbf{T}^j \|} = I - \frac{\mathbf{A}^j}{\| \mathbf{A}^j \|} \Leftrightarrow \| \mathbf{A}^j \| = W^j , \tag{8}$$

which is demonstrated in Appendix, the computation of $\mathbf{T}^j$ can be substituted by $\frac{\mathbf{T}^j}{\|\mathbf{T}^j\|} = I - \frac{\mathbf{A}^j}{\|\mathbf{A}^j\|}$ when $\| \mathbf{A}^j \| = W^j$; nevertheless, since the second order tensor $\hat{\mathbf{T}}^j = I - \frac{\mathbf{A}^j}{\|\mathbf{A}^j\|}$ is generally positive semi-definite, in the first voting pass of the NV algorithm we always compute $\hat{\mathbf{T}}^j$ as the vote at receiver j, also when the condition in Eq. (8) does not hold. This is useful for it not only simplifies the computation of the first voting pass, but also enforces the classified dimensionality d to be $0 < d < D$, and $\forall j, \| \hat{\mathbf{T}}^j \| \leq 1$.

In the NV algorithm, the first voting pass uses as its input a set of data points, and generates as output a set of tokens (point,tensor) describing the manifold's normal space . The voting passes executed after the first are different, since the input of the algorithm is no more a set of points, but a set of tokens, that is a set of (point, tensor) pairs. To exploit this information, each emitter $(\mathbf{p}_i, \mathbf{T}_i)$ generates tensor votes equal to $W^i \mathbf{T}_i$, where W^i is a weight computed according to the unnormalized Gaussian decay function centered in $\mathbf{p}_i$. To face additive noise and to cast strongest votes in the direction tangent to the underlying manifold, the decay function is elongated proportionally to a scale parameter σ in the tangent space, and proportionally to a noise standard deviation parameter γ ($\gamma \leq \sigma$) in the normal space.

To build such a Gaussian function, we at first classify the dimensionality of the underlying manifold by selecting the maximum saliency value, s_h; then, we synthesize the precision matrix $\mathbf{P}_i = \mathbf{X}_i \Lambda_i \mathbf{X}_i^T$, describing the Gaussian function, by setting the eigenvalues in the diagonal matrix Λ_i to $\lambda_k = \frac{1}{\sigma^2}$ for $1 \leq k \leq h$, and $\lambda_k = \frac{1}{\gamma^2}$ for $h < k \leq D$, while $\mathbf{X}_i$ is the matrix of the eigenvectors of the tensor obtained from the previous voting pass. Once the precision matrix is obtained the scaling values W^i are computed and used to cast tensorial votes.

3.1 Statistical Interpretation

In the described NV settings, each emitter token $(\mathbf{p}_i, \mathbf{T}_i)$ can be viewed as a local weak classifier iteratively trained by an information diffusion process. This method can be compared to belief propagation (BP, [1]) and to similar techniques where every node of a dependency graph is iteratively updated using information from neighboring nodes.

After the training step, the vote $W^i\mathbf{T}_i^j$ cast from the emitter i to the receiver position $\mathbf{p}_j$, can be considered as the precision matrix of the following unnormalized Gaussian function:

$$N(\mathbf{x};\mathbf{p}_j,W^i\mathbf{T}_i^j) = e^{-(\mathbf{x}-\mathbf{p}_j)^T W^i \mathbf{T}_i^j (\mathbf{x}-\mathbf{p}_j)/2}\ , \tag{9}$$

which is proportional to the probability density function $P(\mathbf{x} \in \mathscr{M}_{\mathbf{p}_j} | (\mathbf{p}_i, \mathbf{T}_i))$, assigning to each point $\mathbf{x} \in \Re^D$ the probability to be on the manifold $\mathscr{M}_{\mathbf{p}_j}$, conditioned on $(\mathbf{p}_i, \mathbf{T}_i)$.

Since

$$\begin{aligned} P\left(\mathbf{x} \in \mathscr{M}_{\mathbf{p}_j}\right) &= N(\mathbf{x};\mathbf{p}_j,\mathbf{T}^j) = N(\mathbf{x};\mathbf{p}_j,\sum_{\forall i} W^i\mathbf{T}_i^j) \\ = \prod_{\forall i} N(\mathbf{x};\mathbf{p}_j,W^i\mathbf{T}_i^j) &= \prod_{\forall i} P\left(\mathbf{x} \in \mathscr{M}_{\mathbf{p}_j} | (\mathbf{p}_i,\mathbf{T}_i)\right)\ , \end{aligned} \tag{10}$$

the accumulated vote $\mathbf{T}^j$ represents the joint distribution in $\mathbf{p}_j$; therefore, the strong dimensionality classifier obtained by employing $P\left(\mathbf{x} \in \mathscr{M}_{\mathbf{p}_j}\right)$ is an ensemble classifier made by combining the weak classifiers obtained by employing the $P\left(\mathbf{x} \in \mathscr{M}_{\mathbf{p}_j} | (\mathbf{p}_i,\mathbf{T}_i)\right)$ for each i.

4 Applications of the Neighbors Voting Algorithm

In this section we show how NV can be reinterpreted to perform clustering , classification and image inpainting . Experimental results on both synthetic and real data will be reported in Sect. 5.

4.1 Point Clustering

Given a set of points $\mathscr{P} = \{\mathbf{p}_i\}$, the NV algorithm can be used to train an ensemble classifier in order to learn the manifold underlying the data. Each point $\mathbf{p}_i$ can then be classified according to the dimensionality of the inferred manifold going through it; point clustering is therefore realized by grouping points belonging to inferred manifolds with the same dimensionality.

With the aim of manifold dimensionality inference, the NV algorithm starts from a set of pairs $(\mathbf{p}_i, \mathbf{T}_i)$ and iterates voting passes in order to modify (train) the $\mathbf{T}_i$s by means of the information exchanged among neighboring points. This information diffusion process allows to improve the dimensionality and local orientation estimation, encoded in $\mathbf{T}_i$, with respect to the underlying manifold going through $\mathbf{p}_i$.

After each voting pass the inferred dimensionality information is reinforced by applying a nonlinear filtering as follows:

1. the eigensystem $\mathbf{T}_i = \mathbf{X}_i \Lambda_i \mathbf{X}_i^T$ is computed;
2. the saliency values are computed as described in Sect. 2;

3. the maximum saliency value, s_h, is found to select the most likely dimensionality of the underlying manifold;
4. a new eigenvalue matrix $\overline{\Lambda}_i$ is generated by setting to zero the eigenvalues corresponding to the tangent directions and to one the others;
5. the resulting tensor is $\overline{\mathbf{T}}_i = \mathbf{X}_i \overline{\Lambda}_i \mathbf{X}_i^T$.

The described algorithm allows to efficiently classify the dimensionality of the underlying manifold going through each input data point, so that points related to manifolds with different dimensionality can be separated.

4.2 Classification

The general classification problem can be stated as follows: given a set of training data, that is, points in a space $S \subseteq \Re^D$ with associated labels taken from a finite set $\mathscr{L} = \{\lambda_l\}_{l=1}^{L}$, a classifier is trained to learn the unknow classification function $f : S \to \mathscr{L}$, that is, the function that takes points in S and maps them in $\mathscr{L}$. After the training phase, the classification consists in using the classifier instead of the real but unknown function $f(\cdot)$ in order to distinguish data points belonging to different classes.

In this section we describe how the NV algorithm can be employed to train a set of classifiers allowing to:

- recognize points belonging to one class, thus creating a two-class classifier discriminating in-class/off-class points (see Sect. 4.2.1);
- recognize points belonging to different classes, thus creating a multi-class classifier (see Sect. 4.2.2).

4.2.1 Two-Class Classifier

The NV algorithm can be used to train a two-class classifier, since the computed (trained) tokens can be used as an estimation of the unknown function $f(\cdot)$, by reinterpreting them as Gaussian functions as in Eq. (9) and combining them as described in Eq. (10).

More precisely, given a set of points $\mathbf{p}_j \subset S$, taken from a class $\lambda_{\bar{l}} \in \mathscr{L} = \{0, 1\}$, it is possible to map them from their original space S, to a 'better shaped' space $\tilde{S}$, where the given points are interpreted as points drawn from the manifold $\mathscr{M}_{\bar{l}}$. Therefore, the problem of recognizing whether an unknown point $\mathbf{x} \in S$ belongs to $\lambda_{\bar{l}}$, can be restated as the problem of discovering if the point $\tilde{\mathbf{x}} \in \tilde{S}$[4], is part of the manifold $\mathscr{M}_{\bar{l}}$ from which the training points are drawn.

Considering that the real manifold is not known, and that most of the times it is a manifold with boundaries, the solution to this problem cannot be found in practice (in fact the classification problems are generally ill posed); for this reason a probabilistic approach is required.

[4] $\tilde{\mathbf{x}}$ is the projection of $\mathbf{x}$ into $\tilde{S}$.

Therefore $\mathcal{M}_{\bar{l}}$ can be represented by means of a real valued function: $\hat{f}(\cdot|\mathbf{x}) : S \rightarrow [0,1]$ that can be interpreted as the PDF of a Bernoulli random variable, parameterized by the point $\mathbf{x} \in S$ to be classified. The two classes of $\hat{f}(\cdot|\mathbf{x})$ represent the on-manifold/off-manifold classes.

The function $\hat{f}$ is learned by employing the NV algorithm; after the training process each token $(\mathbf{p}_j, \mathbf{T}_j)$ contains a tensor that allows to compute a local estimation of the probability $p_{\bar{l},j}$ of the point $\mathbf{x}$ to be on the same manifold of $\mathbf{p}_j$. This estimation is computed as described in Eq. (10), that is, $p_{\bar{l},j} = P\left(\mathbf{x} \in \mathcal{M}_{\mathbf{p}_j}\right)$. This is a Bernoulli probability distribution that can be employed as $\hat{f}$ for the simplest case where only one trained token $(\mathbf{p}_j, \mathbf{T}_j)$ is used for classification . If N trained tokens are used to perform classification, the function $\hat{f}$ must be computed as the composition of N Bernoulli distributions and must remain a valid Bernoulli distribution. In this work $\hat{f}$ is a Bernoulli distribution whose parameter, $p_{\bar{l}}$, is fixed to the maximum of the N Bernoulli distributions parameters, that is,

$$p_{\bar{l}} = max_{j=1,..,N}\left\{p_{\bar{l},j}\right\} = max_{j=1,..,N}\left\{P\left(\mathbf{x} \in \mathcal{M}_{\mathbf{p}_j}\right)\right\} . \tag{11}$$

4.2.2 Multi-class Classifier

When the training set is composed of points belonging to L different classes, that is, each training point is associated with a label λ_l in a set $\mathcal{L} = \{\lambda_l\}_{l=1}^{L}$, the problem can be solved by combining L two-class classifiers, each trained to recognize the elements of one class, as previously described. With this setting, each manifold can be learned separately, allowing for a simple trained data composition for new classification problems[5].

Given a point $\mathbf{x}$ to be classified, consider the N trained two-class classifiers based on Bernoulli distributions with parameters $p_l = \max_{i=1,..,N}\{p_{l,i}\}$. The classification is performed by choosing the class index $\bar{l}$ depending on the p_l values. In this work, we have experimented with two strategies (see also Sect. 5.2).

- The first strategy chooses the class $\bar{l}$ that maximizes p_l, that is, $\bar{l} = \arg\max_l p_l$; note that, although this choice is not exact from the probability theory point of view, in practice it may achieve good results (see Sect. 5). Furthermore, this simple technique can be viewed as a generalization of the nearest neighbor classifier when all the Gaussian functions are equally scaled and isotropic; indeed, the Gaussian giving the highest probability value is the one whose mean is the training point $\mathbf{p}$ that is the nearest neighbor to $\mathbf{x}$.
- The second strategy is to consider all the 2^L possible events; that is, point $\mathbf{x}$ may be drawn from exactly one manifold, from none, or from more than one manifold at the same time (junction point). To perform classification , given a binary vector with L values, $\mathbf{b} = \{b_l\}_{l=1}^{L}$, where b_l represents the fact that $\mathbf{x}$

[5] As an example consider the practical case where a lot of classes are already trained and can be correctly classified, and a new class must be classified as well. In this case, it may be convenient to learn only one manifold without affecting the already trained classifiers.

belongs or not to the class l, it is possible to compute the probability of the configuration represented by $\mathbf{b}$ as

$$P(\mathbf{b}) = \prod_{l=1}^{L} p_l^{b_l}(1-p_l)^{1-b_l} \ . \tag{12}$$

Among the 2^L possible configurations, the preferred configuration is the one that maximizes $P(\mathbf{b})$, that is, $\overline{\mathbf{b}} = \arg\max_{\mathbf{b}} P(\mathbf{b})$.

4.2.3 Scaling Parameters

The described classification algorithm is based on a set of Gaussian functions whose precision matrix is computed through the NV algorithm. This algorithm generates tensors with a well understood orientation but with an uncontrolled scale.

Therefore, after the training step, the trained tensors must be rescaled to control the $\hat{f}$ function approximation. To this aim, every Gaussian is scaled so that the tangent space uncertainty is fixed to an a priori σ value. To do this we normalize the tensors by dividing them by their spectral norm times σ^2 so that the covariance matrix spectral norm becomes exactly σ^2, that is,

$$\overline{\mathbf{T}}_j = \frac{\mathbf{T}_j}{\sigma^2 \parallel \mathbf{T}_j \parallel} \ . \tag{13}$$

4.3 Image Inpainting with NV

In [6] a TVF-based image inpainting technique was proposed. In this section we describe how to solve this problem in a similar way, by using a NV-based technique.

Given an image region, a set Ω of $(2r+1 \times 2r+1)$-pixels sub-images (training patches) can be extracted; each patch can be seen as a vector in $\Re^{(2r+1)^2}$. The NV algorithm is then applied to Ω to obtain a set of trained tensors $\mathscr{T}$ that describe an estimate of the underlying manifold; therefore, they can be used to recover missing information.

A 'patch vector' $\mathbf{p}$ containing unspecified coefficients identifies the submanifold $\mathscr{M} \subset \Re^{(2r+1)^2}$. To recover the missing information in $\mathbf{p}$, we select from the set of tensors $\mathscr{T} = \{(\mathbf{p}_i, \mathbf{T}_i)\}$ the pair $(\mathbf{p}_h, \mathbf{T}_h)$ so that the distance $\parallel \tilde{\mathbf{p}} - \tilde{\mathbf{p}}_h \parallel$ is minimum with respect to all the training patches[6]. Let d be the estimated dimensionality of the manifold underlying $\mathbf{p}_h$, $\mathscr{T}$ be the tangent space identified by the tensor $\mathbf{T}_h$, and $\mathbf{e}_j$ $(1 \leq j \leq d)$ be its column vectors, the inferred patch $\bar{\mathbf{p}}$ can be computed as

$$\bar{\mathbf{p}} = \frac{(\tilde{\mathbf{p}}_h - \mathbf{p}_h) \cdot \bar{\mathbf{p}}'}{\parallel \tilde{\mathbf{p}}_h - \mathbf{p}_h \parallel^2} \bar{\mathbf{p}}' + \mathbf{p}_h \ , \tag{14}$$

where

[6] We represent with the notation $\tilde{\mathbf{x}}$ the projection of a point $\mathbf{x} \in \Re^{(2r+1)^2}$ on $\mathscr{M}$.

$$\bar{\mathbf{p}}' = \sum_{l=1}^{d} \mathbf{e}_l \cdot (\tilde{\mathbf{p}}_h - \mathbf{p}_h)\mathbf{e}_l \,. \tag{15}$$

These equations compute $\bar{\mathbf{p}}$ as the point nearest to $\tilde{\mathbf{p}}_h$ and constrained on the linear space $\mathrm{span}\left\langle \{\mathbf{e}_j\}_{j=1}^{d} \right\rangle \cap \mathscr{M}$.

Every patch $\bar{\mathbf{p}}$, inferred for a partially specified patch $\mathbf{p}$, is an estimate of the unknown real patch and contains the estimated pixel gray levels $\bar{\mathbf{p}}_{x,y}$. Inferring overlapping patches allows to generate different gray level estimates for the same pixel; their mean value is the maximum likelihood estimation of the unknown pixel gray level. Given an unknown image region, its boundary patches are the easiest to recover since they have less unknown coordinates.

Based on these considerations, the inpainting algorithm proceeds by iterating the following steps:

1. the external edge pixels, p_i, of the unknown region are morphologically identified;
2. for each p_i:
   a. a partially specified patch is centered on p_i and its unknown pixel values are inferred;
   b. the inferred pixel values are accumulated in a working image and a counter image is used to count the number of contributions for each pixel;
3. the maximum value mc in the counter image is found and the gray levels of pixels p_k with at least $\left\lceil \frac{8}{9} mc \right\rceil$ contributions are estimated by averaging[7];
4. the pixels p_k are removed from the unknown region.

These steps are iterated until the unknown region becomes empty. This algorithm stops after few steps and has proved to produce promising results that are described in Sect. 5.

5 Results

In this section we report quantitative and qualitative results obtained by the algorithms presented in the previous section. In Sect. 5.1 the results of clustering experiments performed on synthetic data are reported, while in Sect. 5.2 and Sect. 5.3 we present the performance obtained by testing our classification and inpainting algorithms on real images.

5.1 Clustering Results

To test our point clustering algorithm we perform experiments by creating two manifolds, $\mathscr{M}$ and $\mathscr{N}$, of dimensionality d and e, respectively (with $d \neq e$); $\mathscr{M}$ and $\mathscr{N}$ are embedded in $\Re^D$, and $\mathscr{M} \cap \mathscr{N} \neq \emptyset$. Two sets of points $\mathscr{P}_{\mathscr{M}}$ and $\mathscr{P}_{\mathscr{N}}$, containing approximately 200 points each, are drawn from $\mathscr{M}$ and $\mathscr{N}$, respectively, and

[7] The threshold value $\left\lceil \frac{8}{9} mc \right\rceil$ has been experimentally set.

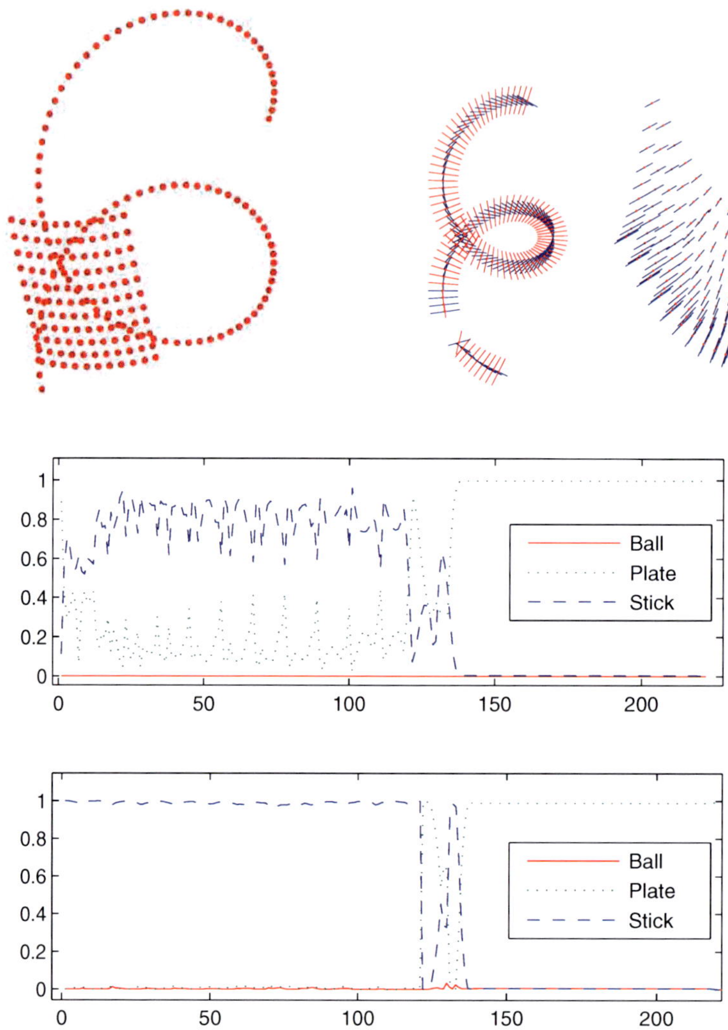

Fig. 2 Top row: the input dataset and the obtained clusters shown with their normal space estimations. Central and bottom plots: saliency values after 1 and 5 voting passes

$\mathscr{P} = \mathscr{P}_{\mathscr{M}} \cup \mathscr{P}_{\mathscr{N}}$ is used as input to the NV algorithm (with 5 voting passes) whose aim is to classify each point as belonging to either $\mathscr{M}$ or $\mathscr{N}$.

As an example, consider the dataset depicted in the left of Fig. 2: it contains a curve intersecting a surface in $\Re^3$. After 5 voting passes of the NV algorithm, tensors are normally oriented with respect to the underlying manifolds; therefore, we compute the saliency values of each tensor and use them to infer the dimensionality of the manifold.

In the plots of Fig. 2 three saliency values for each data point are shown after the first and the last voting passes, respectively; note that after the last voting pass

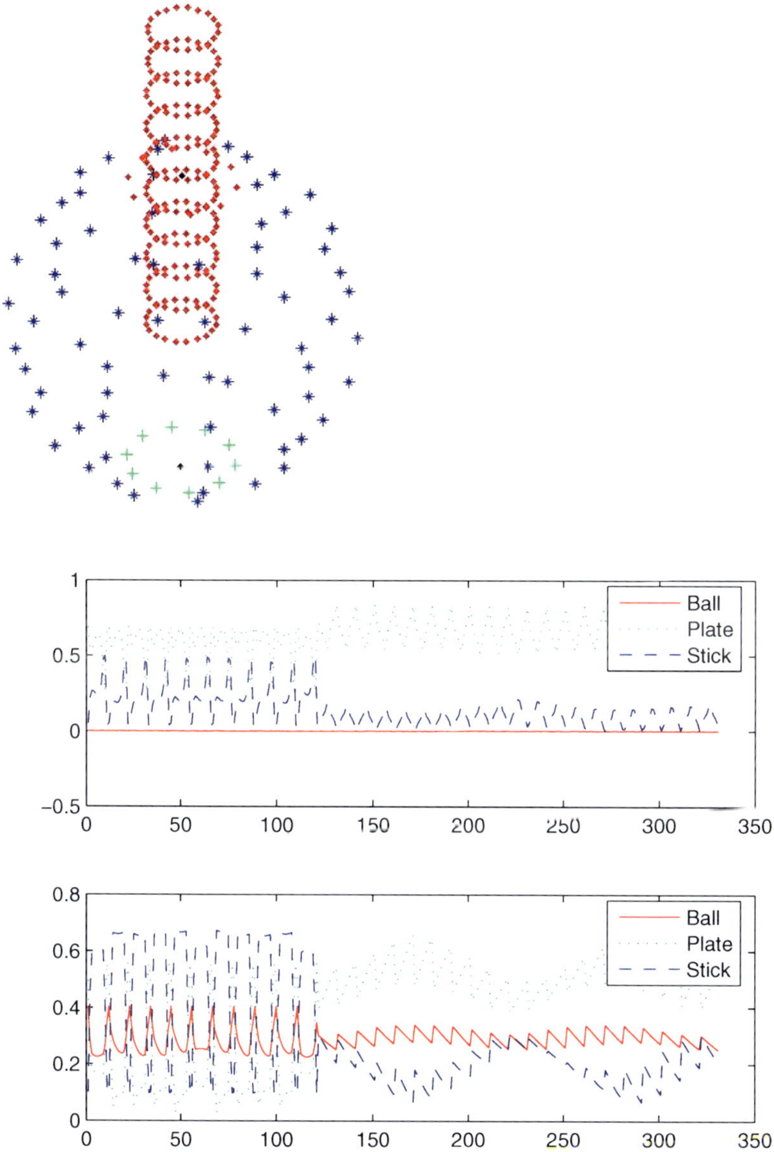

Fig. 3 Top: the input dataset clustered in four clusters: cylinder (points), sphere (asterisks), bottom circle (crosses), central bottom vertex (point). Middle and bottom: saliency values showing that in this case the dimensionality is not well defined; indeed, for the cylinder's points it is possible to choose between several 1D circles (solution selected by NV) and one cylindrical 2D surface

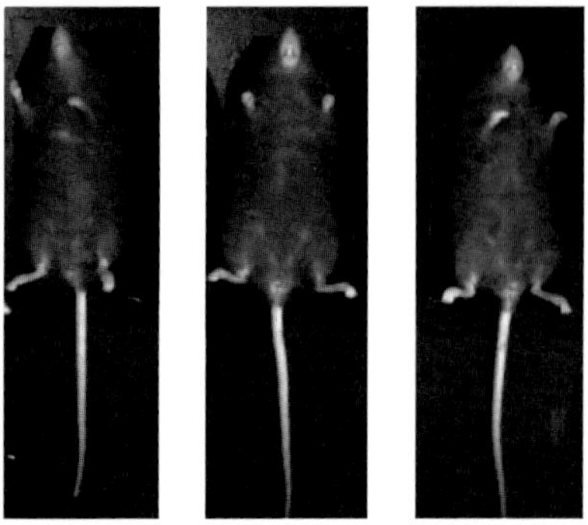

Fig. 4 Examples of mouse images

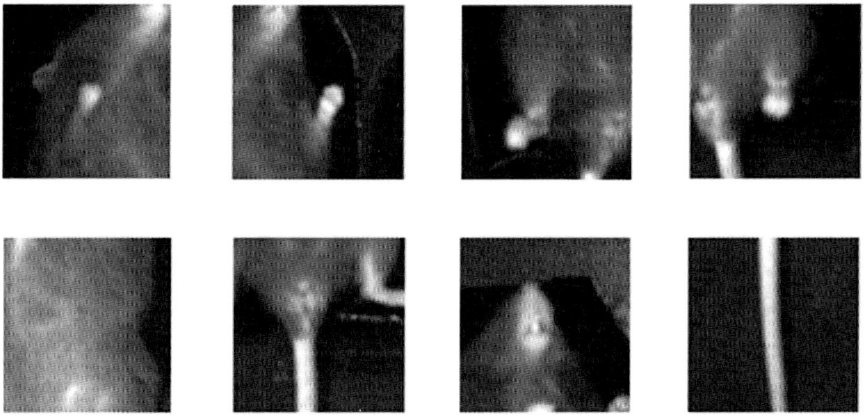

Fig. 5 Mouse sub images. Top row: sub-images of the anterior and posterior limbs. Bottom row: sub-images of the abdominal area, genital area, head, and tail

the saliency values are either ≈ 1, or ≈ 0. Therefore, they clearly identify the dimensionality of the manifold. In the top-right of Fig. 2 the clustered geometrical structures are shown for the dataset shown on the left.

In Fig. 3 another example is shown, where the NV algorithm detects four clusters, for both these tests we used $\sigma = 0.1$ and $\gamma = \sigma/10$. The clustering algorithm has been executed 10 times by varying $\mathscr{M}$ and $\mathscr{N}$.

The clustering code has been implemented in Matlab; when executed on an Intel Centrino Duo 2.0GHz CPU with 2.0GB RAM, it takes about 3 s to partition the set $\mathscr{P}$, composed of 222 points, in two clusters. Notice that a Matlab implementation of TVF, written with respect to the Matlab optimization manual, took about 126 s to process the same dataset. Also notice that a C/C++ implementation of the NV can provide a very efficient TVF approximation[8].

[8] We have published the open-source project OpenTVF containing some TVF-based algorithms; a C++ implementation of the NV algorithm will be soon available in that library.

5.2 Classification Results

The classification algorithm described in Sect. 4.2.2 was employed to solve an image classification problem. More precisely, given mouse images such as those shown in Fig. 4, the goal is to localize five anatomical parts: the head, the anterior limbs, the abdominal area, the posterior limbs, the genital area, and the tail. This application is helpful to pharmacologists who need to measure the response of the animal to the administration of drugs [14].

To automatically solve the problem of mouse image segmentation we have employed our ensemble classifier method to classify sub-images, randomly extracted from the mouse images in order to recognize those centered on the anatomical parts of interest.

More precisely, the image set is composed of 261 mouse images; for each image and for each anatomical part, *ROI*, we extracted the sub-image centered in the center of mass, $R = (x_R, y_R)$, of the *ROI*, and other 4 sub-images centered on the points $R_1 = (x_R - w, y_R)$, $R_2 = (x_R + w, y_R)$, $R_3 = (x_R, y_R - h)$, and $R_4 = (x_R, y_R + h)$, where w and h correspond to $\frac{1}{3}$ of the width and the height of the *ROI*[9]. In this way we have built, for each anatomical part of interest, about 1300 sub-images approximately centered on it. Therefore, our set is composed of about 10400 sub-images; examples of the sub-images are shown in Fig. 5.

To perform classification , we at first encoded the mouse sub-images by applying a filter bank, $Bank_G$, composed of Gabor filters at four different scales and eight different orientations. Each sub-image is therefore represented by a vector of 32

Table 1 Comparison of the mean (standard deviation) accuracy values achieved by the tested classification algorithms. Columns stand for different train/test ratios while rows correspond to classification algorithms

	1/5	1/10	1/20	1/40
NV	98.59% (0.16%)	97.55% (0.22%)	96.04% (0.34%)	93.88% (0.45%)
NN	98.35% (0.21%)	96.81% (0.32%)	95.12% (0.54%)	92.00% (0.55%)
PPNN	98.21% (0.21%)	97.09% (0.30%)	95.37% (0.34%)	92.95% (0.59%)
PCF	87.63% (3.55%)	84.76% (4.25%)	81.12% (4.01%)	83.58% (3.75%)

Table 2 Average training and classification times for NV and TVF

	1/5	1/10	1/20	1/40
training (NV)	58 ms	117 ms	212 ms	406 ms
training (TVF)	291 ms	409 ms	652 ms	1206 ms
testing	2.9 ms	6.5 ms	12.7 ms	25.1 ms

[9] The four sub-images centered in R_i (i=1,..,4) have been extracted to train the classifier in order to recognize sub-images not perfectly centered on the *ROI*.

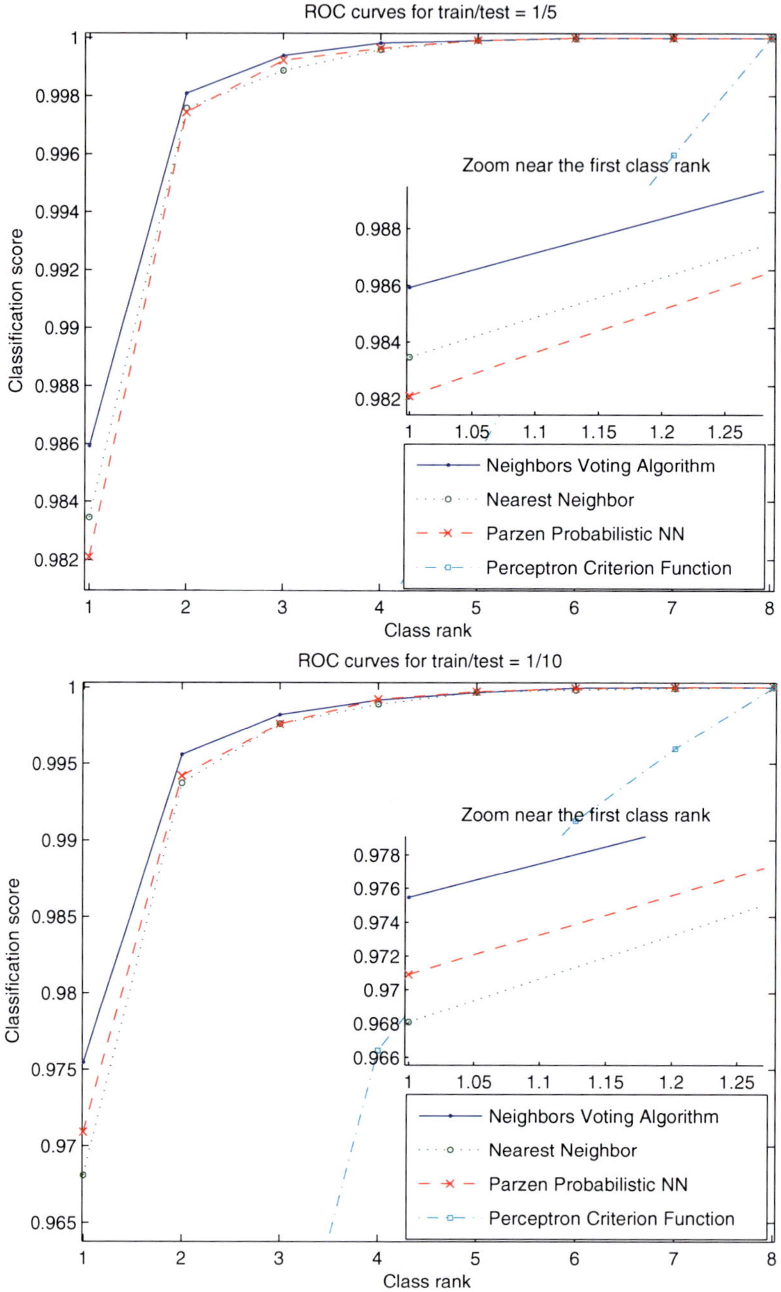

Fig. 6 Mean ROC curves obtained testing the classification algorithms for two train/test ratios: 1/5 and 1/10

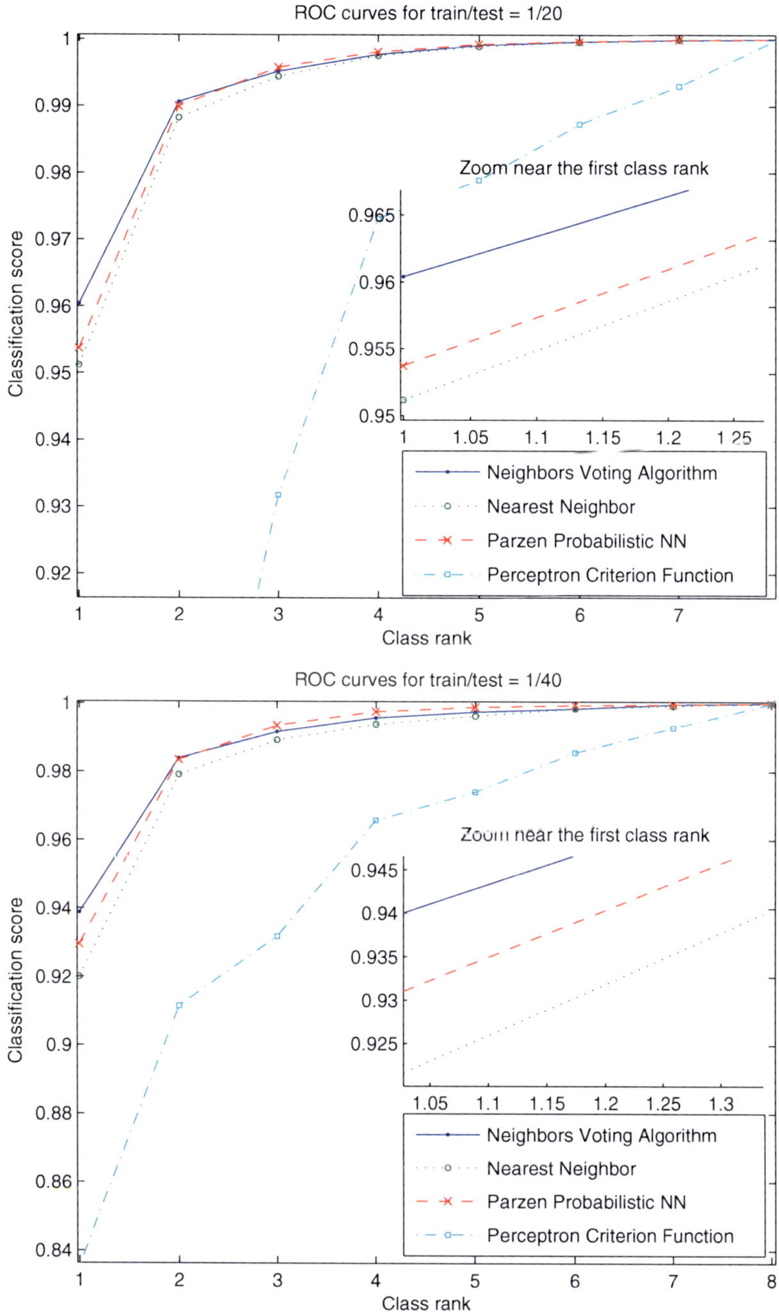

Fig. 7 Mean ROC curves obtained testing the classification algorithms for two train/test ratios: 1/20 and 1/40

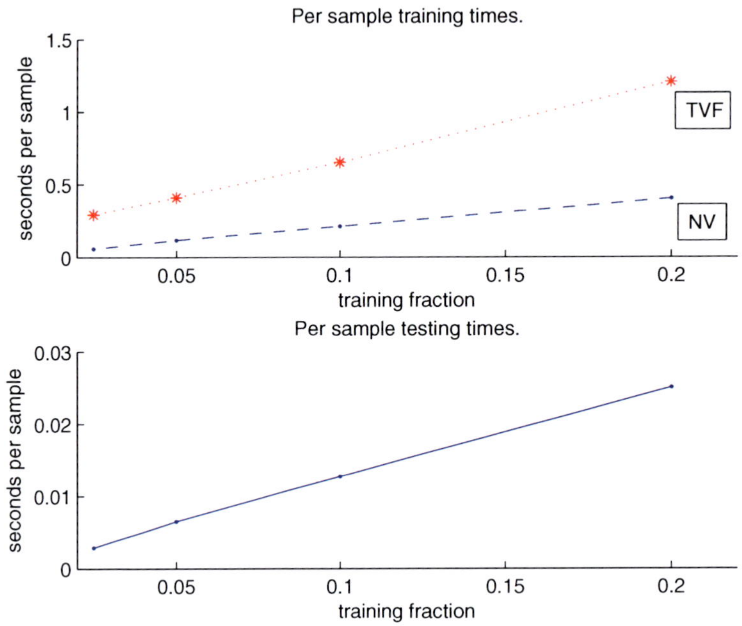

Fig. 8 Comparison between the NV and TVF training times for the same classification algorithm. Columns stand for different train/test ratios while rows correspond to classification algorithms

features corresponding to the response of each filter in the central pixel of the sub-image, $C = (x, y)$.

Since the results obtained with this method were not satisfactory, we looked for a more discriminative feature space. To this aim, on each sub-image we have evaluated the response of $Bank_G$ not only in the pixel $C = (x, y)$, but also in the four pixels $P_1 = (x - w, y)$, $P_2 = (x - w, y)$, $P_3 = (x, y - w)$, $P_4 = (x, y + w)$, where w is the half size of the employed Gabor filter (this size varies at different scales), thus obtaining a feature vector of 160 elements. To reduce the dimensionality of the input vectors while maintaining the most significant features, we applied the Principal Component Analysis (PCA) to the training samples. Keeping the 90% of the variance in the training data, we obtained feature vectors with no more than 40 components[10].

With this dataset we performed four experiments by varying the train/test ratio on the set of values $\left\{\frac{1}{5}, \frac{1}{10}, \frac{1}{20}, \frac{1}{40}\right\}$. To have an unbiased evaluation scheme, for each train/test ratio we repeated the experiment ten times by randomly choosing the training and test sets and then computed the average classification performance.

[10] Note that by varying the training set the projected datasets may have different dimensionalities.

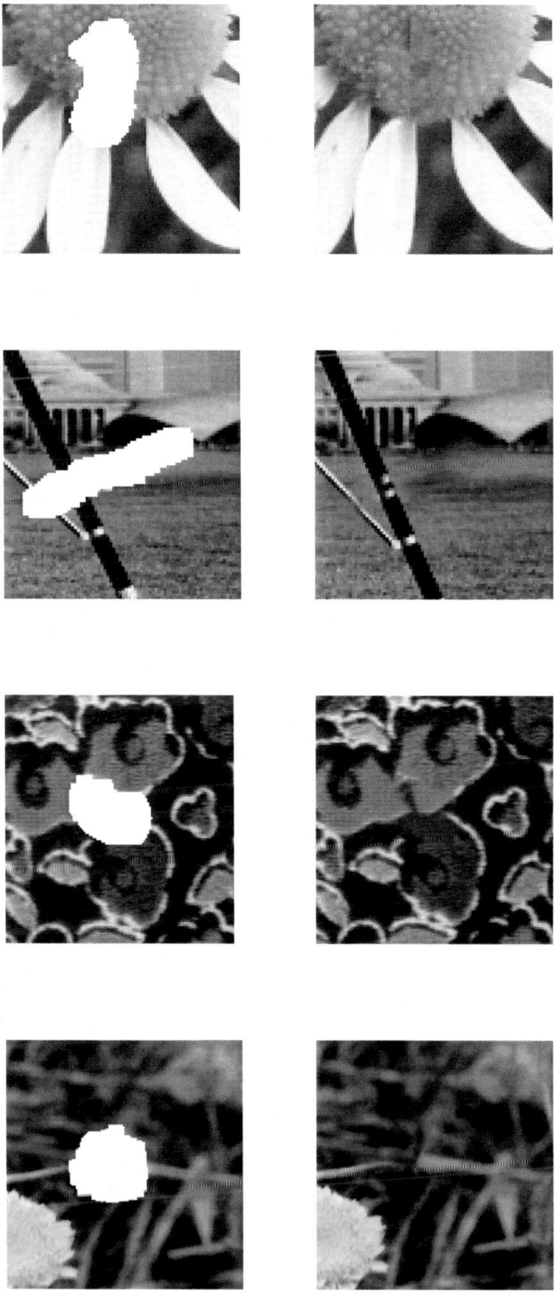

Fig. 9 Four images with unknown regions (left column) and the inpainted result (right column)

The first row of Table 1 reports the mean classification accuracy and the standard deviation achieved in these experiments by fixing the σ parameter[11] to 150 (see Sect. 3) and by employing the first classification strategy. Results obtained by employing the second classification method are not reported here because they are similar. Notice that the achieved results are promising even when the number of training examples is small.

To further prove the efficacy of the NV algorithm for classification tasks, we have compared its performance to three algorithms described in [4]: Nearest Neighbor (NN), Parzen Probabilistic Neural Network (PPNN), and the minimization of the Perceptron Criterion Function (PCF). As can be seen in Table 1, their results are always worse than those obtained by NV, and the decrease in performance is even more significant when the training set becomes smaller.

For clarity of presentation, Figs. 6 and 7 show the mean ROC curves obtained by the four classification algorithms.

The training and classification algorithms have been executed on an Intel Centrino Duo 2.0GHz CPU with 2.0GB RAM by using Matlab. In Table 2 and Fig. 8, the average training and testing times are reported[12]; it can be noticed in Fig. 8 that the computation times are linearly dependent with respect to the train/test ratio.

5.3 Inpainting Results

We have tested our image inpainting algorithm on a set of 50 generic images, where the unknown region and the training region have been manually chosen. We have used 7×7 square patches, $\sigma = 0.2$, and $\gamma = \sigma/10$. Experiments have proved that good inpainting results can be obtained by employing the first voting pass only, thus increasing the efficiency of the overall algorithm. In Fig. 9 examples of inpainting results are shown.

The inpainting algorithm has demonstrated to be very efficient during the reconstruction step; the time complexity is indeed dominated by the training step. Using an Intel Centrino Duo 2.0GHz CPU with 2.0GB RAM, our Matlab implementation has taken on about 18 s to infer, on average, 1171 pixel values per image, starting from a training set of about 550 training patches.

6 Conclusions and Future Work

In this work we have described the theory and applications of the NV algorithm, which is an efficient approximation of the TVF technique, developed to overcome the high time and space complexity that discourage the TVF usage. The NV algorithm is an iterative procedure based on the tensorial vote cast between neighboring points; considering each pair (point, tensor) as a weak classifier, their combination is employed to build a strong classifier.

[11] The value of σ is chosen so that it is approximately equal to one third of the mean intra-class points distance evaluated over the training data.

[12] These times are referred to Matlab implementations of NV and TVF.

To prove the efficacy of the developed algorithm, the NV has been applied to solve three problems on both synthetic and real data: point clustering , image classification , and image inpainting .

The achieved results and the time-space complexity of the developed applications prove the efficacy of the NV algorithm and show that although being an approximation of the TVF technique, it can be successfully applied to solve problems in arbitrary dimensional spaces.

Future work will be aimed at both the improvement of the point clustering , image classification and image inpainting algorithms and at the application of the NV algorithm to solve other problems. Furthermore, we plan to modify the implementation of the NV algorithm in order to make it more efficient, both by computing tensorial votes only when strictly needed and by employing appropriate data structures (e.g., ANN-trees [7]) to use range queries for computing the Gaussian values only for neighboring points.

Finally, considering the promising results achieved by the NV, we are planning to apply it to solve other problems on real data; more precisely, we are working on the spam-classification task. Our preliminary results achieved on that classification problem are very promising.

Acknowledgements. The authors would like to thank Phd. Gianpaolo Rando and Prof. Adriana Maggi (*gianpaolo.rando@unimi.it*) of Dipartimento di Scienze Farmacologiche of the Università degli Studi di Milano for providing the mouse images used for the classification experiments.

Appendix

In this appendix the proofs of the results used in Sect. 3 are reported.

At first, representing the spectral norm with $\| \cdot \|$ and defining $s_i^j = \frac{(\mathbf{p}_j - \mathbf{p}_i)}{\|\mathbf{p}_j - \mathbf{p}_i\|}$, we demonstrate that $\| W^j \mathbf{I} \| = W^j \geq \| \mathbf{A}^j \|$:

$$\begin{aligned} \| \mathbf{A}^j \| &= \| \sum_{\forall i} W_i^j \frac{(\mathbf{p}_j - \mathbf{p}_i) \cdot (\mathbf{p}_j - \mathbf{p}_i)^T}{\| \mathbf{p}_j - \mathbf{p}_i \|^2} \| = \\ &= \| \sum_{\forall i} W_i^j s_i^j (s_i^j)^T \| \leq \sum_{\forall i} W_i^j \underbrace{\| s_i^j (s_i^j)^T \|}_{1} = W^j . \end{aligned} \quad (16)$$

Next, we prove that:

$$\frac{\mathbf{T}^j}{\| \mathbf{T}^j \|} = I - \frac{\mathbf{A}^j}{\| \mathbf{A}^j \|} \Leftrightarrow \| \mathbf{A}^j \| = W^j . \quad (17)$$

At first, we note that if $\| \mathbf{A}^j \| = W^j$, it must be because

$$\| \sum_{\forall i} W_i^j s_i^j (s_i^j)^T \| = \sum_{\forall i} W_i^j \| s_i^j (s_i^j)^T \| . \quad (18)$$

Since $\forall i, j$ we have that $W_i^j > 0 \wedge \| s_i^j \| = 1$, from Eq. (18) it follows that $\forall i, j, k; \| s_i^j s_k^j \| = 1$. Therefore $rank(\mathbf{A}^j) = 1$ and $\min\left(\lambda\left(\frac{A^j}{\|A^j\|}\right)\right) = 0$, where $\lambda(\cdot)$ is the function that computes the eigenvalues of a given matrix. Thanks to this demonstration we can write:

$$\mathbf{T}^j = \mathbf{W}^j I - \mathbf{A}^j = \mathbf{W}^j \left(I - \frac{\mathbf{A}^j}{\mathbf{W}^j}\right) \Rightarrow \frac{\mathbf{T}^j}{\mathbf{W}^j} = I - \frac{\mathbf{A}^j}{\mathbf{W}^j} \Rightarrow \frac{\mathbf{T}^j}{\| \mathbf{A}^j \|} = I - \frac{\mathbf{A}^j}{\| \mathbf{A}^j \|} \tag{19}$$

and

$$\| \mathbf{T}^j \| = \| \mathbf{A}^j \| \| I - \frac{\mathbf{A}^j}{\| \mathbf{A}^j \|} \| = \| \mathbf{A}^j \| \left(1 - \underbrace{\min\left(\lambda\left(\frac{A^j}{\| A^j \|}\right)\right)}_{0} \right) = \| \mathbf{A}^j \| . \tag{20}$$

Thus, we have demonstrated the 'if' part of Eq. (17); the 'only if' part can be demonstrated in a similar way.

References

1. Bishop, C.M.: Pattern Recognition and Machine Learning. Springer Science+Business Media, New York (2006)
2. Campadelli, P., Lombardi, G.: Tensor voting fields: direct votes computation and new saliency functions. In: Cucchiara, R. (ed.) Proc. 14th Int. Conf. Image Analysis and Process, Modena, Italy, pp. 677–684. IEEE Comp. Soc., Los Alamitos (2007)
3. Campadelli, P., Casiraghi, E., Lombardi, G.: The neighbors voting algorithm. In: Okun, O., Valentini, G. (eds.) Proc. 2nd Supervised and Unsupervised Ensemble Methods and their Appl., Patras, Greece (2008)
4. Duda, R.O., Hart, P.E., Stork, D.G.: Pattern Classification. John Wiley & Sons, Hoboken (2001)
5. Guy, G., Medioni, G.: Inferring global perceptual contours from local features. Int. J. Comp. Vis. 20(1-2), 113–133 (1996)
6. Jia, J., Tang, C.-K.: Image repairing: robust image synthesis by adaptive ND tensor voting. In: Proc. IEEE Conf. Comp. Vis. Patt. Recogn., Madison, WI, pp. 643–650. IEEE Comp. Soc., Los Alamitos (2003)
7. Lin, K.-I., Yang, C.: The ANN-tree: an index for efficient approximate nearest neighbor search. In: Proc. 7th Int. Conf. Database Syst. for Advanced Appl., Hong Kong, China, pp. 174–181. IEEE Comp. Soc., Los Alamitos (2001)
8. Medioni, G., Kang, S.B.: Emerging Topics in Computer Vision. Prentice Hall, Upper Saddle River (2004)
9. Mordohai, P., Medioni, G.: Dense multiple view stereo with general camera placement using tensor voting. In: Proc. 2nd int. Symp. 3D Data Processing, Visualization and Transmission, Thessaloniki, Greece, pp. 725–732. IEEE Comp. Soc., Los Alamitos (2004)
10. Mordohai, P., Medioni, G.: Dimensionality estimation and manifold learning using tensor voting (2005), `http://iris.usc.edu/~medioni/download/ND_manual1.pdf`

11. Mordohai, P., Medioni, G.: Tensor Voting: A Perceptual Organization Approach to Computer Vision and Machine Learning: Synthesis Lectures on Image, Video, and Multimedia Processing. Morgan and Claypool Publ., San Rafael (2006)
12. Mordohai, P., Medioni, G.: Stereo using monocular cues within the tensor voting framework. IEEE Trans. Patt. Analysis Mach. Intell. 28(6), 968–982 (2006)
13. Nicolescu, M., Medioni, G.: Motion segmentation with accurate boundaries: a tensor voting approach. In: Proc. IEEE Conf. Comp. Vis. Patt. Recogn., Madison, WI, pp. 382–389. IEEE Comp. Soc., Los Alamitos (2003)
14. Rando, G., Arca, S., Casiraghi, E., Maggi, A.: Automatic segmentation of mouse images. In: Proc. 10th European Congress Stereology and Image Analysis, Milan, Italy (2009)
15. Tang, C.-K., Medioni, G.: Curvature-augmented tensor voting for shape inference from noisy 3D data. IEEE Trans. Patt. Analysis Mach. Intell. 24(6), 858–863 (2002)
16. Tong, W.-S., Tang, C.-K., Mordohai, P., Medioni, G.: First order augmentation to tensor voting for boundary inference and multiscale analysis in 3D. IEEE Trans. Patt. Analysis Mach. Intell. 26(5), 858–863 (2004)

Clustering Ensembles with Active Constraints

Muna Al-Razgan and Carlotta Domeniconi

Abstract. In this work we combine clustering ensembles and semi-supervised clustering to address the ill-posed nature of clustering. We introduce a hybrid approach that extends our previous work on clustering ensembles to situations where some knowledge from the end user is available, by enforcing constraints during the partitioning process. The experimental results show that our constrained ensemble technique is capable of producing a partition that is as good as, or better, than those computed by other semi-supervised clustering approaches.

1 Introduction

Clustering is the process of discovering homogeneous groups or clusters according to a given similarity measure. Clustering is well suited for data analysis. However, clustering is susceptible to several difficulties. It is well known that off-the-shelf clustering methods may discover different structures in a given set of data. This is because each clustering algorithm has its own bias resulting from the optimization of different criteria. Furthermore, there is no ground truth against which the clustering result can be validated. Thus, no cross-validation technique can be carried out to tune input parameters involved in the clustering process. Recently, clustering ensembles have emerged as a technique for overcoming problems with clustering algorithms. A clustering ensemble consists of different partitions. These partitions can be obtained from multiple applications of any single algorithm with different initializations, from various bootstrap samples of the available data, or from the application of different algorithms to the same dataset. Clustering ensembles offer a solution to challenges inherent to clustering arising from its ill-posed nature: they can provide more robust and stable solutions by making use of the consensus across multiple clustering results, while averaging out emergent spurious structures that arise due to the various biases to which each participating algorithm is tuned.

Muna Al-Razgan · Carlotta Domeniconi
George Mason University, 4400 University Drive, MS 4A5, Fairfax VA 22030, USA
e-mail: malrazga@gmu.edu, carlotta@cs.gmu.edu

O. Okun & G. Valentini (Eds.): Appli. of Supervised & Unsuper. Ensemble Meth., SCI 245, pp. 175–189.
springerlink.com © Springer-Verlag Berlin Heidelberg 2009

Although clustering is traditionally an unsupervised learning problem, in some applications the end user can provide limited information about the data. Semi-supervised clustering uses prior knowledge to guide the clustering process, and to provide results that adhere to the user's preference. As such, semi-supervised clustering techniques promise an ease of use and a natural approach along with accurate results.

In this work, we extend our previous clustering ensemble technique [1] to situations where some knowledge is available from the end user. Our newly proposed method combines the clustering ensemble's ability to overcome the ill-posed nature of clustering with the semi-supervised clustering's ability to leverage an end user's knowledge. Our technique enforces knowledge-based constraints during the partitioning process of each ensemble's component to improve the quality of the overall ensemble. To generate useful domain knowledge, our method takes an active approach, and capitalizes on those points that are not easily clustered by the ensemble. By enforcing the resulting constraints at the components' level, we ensure that the corresponding structure they represent is carried into the consensus function, and therefore into the consensus partition.

Our method of generating clustering components relies on a locally adaptive clustering algorithm (called LAC) [6, 7] that depends on two input parameters. The first one is common to all clustering algorithms: the number of clusters k to be discovered in the data. The second one (called h) controls the strength of the incentive to cluster on more features (more details are provided in Sect. 3). Our technique embeds constraints into the individual partitionings of each LAC component, when different values of h are used. We assume that the number of clusters k is fixed. Our constrained clustering ensemble approach can be easily extended to any clustering algorithm with prior knowledge. The major challenge we face is finding a method of incorporating constraints to improve upon the demonstrated reliability of weighted clustering ensembles. Our solution proposes a consensus function that is then mapped to a graph partitioning problem.

2 Related Work

Our approach is motivated by the work in [5], where the authors use the ensemble methodology to produce a consensus partition, and then apply labeled data to assign clusters to classes. The authors in [5] aim at labeling points more accurately, whereas our approach aims at enhancing clustering ensembles by enforcing constraints during the partitioning process. In [8] the authors combine different clusterings obtained via the k-means algorithm. The clusterings produced by k-means are mapped into a co-association matrix, which provides a measure of similarity between pairs of points. Kuncheva et al. [13] extend the work in [8] by choosing at random the number of clusters for each ensemble member. The authors in [16] introduce a meta-clustering procedure: first, each clustering is mapped into a distance matrix; then, the multiple distance matrices are combined; and finally, a hierarchical clustering method is introduced to compute a consensus clustering. In [10] the author proposes a similar approach, where a graph-based partitioning algorithm is used to generate the combined clustering.

The authors in [15] propose the COP-Kmeans algorithm as a variation of k-means, where constraints are embedded during the clustering process: each point is assigned to the closest cluster that will enact the least violation of constraints. The algorithm will not assign the point if no such cluster can be found. Two additional constraint-based variants of k-means are Seeded-KMeans and Constrained-KMeans [4]. In both algorithms, the given labeled data are used to initialize a seeded set; the constraints obtained from this labeled set are then used to guide the k-means algorithm. Seeded-KMeans allows its constraints to be violated in successive iterations, while Constrained-KMeans enforces the constraints in each iteration. The approach proposed in [9] imputes information from the co-association values generated from a clustering ensemble. Such imputed constraints have proven to achieve more accurate results than random constraints. The technique discussed in this Chapter uses this method for selecting constraints as its basis.

3 Locally Adaptive Clustering

In this section we briefly describe the Locally Adaptive Clustering algorithm [6, 7] used to generate the clustering components of our ensembles. Locally Adaptive Clustering (LAC) is a *soft* feature selection procedure that assigns weights to features according to the local variance of data along each dimension. Dimensions along which data are loosely clustered receive a small weight, which has the effect of elongating distances along that dimension. Features along which data manifest a small variance receive a large weight, which has the effect of constricting distances along that dimension. Thus the learned weights perform a directional local reshaping of distances which allows a better separation of clusters, and therefore the discovery of different patterns in different subspaces of the original input space.

Let us consider a set of n points in some space of dimensionality D. A *weighted cluster* is a subset of data points, together with a vector of weights $\mathbf{w} = (w_1, \ldots, w_D)^t$, such that the points in the cluster are close to each other according to the L_2 norm distance weighted using $\mathbf{w}$. The component w_j measures the degree of participation of feature j to the cluster. The problem is how to estimate the weight vector $\mathbf{w}$ for each cluster in the dataset.

In traditional clustering, the partition of a set of points is induced by a set of *representative* vectors, also called *centroids* or *centers*. The partition induced by discovering weighted clusters is formally defined as follows.

Definition 0.1. : Given a set S of n points $\mathbf{x} \in \Re^D$, a set of k centers $\{\mathbf{c}_1, \ldots, \mathbf{c}_k\}$, $\mathbf{c}_j \in \Re^D$, $j = 1, \ldots, k$, coupled with a set of corresponding weight vectors $\{\mathbf{w}_1, \ldots, \mathbf{w}_k\}$, $\mathbf{w}_j \in \Re^D$, $j = 1, \ldots, k$, partition S into k sets:

$$S_j = \{\mathbf{x} | (\sum_{s=1}^{D} w_{js}(x_s - c_{js})^2)^{1/2} < (\sum_{s=1}^{D} w_{li}(x_s - c_{ls})^2)^{1/2}, \forall l \neq j\}, j = 1, \ldots, k, \tag{1}$$

where w_{js} and c_{js} represent the sth components of vectors $\mathbf{w}_j$ and $\mathbf{c}_j$ respectively (ties are broken randomly).

The set of centers and weights is *optimal* with respect to the Euclidean norm, if they minimize the error measure:

$$E_1(P,W) = \sum_{j=1}^{k}\sum_{s=1}^{D}(w_{js}\frac{1}{|S_j|}\sum_{\mathbf{x}\in S_j}(c_{js} - x_s)^2), \tag{2}$$

subject to the constraints $\forall j$, $\sum_s w_{js} = 1$. $\mathbf{P}$ and $\mathbf{W}$ are $(D \times k)$ matrices whose columns are $\mathbf{c}_j$ and $\mathbf{w}_j$ respectively, i.e., $\mathbf{P} = [\mathbf{c}_1 \ldots \mathbf{c}_k]$ and $\mathbf{W} = [\mathbf{w}_1 \ldots \mathbf{w}_k]$. For shortness of notation, we set $X_{js} = \frac{1}{|S_j|}\sum_{\mathbf{x}\in S_j}(c_{js} - x_s)^2$, where $|S_j|$ is the cardinality of set S_j. X_{js} represents the average distance from the centroid $\mathbf{c}_j$ of points in cluster j along dimension s. The solution

$$(\mathbf{P}^*, \mathbf{W}^*) = \arg\min_{(\mathbf{P},\mathbf{W})} E_1(\mathbf{P}, \mathbf{W})$$

will discover one dimensional clusters: it will put maximal (unit) weight on the feature with smallest dispersion X_{js} within each cluster j, and zero weight on all other features. Our objective, instead, is to find weighted multidimensional clusters, where the unit weight gets distributed among all features according to the respective dispersion of data within each cluster. One way to achieve this goal is to add the regularization term $\sum_{s=1}^{D} w_{js} log w_{js}$, which represents the negative entropy of the weight distribution for each cluster. It penalizes solutions with maximal weight on the single feature with smallest dispersion within each cluster. The resulting error function is

$$E_2(\mathbf{P}, \mathbf{W}) = \sum_{j=1}^{k}\sum_{s=1}^{D}(w_{js}X_{js} + h w_{js} log w_{js}), \tag{3}$$

subject to the same constraints $\forall j$, $\sum_s w_{js} = 1$. The coefficient $h \geq 0$ is a parameter of the procedure; it controls the strength of the incentive for clustering on more features. Increasing (decreasing) its value will encourage clusters on more (less) features. This constrained optimization problem can be solved by introducing the Lagrange multipliers. It gives the solution

$$w_{js}^* = \frac{exp(-X_{js}/h)}{\sum_{s=1}^{D} exp(-X_{js}/h)}, \tag{4}$$

$$c_{js}^* = \frac{1}{|S_j|}\sum_{\mathbf{x}\in S_j} x_s. \tag{5}$$

Solution (4) puts increased weights on features along which the dispersion X_{js} is smaller, within each cluster. The degree of this increase is controlled by the value h. Setting $h = 0$ places all weight on the feature s with smallest X_{js}, whereas setting $h = \infty$ forces all features to be given equal weight for each cluster j.

A search strategy needs to be designed to find a partition P that identifies the solution clusters. We proposed the approach [7] that progressively improves the quality of initial centroids and weights, by investigating the space near the centers to estimate the dimensions that matter the most. *Well-scattered* points in S are first chosen as the k centroids. All weights are initially set to $1/D$. Given the initial centroids $\mathbf{c}_j$, for $j = 1, \ldots, k$, the corresponding sets S_j are computed as previously defined. The average distance X_{js} along each dimension from the points in S_j to $\mathbf{c}_j$ is then computed. The smaller X_{js}, the stronger is the degree of participation of feature s to cluster j. The value X_{js} is used in an exponential weighting scheme to credit weights to features (and to clusters), as given in Eq. (4). The computed weights are used to update the sets S_j, and therefore the centroids' coordinates as given in Eq. (5). The procedure is iterated until convergence is reached.

LAC has shown a highly competitive performance with respect to other state-of-the-art subspace clustering algorithms [6, 7]. Despite its strong performance, LAC's dependence on the setting of h is a liability. Improving upon this aspect of LAC's performance is desirable, and we have sought such improvement through the development of a clustering ensemble technique that makes use of pairwise constraints. Informative constraints are actively identified by leveraging different runs of the LAC algorithm, each using a different value of the h parameter. We describe the details of this process in the following section.

4 Selecting Informative Constraints

A number of semi-supervised clustering algorithms have been proposed [4, 15]. However, most of these techniques construct *must-link* and *cannot-link* constraints by first randomly selecting pairs of points, and then querying an oracle expert for information about their relationship (i.e., whether the two points belong to the same cluster or not). Although this method is relied on by many researchers, it has the liability of not improving the clustering process to its fullest potential. Because the selection process is random, and does not seize on associations available in the raw data, this method neglects a very important source of information.

To avoid these limitations, we follow the approach described in [9] to generate constraints. The authors base their method on the relationships between intra-cluster and inter-cluster points. The authors begin by mapping the results of k-means to a co-association matrix, whose element of position (i, j) is equal to one if points $\mathbf{x}_i$ and $\mathbf{x}_j$ are placed in the same cluster, and zero otherwise. The algorithm is then run v times, and the resulting v matrices are averaged into a final matrix $\mathbf{T}$. With a sufficient number of base clusterings, the relationship between two data points becomes apparent in the final matrix $\mathbf{T}$. The entry T_{ij} indicates the portion of the v clusterings in which two data points $\mathbf{x}_i$ and $\mathbf{x}_j$ were assigned to the same cluster. A value of $T_{ij} = 1$ indicates that the points were assigned to the same cluster in each of the v matrices, and therefore represents a very high probability that the points belong to the same class. A value of $T_{ij} = 0$, on the other hand, indicates that the points were never assigned to the same cluster, and therefore represents a very low probability that the points belong to the same class.

However, it is possible that the value of T_{ij} is neither $T_{ij} \approx 1$ nor $T_{ij} \approx 0$. In particular, if $T_{ij} \approx 0.5$, there will be a great deal of uncertainty about the placement of the corresponding two points. To solve this ambiguity, the authors in [9] set two threshold values, t_c and t_m, to identify the pairs of points that were easily clustered or separated. Then, they select the pairs $(\mathbf{x}_i, \mathbf{x}_j)$ for which the entry value is such that $t_c \leq T_{ij} \leq t_m$, and query the oracle expert about the placement of such pairs. If the points $(\mathbf{x}_i, \mathbf{x}_j)$ are to be clustered together, a must-link constraint is generated; otherwise, a cannot-link constraint is generated.

Since this method proved to be effective, we adopted it to select a set of constraints. We run the LAC algorithm m times. Each partition v gives a co-association matrix $\mathbf{T}_v$, from which we compute the average co-association matrix $\mathbf{T} = \frac{1}{m}\sum_{v=1}^{m} T_v$. We then select all pairs of points $(\mathbf{x}_i, \mathbf{x}_j)$ such that $T_{ij} \in [t_c, t_m]$. This selection mechanism allows to identify pairs of points with great uncertainty regarding their clustering placement. Thus, querying an oracle about their underlying relationships adds valuable information which cannot be derived from the data alone.

This process generates two constraint sets (M, C), where M corresponds to must-link constraints, and C corresponds to cannot-link constraints. We use these constraints to form a chunklet graph as discussed in the following section. The pseudo code for the imputation of constraints is given in Algorithm 3.

Algorithm 3. Imputation of Constraints (IC) Algorithm

Input: m partitions of n data points

1. For each partition $v = 1, \ldots, m$:
   - Build the co-association matrix $\mathbf{T}_v$ of size $n \times n$

$$T_{v_{ij}} = \begin{cases} 1 & \text{if } \mathbf{x}_i, \mathbf{x}_j \text{ are in the same cluster in partition } v \\ 0 & \text{if } \mathbf{x}_i, \mathbf{x}_j \text{ are in different clusters in partition } v \end{cases}$$

2. Compute the final co-association matrix $\mathbf{T}$ from all $\mathbf{T}_v$ where $v = 1, \ldots, m$:

$$\mathbf{T} = \frac{1}{m}\sum_{v=1}^{m} \mathbf{T}_v$$

3. Select all pairs (or a random sample) $(\mathbf{x}_i, \mathbf{x}_j)$ such that $T_{ij} \in [t_c, t_m]$
4. Query the oracle for the selected pairs $(\mathbf{x}_i, \mathbf{x}_j)$
5. Construct constraint sets (M, C)

Output: The resulting constraint sets (M, C)

---

5 Chunklet Graph

In order to embed constraints into subspace clusterings, we organize the constraints into a graph called *chunklet graph*. In the following, we describe the procedure to construct such graph.

A chunklet is a group of points that belong to the same cluster, although the identity of the cluster is unknown [3]. The size of the chunklet is equal to the number of points it contains, e.g., the chunklet $\triangle = (\mathbf{x}_1, \mathbf{x}_2)$ has size $s(\triangle) = 2$.

Each chunklet is formed by must-link constraints (M) obtained from the oracle expert. If the oracle imposes a must-link between points $\mathbf{x}_1$ and $\mathbf{x}_2$, then the chunklet $\triangle_1 = (\mathbf{x}_1, \mathbf{x}_2)$ is formed. Following the formation of chunklets through must-link constraints, a transitive closure process, in which chunklets are merged, is initiated. For example, if there is a must-link constraint between $(\mathbf{x}_1, \mathbf{x}_2)$ and between $(\mathbf{x}_1, \mathbf{x}_3)$, then by transitive closure the chunklet $\triangle_2 = (\mathbf{x}_1, \mathbf{x}_2, \mathbf{x}_3)$ is formed.

Once chunklets are formed and all transitive closures completed, a graph is created using the cannot-link constraints imposed by the oracle expert. These constraints prevent the assignment of some chunklets to the same cluster. For example, if there is a cannot-link constraint between the pair $(\mathbf{x}_3, \mathbf{x}_5)$ and we have the chunklet $\triangle_3 = (\mathbf{x}_4, \mathbf{x}_5)$, then our previously cited chunklet $\triangle_2 = (\mathbf{x}_1, \mathbf{x}_2, \mathbf{x}_3)$ will be prevented from the assignment to the same cluster as chunklet $\triangle_3$.

A cannot-link constraint is represented on the graph as an edge between two vertices, where each vertex corresponds to one chunklet. In this way the entire chunklet graph is obtained, where a chunklet is a vertex and each cannot-link constraint is an edge. Edges indicate that the corresponding vertices (chunklets) should be assigned to different clusters.

Thus, this procedure generates a chunklet graph $G_{ch} = (V, E)$, where V is a set of vertices (or chunklets) constructed from the must-link constraints M, and $|V|$ is the total number of chunklets. E is the set of edges, and an edge E_{ij} exists between vertices (chunklets) v_i and v_j *iff* there exist $\mathbf{x}_i \in v_i$, $\mathbf{x}_j \in v_j$ such that $(\mathbf{x}_i, \mathbf{x}_j) \in C$. The pseudo code of the chunklet graph construction is presented in Algorithm 4.

Algorithm 4. Chunklet Graph (CG) Algorithm

Input: Constraint sets (M, C)

1. Compute the transitive closure for all the must-link constraints M
2. Build chunklets v_i using the must-link constraints
3. Construct the chunklet graph $G_{ch} = (V, E)$, where V corresponds to the set of chunklets and $E_{ij} = 1$ *iff* there exist $\mathbf{x}_i \in v_i$, $\mathbf{x}_j \in v_j$ such that $(\mathbf{x}_i, \mathbf{x}_j) \in C$

Output: The resulting graph G_{ch}

6 Chunklet Assignment

Chunklet assignment is the process of assigning vertices (chunklets) in the graph to the appropriate centroid without violating any of the must-link or cannot-link constraints. We consider each chunklet as a group of points, and assign all points in the chunklet to the closest centroid which does not violate any constraint. LAC provides a set of k centroids $\{\mathbf{c}_1, \ldots, \mathbf{c}_k\}$, and a set of k weight vectors $\{\mathbf{w}_1, \ldots, \mathbf{w}_k\}$, where each weight vector reflects the relevance of features within the corresponding cluster. Given a vertex (chunklet) v_i, we calculate the weighted Euclidean distances between all the points $\mathbf{x}_j \in v_i$ and each centroid $\mathbf{c}_l$, where $l = 1, \ldots, k$, and look for the centroid $\mathbf{c}_t$ that satisfies

$$\mathbf{c}_t = \arg\min_l (d(v_i, \mathbf{c}_l)), \tag{6}$$

where $d(v_i, \mathbf{c}_l) = \sum_{j=1}^{|v_i|} \sqrt{\sum_{s=1}^{D} w_{ls}(x_{js} - c_{ls})^2}$, D is the dimensionality of the data, and $\mathbf{w}_l$ is the weight vector associated with centroid $\mathbf{c}_l$. To assign a chunklet to a centroid, we need to consider possible cannot-link constraints involving the chunklet. Three cases which require different centroid assignment strategies are given below.

Case 1. v_i is an isolated chunklet that does not have an edge in the graph G_{ch}. We assign this chunklet to the centroid $\mathbf{c}_t = \arg\min_l(d(v_i, \mathbf{c}_l))$.

Case 2. v_i is a chunklet that has at least one neighbor in the graph G_{ch}, and none of its neighbors has been assigned to a centroid. As before, we assign v_i to the centroid $\mathbf{c}_t = \arg\min_l(d(v_i, \mathbf{c}_l))$.

Case 3. v_i ia a chunklet that has at least one neighbor in the graph G_{ch} that has been assigned to a centroid. We construct the set of centroids R_i to which the neighboring nodes of v_i have been assigned. We then assign v_i to the centroid $\mathbf{c}_t = \arg\min_l(d(v_i, \mathbf{c}_l))$ such that $\mathbf{c}_t \notin R_i$.

This procedure assigns chunklets to centroids according to the local similarities being discovered by the subspace clustering algorithm. The pseudo code of the chunklet assignment procedure is given in Algorithm 5.

7 Constrained-Weighted Bipartite Partitioning Algorithm (C-WBPA)

Our aim is to generate robust and stable solutions via a consensus clustering method that makes use of prior knowledge under the form of must-link and cannot-link constraints. We generate contributing clusterings by running the LAC algorithm multiple times by changing the h parameter. In details, the overall process works as follows.

We generate imputed constraints using the approach described in Sect. 4. Once we have identified the proper constraints, we build the chunklet graph as illustrated in Sect. 5. We then incorporate the graph into the clustering ensemble

Algorithm 5. Chunklet Assignment (CA) Algorithm

Input: Chunklet graph G_{ch}, centroids $\{\mathbf{c}_1^v, \ldots, \mathbf{c}_k^v\}$, and weights $\{\mathbf{w}_1^v, \ldots, \mathbf{w}_k^v\}$

1. Create a matrix $\mathbf{O} = zeros(u,k)$, where u is total number of points in all chunklets of G_{ch}
2. For each vertex $V_i \in G_{ch}$, we consider three cases:

   a. Chunklet v_i does not have a neighbor in the graph G_{ch}. Assign chunklet v_i to the closest centroid:

      i. $\mathbf{c}_t = \arg\min_l(d(v_i, \mathbf{c}_l))$, such that $d(v_i, \mathbf{c}_l) = \sum_{j=1}^{|v_i|} \sqrt{\sum_{s=1}^{D} w_{ls}(x_{js} - c_{ls})^2}$
      ii. $\forall \mathbf{x}_j \in v_i$, $O_{j,t} = 1$

   b. Chunklet v_i does not have a neighbor that has been assigned to a centroid.

      i. Assign chunklet v_i to closet centroid $\mathbf{c}_t$ as above
      ii. $\forall \mathbf{x}_j \in v_i$, $O_{j,t} = 1$

   c. Chunklet v_i has at least one neighbor that has been assigned to a centroid.

      i. Construct the set of centroids R_i to which the neighboring chunklets of v_i have been assigned
      ii. Find the closest centroid $\mathbf{c}_t$ to chunklet v_i as described above, satisfying the condition $\mathbf{c}_t \notin R_i$
      iii. $\forall \mathbf{x}_j \in v_i$, $O_{j,t} = 1$

Output: The resulting matrix $\mathbf{O}$

components without violating any cannot-link constraint using the procedure described in Sect. 6. The assignment strategy is applied to each component of the ensemble.

Since LAC produces weighted clusters, we use this information to assign the vertices of the chunklet graph to the closest centroid. For each partition produced by the LAC algorithm we embed the given constraints, by assigning each chunklet to the centroid that minimizes the sum of the weighted distances between all the points in each chunklet. This assignment will produce a matrix $\mathbf{O}$ of 0 and 1 values, where the number of rows is equal to the number of points in the chunklet graph, and the number of columns is equal to k (the number of clusters) (see Algorithm 5). An entry $O_{i,j} = 1$ means that the point i is highly likely to belong to cluster j. We also ensure the assignment of chunklets connected by an edge to different centroids.

For points $\mathbf{x}_i$ not involved in the constraint sets, and for each clustering $v = 1, \ldots, m$, we follow our consensus clustering approach WBPA [1] as described in the following. The weighted distance of $\mathbf{x}_i$ from cluster C_l is given by $d_{il} = \sqrt{\sum_{s=1}^{D} w_{ls}(x_{is} - c_{ls})^2}$. Let $D_i = max_l\{d_{il}\}$ be the largest distance of $\mathbf{x}_i$ from any cluster. We want to define the probability associated with cluster C_l given that we have observed $\mathbf{x}_i$. Given a point $\mathbf{x}_i$, the cluster label C_l is assumed to be a random variable from a distribution with probabilities $\{P(C_l|\mathbf{x}_i)\}_{l=1}^{k}$. We provide a nonparametric estimation of such probabilities based on the data and on the clustering result. We do not make any assumption about the specific form (e.g., Gaussian) of the underlying data distributions, thereby avoiding parameter estimation of models, which is problematic in high dimensions when the available data are limited.

In order to embed the clustering result in our probability estimations, the smaller the distance d_{il} is, the larger the corresponding probability credited to C_l should be. Thus, we can define $\{P(C_l|\mathbf{x}_i)\}$ as follows:

$$P(C_l|\mathbf{x}_i) = \frac{D_i - d_{il} + 1}{kD_i + k - \Sigma_l d_{il}}, \quad (7)$$

where the denominator serves as a normalization factor in order to guarantee $\Sigma_{l=1}^{k} P(C_l|\mathbf{x}_i) = 1$. We observe that $\forall l = 1,\ldots,k$ and $\forall i = 1,\ldots,n$ $P(C_l|\mathbf{x}_i) > 0$. In particular, the added value of 1 in Eq. (7) allows for a non-zero probability $P(C_L|\mathbf{x}_i)$ when $L = \arg\max_l d_{il}$[1]. In this last case $P(C_l|\mathbf{x}_i)$ assumes its minimum value $P(C_L|\mathbf{x}_i) = 1/(kD_i + k - \Sigma_l d_{il})$. For smaller distance values d_{il}, $P(C_l|\mathbf{x}_i)$ increases proportionally to the difference $D_i - d_{il}$: the larger the deviation of d_{il} from D_i, the larger the increase. As a consequence, the corresponding cluster C_l becomes more likely, as it is reasonable to expect based on the information provided by the clustering process. Thus, Eq. (7) provides a nonparametric estimation of the posterior probability associated to each cluster C_l.

We can now construct the vector P_i of posterior probabilities associated with $\mathbf{x}_i$:

$$P_i = (P(C_1|\mathbf{x}_i), P(C_2|\mathbf{x}_i), \ldots, P(C_k|\mathbf{x}_i))^t, \quad (8)$$

where t denotes the transpose of a vector.

We initialize a matrix $\mathbf{N}$ to zero values, where the number of rows are equal to n (total number of points), and the number of columns are equal to k (the number of clusters). We then start filling the matrix $\mathbf{N}$ for each point $\mathbf{x}$. For the points participating in the chunklet graph, we retrieve their row value from the i-th row of the matrix $\mathbf{O}$ and assign it to $\mathbf{N}$ (i-th row of $\mathbf{N}$). For points not involved in the constraint set, we extract their probability vectors P_i and assign it to N_i. We follow the above procedure for each partition. Finally, we construct the following $\mathbf{AN}$ matrix:

$$\mathbf{AN} = (\mathbf{N}^1 \mathbf{N}^2 \ldots \mathbf{N}^m), \quad (9)$$

where the dimensionality of $\mathbf{AN}$ is $(n \times km)$.

Based on $\mathbf{AN}$ we can now define a bipartite graph to which our consensus partition problem maps. Consider the graph $G = (V,E)$ with V and E constructed as follows. $V = V^C \cup V^I$, where V^C contains km vertices, each representing a cluster of the ensemble, and V^I contains n vertices, each representing an input data point. Thus $|V| = km + n$. The edge E_{ij} connecting the vertices V_i and V_j is assigned a weight value defined as follows. If the vertices V_i and V_j represent both clusters or both instances, then $E(i,j) = 0$; otherwise, if vertex V_i represents an instance $\mathbf{x}_i$ and

[1] Any small positive constant achieves this goal, with the normalization factor properly adjusted.

vertex V_j represents a cluster C_j^v (or vice versa) then the corresponding entry of E is $AN(i,k(v-1)+j)$. We run METIS [12] on the resulting graph to compute a k-way partitioning of the vertices that minimizes the edge weight-cut. The partition of the instance vertices gives the consensus clustering we seek.

The steps of the algorithm, which we call C-WBPA (Constrained-Weighted Bipartite Partitioning Algorithm), are summarized in Algorithm 6.

Algorithm 6. Constrained-Weighted Bipartite Partitioning Algorithm (C-WBPA)

Input: n points $\mathbf{x} \in R^D$, and k

1. Run LAC m times with different h values. Obtain the m partitions: $\{\mathbf{c}_1^v,\ldots,\mathbf{c}_k^v\},\{\mathbf{w}_1^v,\ldots,\mathbf{w}_k^v\}$, $v=1,\ldots,m$
2. (M,C)= IC(m partitions of n points) (Algorithm 3)
3. G_{ch}= CG(M,C) (Algorithm 4)
4. For each partition $v=1,\ldots,m$:

   a. O^v= CA($G_{ch},\{\mathbf{c}_1^v,\ldots,\mathbf{c}_k^v\},\{\mathbf{w}_1^v,\ldots,\mathbf{w}_k^v\}$) (Algorithm 5)
   b. $\forall\, \mathbf{x}_i$ not involved in any constraint:

      i. Compute $d_{il}^v = \sqrt{\sum_{s=1}^D w_{ls}^v (x_{is}-c_{ls}^v)^2}$
      ii. Set $D_i^v = \max_l\{d_{il}^v\}$
      iii. Compute $P(C_l^v|\mathbf{x}_i) = \frac{D_i^v - d_{il}^v + 1}{kD_i^v + k - \sum_l d_{il}^v}$
      iv. Set $P_i^v = (P(C_1^v|\mathbf{x}_i), P(C_2^v|\mathbf{x}_i),\ldots,P(C_k^v|\mathbf{x}_i))^t$

   c. Initialize $N^v(n,k) = zeros$

      i. $\forall \mathbf{x}_i$ not involved in constraints, $N_i^v = P_i^v$, [i-th row of N^v is set equal to P_i^v]
      ii. $\forall \mathbf{x}_l$ in constraints, $N_i^v = O_i^v$, [i-th row of N^v is set equal to O_i^v]

5. Construct the matrix **AN** as in Eq. (9)
6. Construct the bipartite graph $G=(V,E)$, where $V=V^C \cup V^I$, $|V^I|=n$ and $V_i^I \equiv \mathbf{x}_i$, $|V^C|=km$ and $V_j^C \equiv C_j$ (a cluster of the ensemble). Set $E(i,j)=0$ if V_i and V_j are both clusters or both instances. Set $E(i,j)=AN(i,k(v-1)+j)$ if V_i represents an instance $\mathbf{x}_i$, and V_j represents a cluster C_j^v (or vice versa)
7. Run METIS [12] on the resulting graph G

Output: The resulting k-way partition of the n vertices in V^I

---

8 Empirical Evaluation

In our experiments, we used eight real datasets. The characteristics of all datasets are given in Table 1. Iris, Breast, Letter(A,B), Wine and Ionosphere are from the UCI Machine Learning Repository [2]. WDBC is the Wisconsin Diagnostic Breast Cancer dataset [14]. Ling-Spam and 20 Newsgroups are two high dimensional text datasets. The documents in each dataset were preprocessed by eliminating stop words (based on a stop words list) and stemming words to their root source. As

Table 1 Characteristics of the datasets

Dataset	k	D	n (points-per-class)
Iris	3	4	150 (50-50-50)
WDBC	2	31	424 (212-212)
Breast	2	9	478 (239-239)
Letter(A,B)	2	16	1555 (789-766)
Wine	3	13	178 (59-71-48)
Ionosphere	2	33	239 (126-113)
Newsgroups(ele-med)	2	321	1971 (981-990)
Ling-Spam	2	350	906 (453-453)

feature values in the vector space model we have used the frequency of the terms in the corresponding document. To reduce the dimensionality of the data, we followed the procedure presented in [11]. In particular, a global unsupervised feature selection procedure, based on frequent itemset mining, was applied. The objective of this step is to identify sets of terms that co-occur frequently in the given corpus of documents. Such terms become the features used in the final representation of documents.

Ling-Spam is a mixture of spam messages (453) and valid messages (561) sent via the linguist list, a moderated (hence, spam-free) list about the profession and science of linguistics. The original size of the dictionary is 24627. After processing the data as described above, the dictionary size was reduced to 350. 20 Newsgroups is a collection of 20,000 messages collected from 20 different netnews newsgroups. One thousand messages from each of the 20 newsgroups were chosen at random and partitioned by newsgroups name. In our experiments we consider the categories medical (990) and electronics (981). The original size of the dictionary is 24546; after processing the data, the dictionary size was reduced to 321.

Since METIS [12] requires balanced datasets, we performed random sampling on Breast, WDBC, Ionosphere, and Ling-Spam. In each case, we sub-sampled the most populated class: from 357 to 212 for WDBC, from 444 to 239 for Breast , from 225 to 113 for Ionosphere, and from 561 to 453 for LingSpam. Also for the Letter dataset, we used the classes “A” and “B” (balanced).

We compare our constrained clustering ensemble techniques (C-WBPA) with other semi-supervised clustering approaches: COP-Kmeans [15] and Seeded-COP-Kmeans. We run COP-Kmeans ten times with random initialization. Seeded-COP-Kmeans is initialized using the chunklet graph constructed as in Sect. 5. In particular, the centroids are initialized using the vertices with cannot-link constraints among them. We compute the mean vectors of the points contained in each corresponding chunklet. These mean vectors are the initial centroids. If the number of selected chunklets is less than k, we choose as additional centroids the points that are the farthest from the already chosen centroids.

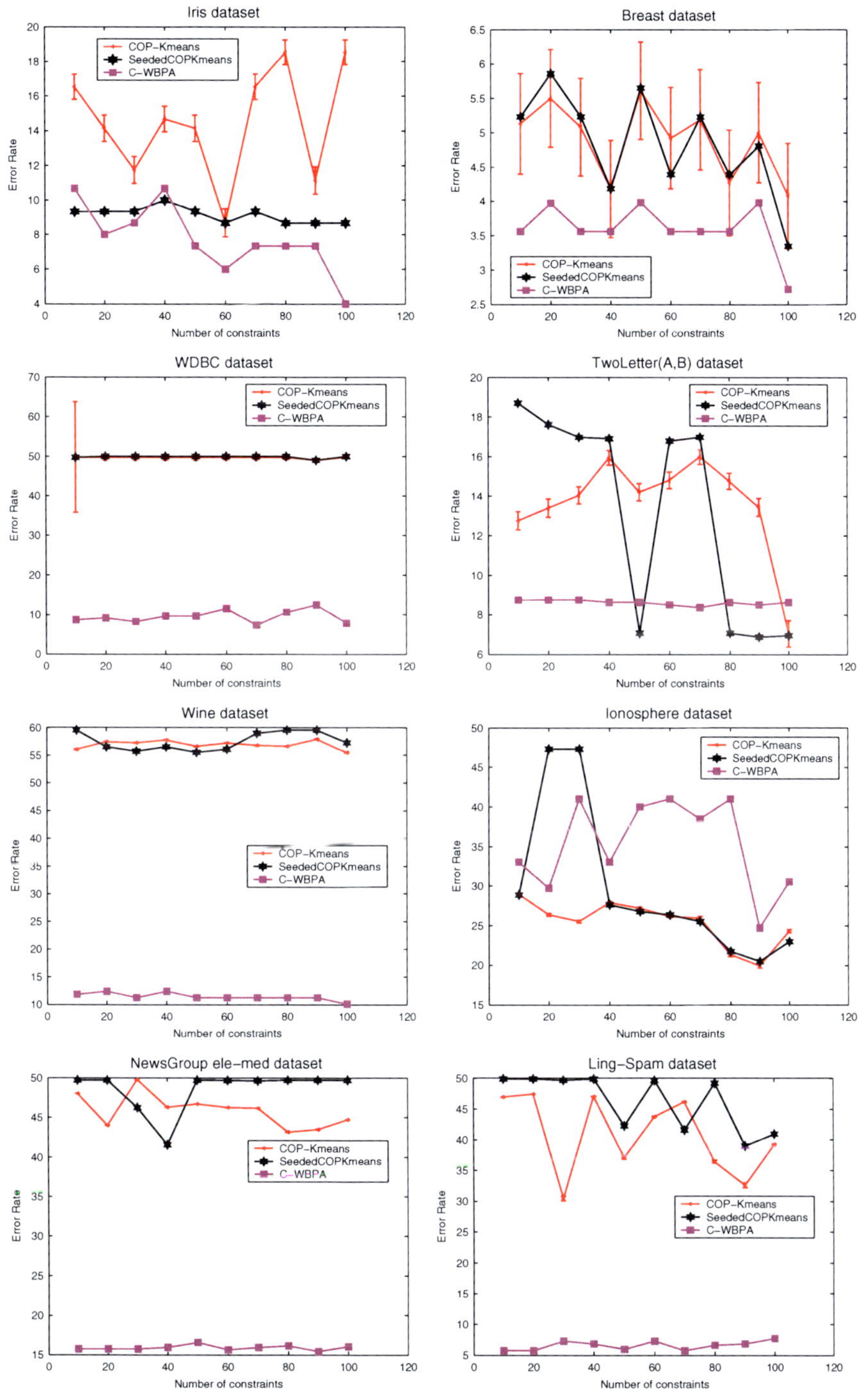

Fig. 1 Constrained clustering ensembles results

Evaluating the quality of clustering is in general a difficult task. Since class labels are available for the datasets used here, we evaluate the results by computing the error rate according to the confusion matrix.

8.1 Analysis of the Results

The same set of constraints is used for C-WBPA, COP-Kmeans, and Seeded-COP-Kmeans. As value of k, the actual number of classes in the data is used. Figure 1 plots the error rates (%) and standard deviations obtained for an increasing number of constraints for each dataset. Each figure clearly shows the improvement of our algorithm C-WBPA with respect to the other techniques. The trends of the error rate clearly depend on the data distribution. We notice the smooth trends of our technique (C-WBPA) with respect to the other techniques and an increasing number of constraints.

Our technique achieves the lowest error rate with a small number of constraints in most cases. Under these conditions, the clustering components are likely to be diverse; thus, the ensemble method is able to filter out the uncorrelated errors made by the individual clusterings. As a consequence, if limited information is available, C-WBPA can achieve a good result. This characteristic makes our approach a valuable asset to be used with very limited knowledge, with no need to query the oracle for more information.

Increasing the number of constraints made available to each component may induce a high degree of correlation between them, causing diversity to decrease. This phenomenon might be the reason for the stable error rate of C-WBPA for increasing constraints. We emphasize the large improvements obtained by C-WBPA for the high dimensional datasets (NewsGroup, Ling-Spam) we tested. This is because the LAC technique is designed to handle data with high dimensionality, while any variation of k-means tends to break down for datasets with high dimensionality.

9 Conclusions

We have introduced a new constrained ensemble technique for clustering. Constrains are bootstrapped in an active fashion through the ensemble, and embedded during the partitioning process carried out by each component. The experimental results show that our constrained ensemble can provide solutions that are superior to other semi-supervised clustering approaches. In our future work, we will consider the design of a semi-supervised clustering method that embeds constraints directly in the final consensus function. In addition, we will investigate techniques to embed constraints across different ensemble components.

Acknowledgements. This work was in part supported by NSF CAREER Award IIS-0447814.

References

1. Al-Razgan, M., Domeniconi, C.: Weighted clustering ensembles. In: Proc. 2006 SIAM Int. Conf. Data Mining, Bethesda, MD, pp. 258–269. SIAM, Philadelphia (2006)
2. Asuncion, A., Newman, D.J.: UCI Machine Learning Repository. University of California, Irvine, School of Information and Computer Sciences (2007)
3. Bar-Hillel, A., Hertz, T., Shental, N., Weinshall, D.: Learning distance functions using equivalence relations. In: Fawcett, T., Mishra, N. (eds.) Proc. 20th Int. Conf. Mach. Learn., pp. 11–18. AAAI Press, Menlo Park (2003)
4. Basu, S., Banerjee, A., Mooney, R.: Semi-supervised clustering by seeding. In: Sammut, C., Hoffmann, A.G. (eds.) Proc. 19th Int. Conf. Mach. Learn., Sydney, NSW, Australia, pp. 27–34. Morgan Kaufmann, San Francisco (2002)
5. Dimitriadou, E., Weingessel, A., Hornik, K.: A mixed ensemble approach for the semi-supervised problem. In: Dorronsoro, J.R. (ed.) ICANN 2002. LNCS, vol. 2415, pp. 571–576. Springer, Heidelberg (2002)
6. Domeniconi, C., Papadopoulos, D., Gunopulos, D., Ma, S.: Subspace clustering of high dimensional data. In: Proc. 2004 SIAM Int. Conf. Data Mining, Arlington, VA, pp. 517–521. SIAM, Philadelphia (2004)
7. Domeniconi, C., Gunopulos, D., Ma, S., Yan, B., Al-Razgan, M., Papadopoulos, D.: Locally adaptive metrics for clustering high dimensional data. Data Mining and Knowl. Discovery J. 14(1), 63–97 (2007)
8. Fred, A., Jain, A.: Data clustering using evidence accumulation. In: Proc. 16th Int. Conf. Patt. Recogn., Quebec, QB, pp. 276–280. IEEE Comp. Soc., Los Alamitos (2002)
9. Greene, D., Cunningham, P.: An ensemble approach to identifying informative constraints for semi-supervised clustering. In: Kok, J.N., Koronacki, J., Lopez de Mantaras, R., Matwin, S., Mladenič, D., Skowron, A. (eds.) ECML 2007. LNCS (LNAI), vol. 4701, pp. 140–151. Springer, Heidelberg (2007)
10. Hu, X.: Integration of cluster ensemble and text summarization for gene expression analysis. In: Proc. 4th IEEE Symp. Bioinformatics and Bioengineering, Taichung, Taiwan, pp. 251–258. IEEE Comp. Soc., Los Alamitos (2004)
11. Kang, N., Domeniconi, C., Barbará, D.: Categorization and keyword identification of unlabeled documents. In: Proc. 5th IEEE Int. Conf. Data Mining, Houston, TX, pp. 677–680. IEEE Comp. Soc., Los Alamitos (2005)
12. Karypis, G., Kumar, V.: A fast and high quality multilevel scheme for partitioning irregular graphs. SIAM J. Scientific Computing 20(1), 359–392 (1998)
13. Kuncheva, L., Hadjitodorov, S.: Using diversity in cluster ensembles. In: Proc. 2004 Int. Conf. Syst. Man and Cybern., The Hague, The Netherlands, pp. 1214–1219. IEEE Comp. Soc., Los Alamitos (2004)
14. Mangasarian, O., Wolberg, W.: Cancer diagnosis via linear programming. SIAM News 23(5), 1–18 (1990)
15. Wagstaff, K., Cardie, C., Rogers, S., Schrödl, S.: Constrained k-means clustering with background knowledge. In: Brodley, C., Pohoreckyj-Danyluk, A. (eds.) Proc. 18th Int. Conf. Mach. Learn., Williamstown, MA, pp. 577–584. Morgan Kaufmann, San Francisco (2001)
16. Zeng, Y., Tang, J., Garcia-Frias, J., Gao, G.: An adaptive meta-clustering approach: combining the information from different clustering results. In: Proc. 1st IEEE Comp. Soc. Conf. Bioinformatics, Stanford, CA, pp. 276–287. IEEE Comp. Soc., Los Alamitos (2002)

Verifiable Ensembles of Low-Dimensional Submodels for Multi-class Problems with Imbalanced Misclassification Costs

Sebastian Nusser, Clemens Otte, and Werner Hauptmann*

Abstract. In this chapter, we discuss different strategies of extending an ensemble approach based on local binary classifiers to solve multi-class problems. The ensembles of binary classifiers were developed with the objective of providing interpretable submodels s for use in safety-related application domains. The ensembles assume highly imbalanced misclassification costs between the two classes. The extension to multi-class problems is not straightforward because common multi-class extensions might induce inconsistent decisions. We propose a solution of this problem that avoids such inconsistencies by introducing a hierarchy of misclassification costs. We show that by following such a hierarchy it becomes feasible to extend the binary ensemble, to maintain the desirable properties (that is, the good interpretability) of the binary ensemble, and to achieve a good predictive performance.

1 Introduction

Safety-related systems are systems whose malfunction or failure may lead to death or serious injury of people, loss or severe damage of equipment, or environmental harm. They are deployed, for instance, in aviation, automotive industry, medical systems and process control. In [16] we proposed a binary ensemble framework for use in safety-related domains. The main design criterion of this approach is to provide an ensemble of binary classification models that only uses small subspaces of

Sebastian Nusser
School of Computer Science, Otto-von-Guericke-University, Universitätsplatz 2, 39106 Magdeburg, Germany
e-mail: mail@seb-nusser.de

Clemens Otte · Werner Hauptmann
Siemens AG, Corporate Technology, Information & Communications, Learning Systems, Otto-Hahn-Ring 6, 81730 Munich, Germany
e-mail: {clemens.otte,werner.hauptmann}@siemens.com

* This work was done when the first author was with Siemens AG, Corporate Technology.

O. Okun & G. Valentini (Eds.): Appli. of Supervised & Unsuper. Ensemble Meth., SCI 245, pp. 191–211.
springerlink.com © Springer-Verlag Berlin Heidelberg 2009

the complete input space enabling the visual interpretation of the models. Because machine learning approaches are regarded with suspiciousness in the field of safety-related domains, the possibility to visualize each submodel greatly facilitates the domain experts' acceptance of the data-driven generated models. An interpretable solution is often required for applications where the available training data is too sparse and the number of input dimensions is too large to sufficiently apply statistical risk estimation methods in practical application tasks.

In most cases, high-dimensional models are required to solve a given problem. Unfortunately, such high-dimensional models are hard to verify (*curse of dimensionality*), may tend to overfitting, and the interpolation and extrapolation behavior is often unclear. An example of such counterintuitive and unintended behavior is illustrated in Fig. 1, where the prediction of the model changes in a region not covered by the given data set. Such behavior becomes even more likely and much more difficult to discover in the high-dimensional case. Our ensemble approach provides an insight into each submodel, which can be evaluated according domain knowledge and, thus, the correct interpolation and extrapolation behavior of the model can be guaranteed.

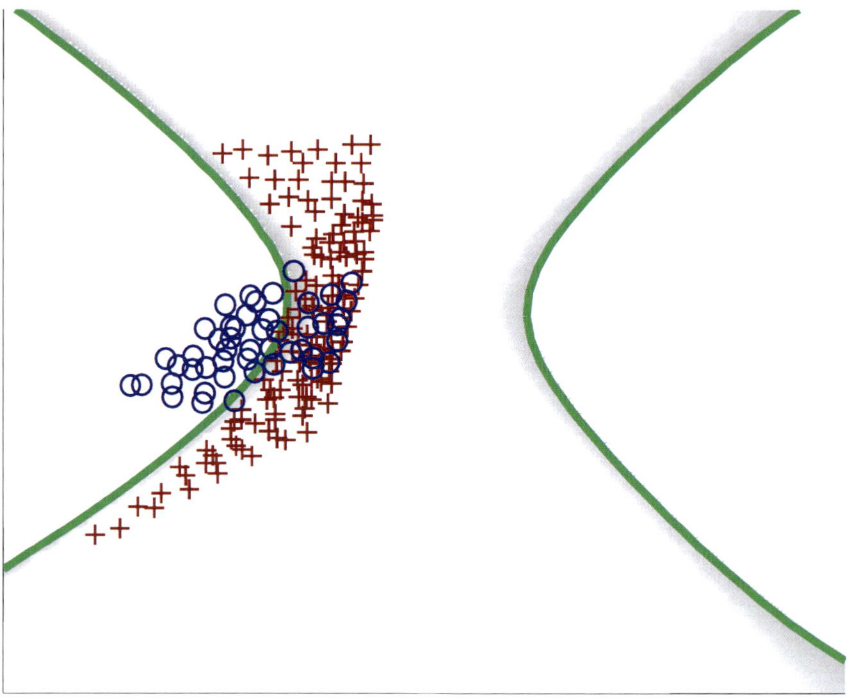

Fig. 1 Counterintuitive extrapolation behavior in a region not covered by the given data set. This two-class problem is solved by a support vector machine (SVM) with an acceptable classification performance on the given data. However, in a region not covered by any data the decision of the SVM changes arbitrarily

The extension of our binary classification ensemble to multi-class problems is not straightforward since commonly used methods like one-against-one or one-against-rest voting [6, 12] may introduce inconsistencies. We will show that such inconsistencies can be avoided by introducing a hierarchy of misclassification costs.

The crucial aspect is to find a suitable trade-off between generation of an interpretable and verifiable model and realization of a high predictive accuracy. In most situations, more complex models will be able to achieve a better predictive performance on the available data compared to simpler models. However, a higher complexity of the model will usually lead to an increased effort for model verification.

This contribution is organized as follows: in Sect. 2 we recall our binary ensemble approaches for use in safety-related domains. In Sect. 3, two commonly used approaches of extending binary classifiers to multi-class problems are briefly discussed and the inconsistencies are illustrated that might arise when applying them. Section 4 extends the binary classification framework to also solve multi-class problems. Experiments on well-known benchmark problems are discussed in Sect. 5 and Sect. 6 concludes this chapter.

2 The Binary Ensemble Framework

This section introduces the basic concepts of our ensemble approach that was first introduced in [16]. In Sect. 4, these algorithms are extended to solve multi-class problems. The original algorithms are designed to solve a binary classification problem. The task is to find an appropriate estimate of the unknown function $f : V^n \rightarrow Y$, where $V^n = X_1 \times X_2 \times \ldots \times X_n = \times_{i=1}^{n} X_i$ with $X_i \subseteq \mathbb{R}$ is the input space and Y is the target value, given an observed data set: $\mathfrak{D} = \{(\mathbf{v}_1, y_1), \ldots, (\mathbf{v}_m, y_m)\} \subset V^n \times Y$.

Basic Idea. Our ensemble framework is motivated by Generalized Additive Models [11, 17] and separate-and-conquer approaches [8]. It can be interpreted as a variant of the projection pursuit [7, 13]. Both approaches (cf. Sect. 2.1 and Sect. 2.2) are designed to find an estimate of the unknown function $f : V^n \rightarrow \mathbb{K}$, where $\mathbb{K} = \{0, 1\}$ is the set of class labels. Our approaches are based on the projection of the high-dimensional data to low-dimensional subspaces. Submodels g_j are trained on these subspaces. By regarding only low-dimensional subspaces a visual interpretation becomes feasible and, thus, the avoidance of unintended extrapolation behavior is possible. The ensemble of submodels boosts the overall predictive accuracy and overcomes the limited predictive performance of each single submodel , while the global model remains interpretable.

Projection of High-Dimensional Data Set. The projection π maps the n-dimensignal input space V^n to an arbitrary subspace of V^n. This mapping is determined by a given index set $\beta \subset \{1, \ldots, n\}$. The index set defines the dimensions of V^n that will be included in the subspace V_β. Thus, the projection π on the input space V^n given the index set β is defined as:

$$\pi_\beta(V^n) = V_\beta = \times_{i\in\beta} X_i\,. \tag{1}$$

Submodels. The j-th submodel is defined as:

$$g_j : \pi_{\beta_j}(V^n) \rightarrow \mathrm{I\!K}, \tag{2}$$

where β_j denotes the index set of the subspace where the classification error of the submodel g_j is minimal. The best projections can be determined, for instance, by a wrapper method for feature selection [15] that performs an exhaustive search through all possible feature combinations. For high-dimensional problems we advise to perform a preceding feature selection [9] in order to reduce the computational costs. The final function estimate $\hat{f}$ of the global model is determined by the aggregation of the results of all submodels $g_j(\pi_{\beta_j}(\mathbf{v}))$.

2.1 DecisionTree-Like Ensemble Model

This approach replaces the classification nodes in a common decision tree by strong classifiers. The classification nodes in the tree are restricted to two input dimensions. This facilitates the visualization of the relevant decision region and avoids overfitting. The best submodel g_j is used to divide the training set into new subsets $\mathfrak{D}_\theta^{new} := \{(\mathbf{v},y)|g_j(\pi_{\beta_j}(\mathbf{v})) = \theta\}$, where $\theta \in \mathrm{I\!K}$. The submodels for these subsets are built recursively until an appropriate termination criterion is fulfilled. The leaf nodes of the tree represent the final classification labels.

2.2 Non-Hierarchical Ensemble Model

This method incorporates prior knowledge about the subgroups of the given problem and avoids hierarchical dependencies of the submodels as in the DecisionTree-like Ensemble Model approach. It is required that the so-called *preferred class* c_{pref} must not be misclassified by any of the trained submodels: $\forall y = c_{pref} : |y - g(\pi_\beta(\mathbf{v}))| = 0$. This requirement typically leads to imbalanced misclassification costs.

The submodels are trained on low dimensional projections of the high dimensional input space with the objective to avoid the misclassification of the preferred class c_{pref}. The submodels greedily separate the samples of the other class from the preferred class samples. Missed samples of the other class are used to build further submodels.

The final function estimate is the disjunctive combination of all learned submodels

$$\hat{f}(\mathbf{v}) = \bigvee_{(g_j,\beta_j)\in Exp} g_j\left(\pi_{\beta_j}(\mathbf{v})\right)\,, \tag{3}$$

where Exp is the set of all learned submodels and their corresponding index sets. For the sake of simplicity it is defined that the preferred class c_{pref} is always encoded as 0 (interpreted as boolean *false*) and the other class is always encoded as 1 (interpreted as boolean *true*) by the submodels g_j.

2.3 An Illustrative Example

The *CUBES* data set is generated from four Gaussian components in a three-dimensional space. This data set is illustrated in Fig. 2. For each *Class 1* cluster 50 samples are drawn from $N(\mathbf{e}_i, 0.2\mathbf{I})$, where $\mathbf{e}_i$ is a unit vector and $\mathbf{I}$ is the identity matrix. 100 samples of the *Class 0* cluster are scattered around the origin, drawn from $N((0,0,0)^T, 0.2\mathbf{I})$. All submodels are trained as SVMs with a Gaussian kernel and the parameter set $\gamma = 0.2$ and $C = 5$.

DecisionTree-like Ensemble Model. This method does not require a predefined default class. However, if a default class is given by the application, it can be considered by different misclassification costs when learning the submodels of the tree-like model. In this toy example, we ignore information about the default class. At the initial state, all two-dimensional projections of the *CUBES* data set are very similar. The best two-dimensional submodel g_1 is depicted in Fig. 3(a). It uses the projection

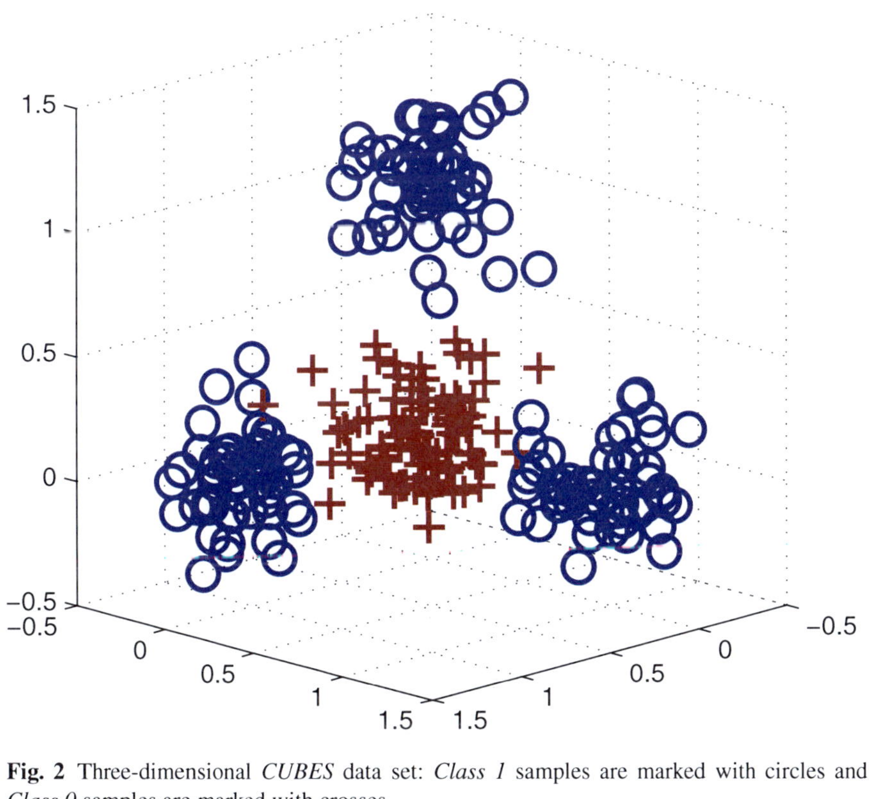

Fig. 2 Three-dimensional *CUBES* data set: *Class 1* samples are marked with circles and *Class 0* samples are marked with crosses

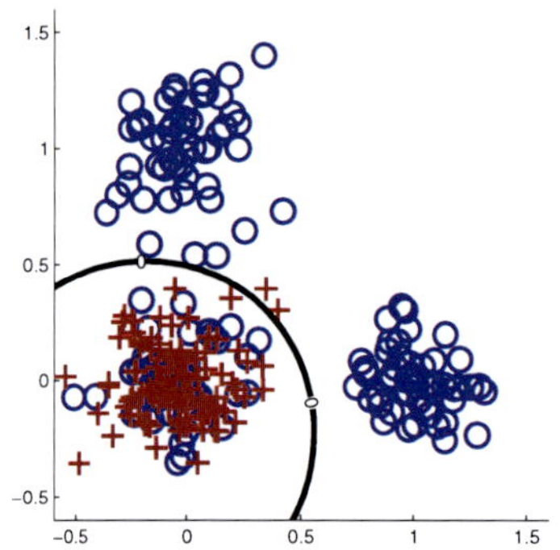

(a) 1st submodel g_1 with index set $\beta_1 = \{2,3\}$

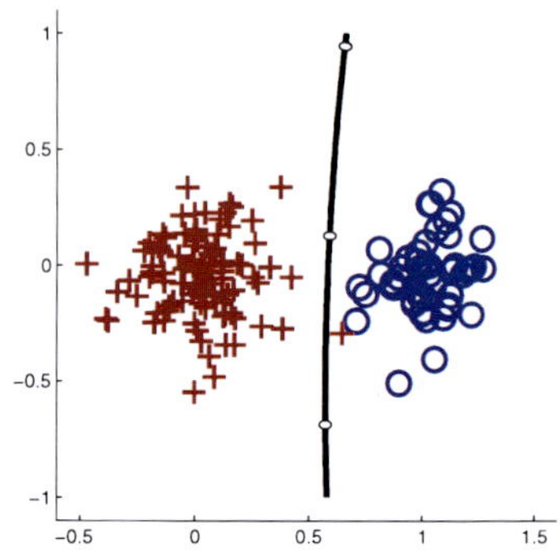

(b) 2nd submodel g_2 with index set $\beta_2 = \{1,2\}$

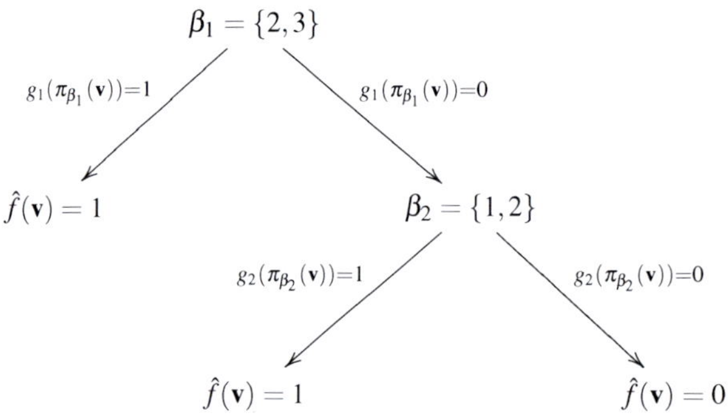

(c) Tree structure of the DecisionTree-like Ensemble Model

Fig. 3 DecisionTree-like Ensemble Model and the *CUBES* data set: *Class 1* samples are marked with circles and *Class 0* samples are marked with crosses. The decision boundaries are drawn as solid lines

$\pi_{\beta_1}(\mathbf{v})$ with $\beta_1 = \{2,3\}$. This submodel assigns 103 data points to *Class 1* (2 errors) and 147 data points to *Class 0* (49 errors). It is not possible to build other submodels for the data points that are assigned to *Class 1*, but for *Class 0* a second submodel can be built. This model, g_2 with $\beta_2 = \{1,2\}$, is depicted in Fig. 3(b). It assigns 50 data points to *Class 1* (1 error) and 97 data points to *Class 0* (0 errors). Further improvements are not possible. The final model, which is depicted in Fig. 3(c), misclassifies three *Class 0* samples. All other data points are correctly assigned to their corresponding class label. The confusion matrix of the final DecisionTree-like Ensemble Model is shown in Table 1(a).

Non-Hierarchical Ensemble Model. *Class 0* is chosen as the default class, $c_{pref} = 0$. That is, *Class 0* must not be misclassified by any learned submodel. This can be achieved, for instance, by using imbalanced misclassification costs for *Class 1* and *Class 0*. The best submodel g_1, see Fig. 4(a), uses the projection $\pi_{\beta_1}(\mathbf{v})$ with $\beta_1 = \{1,2\}$. 53 data points from *Class 1* are misclassified by this submodel. Thus, in

Table 1 Confusion matrices of the *CUBES* data set

(a) DecisionTree-like Ensemble Model

true class	predicted class *Class 0*	*Class 1*
Class 0	97	3
Class 1	0	150

(b) Non-Hierarchical Ensemble Model

true class	predicted class *Class 0*	*Class 1*
Class 0	100	0
Class 1	4	146

the next iteration new submodels are trained only on samples, which are predicted as *Class 0* by the first submodel: $\mathfrak{D}_{\text{new}} = \{(\mathbf{v}, y) | g_1(\pi_{\beta_1}(\mathbf{v})) = 0\}$. In Fig. 4(b) the projection $\pi_{\beta_2}(\mathbf{v})$ with $\beta_2 = \{2,3\}$ of the data set $\mathfrak{D}_{\text{new}}$ and the corresponding submodel g_2 are shown. This submodel misclassifies four *Class 1* samples. Given the chosen parameter set, no further improvements are possible. The final predictive model is $\hat{f}(\mathbf{v}) = g_1(\pi_{\beta_1}(\mathbf{v})) \vee g_2(\pi_{\beta_2}(\mathbf{v}))$. The overall performance of the Non-Hierarchical Ensemble Model is shown in Table 1(b): avoiding the misclassification of the default class $c_{pref} = 0$ leads to four misclassified *Class 1* samples.

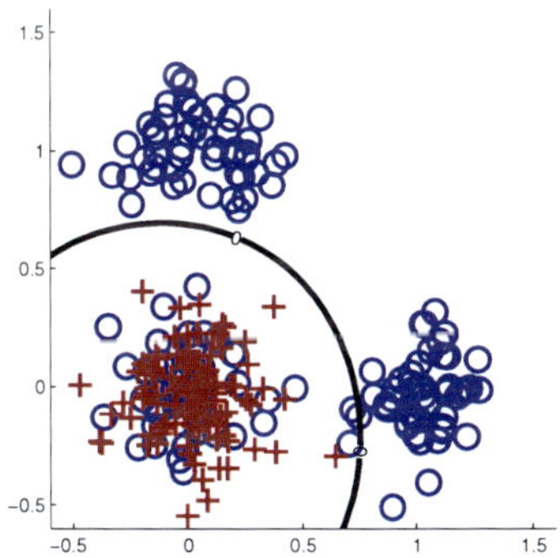

(a) 1st submodel g_1 with index set $\beta_1 = \{1,2\}$

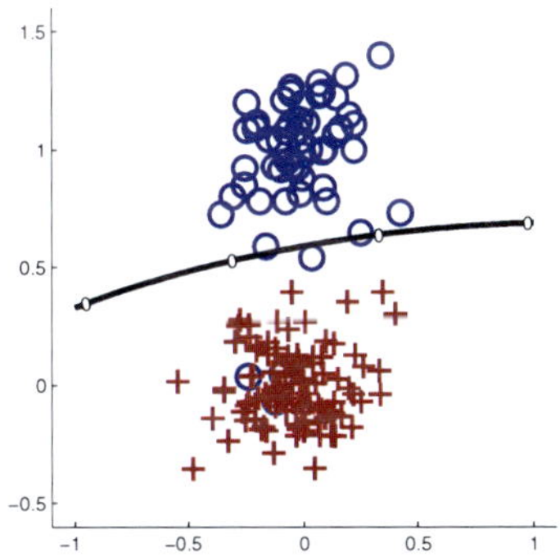

(b) 2nd submodel g_2 with index set $\beta_2 = \{2,3\}$

Fig. 4 Non-Hierarchical Ensemble Model and the *CUBES* data set: *Class 1* samples are marked with circles and *Class 0* samples are marked with crosses. The decision boundaries are drawn as solid lines

3 Multi-class Extensions of Binary Classifiers

There are two commonly used approaches to extend binary classifiers to solve multi-class problems: (1) a one-against-one extension and (2) a one-against-rest extension. A detailed comparison of these methods and an experimental evaluation for support vector machines is given in [12]. Figure 5 illustrates both approaches.

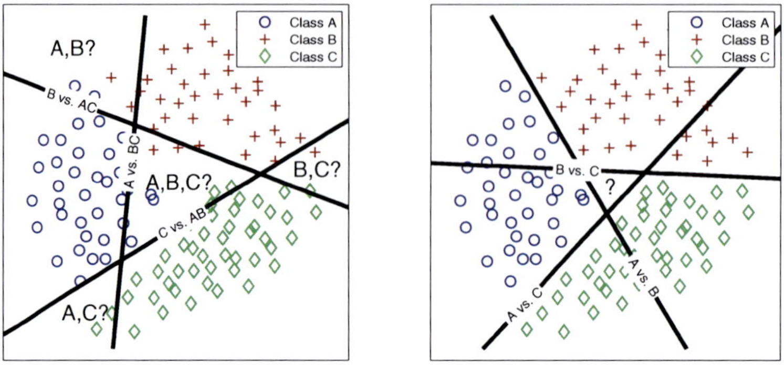

(a) One-against-rest (b) One-against-one

Fig. 5 Illustration of multi-class extensions based on binary classifiers. There are three classes: A, B, C. The discriminant functions are given as solid lines. Regions with possible inconsistent decisions are labeled with question marks

One-against-rest multi-class extension. This method constructs k classifiers ($k = |\mathbb{K}|$ is the number of class labels, $\mathbb{K} = \{c_1, c_2, \ldots, c_k\}$). The model $f_{c_\mathfrak{k}}$ for class $c_\mathfrak{k} \in \mathbb{K}$ is trained on all samples of class $c_\mathfrak{k}$ against all samples from the remaining classes which are combined to a new class $c_\mathfrak{k}^* = \mathbb{K} \setminus c_\mathfrak{k}$ – for the sake of simplicity the class label of $c_\mathfrak{k}^*$ is set to -1. A new data point $\mathbf{v}$ is assigned according to:

$$f(\mathbf{v}) = \arg\max_{c \in \mathbb{K}} f_c(\mathbf{v}) \, .$$

One-against-one multi-class extension. This method builds $k(k-1)/2$ classifiers, each for the pair-wise combination of the classes $c_\mathfrak{k}, c_\mathfrak{l} \in \mathbb{K}$, $\mathfrak{k} \neq \mathfrak{l}$. The final classification is performed by majority voting – that is the most frequent predicted class label is returned as prediction of the multi-class model.

Risk of inconsistent decisions. The issue of inconsistent decisions of combining binary classifiers to multi-class classifiers is addressed in [19], for instance. As illustrated in Fig. 5, there can be regions of the input space where the decision of the multi-class models might be inconsistent. Those regions are marked with question marks in each figure. For the *one-against-rest method*, there are two possibilities of an inconsistent decision: (1) there are several binary classifiers predicting different class labels for one given data point. Such regions are (A,B ?), (A,C ?), (B,C ?). (2) there are regions, where all classifiers are predicting the "rest" class, (A,B,C ?). For the *one-against-one method*, there is only one kind of inconsistent decisions possible: several binary classifiers are predicting different class label for one given data point. The problem of several classifiers predicting different class labels can be solved by assigning the class label at random [12] or to assign the data point to the class with the highest posterior probability [19]. The second kind of inconsistent decisions of the one-against-rest method can be acceptable for some problems, where "no decision" might be better than a "wrong decision". Otherwise, one can use the same strategy as for the other kind of inconsistent decisions.

4 The Multi-class Ensemble Framework

Mainly, there are two possibilities of extending the binary ensembles of low-dimensional submodels, cf. Sect. 2, in order to deal with multi-class problems:

1. The multi-class decision is made on the level of the submodels (Ensemble of Multi-Class Submodels, cf. Sect. 4.1).
2. Submodels are binary classifiers, the multi-class classification task is performed by the ensemble. There are two variants:
   a. the Hierarchical Separate-and-Conquer Ensemble (cf. Sect. 4.2) and
   b. the One-versus-Rest Ensemble (cf. Sect. 4.3).

Another algorithm based on one-against-one classifier combination is given in [18]. This algorithm uses a one-against-one approach to extend binary classifiers to solve multi-class problems. The important similarity of this approach compared to our framework is that it is also based on a reduction of the dimensionality on the level of the submodels. In contrast to our approach, the number of dimensions of the submodels is not limited – all input dimensions that provide statistically sufficient information are included in the training set to build a single submodel to separate the pair of classes. Our approach may use several submodels with limited dimensionality to solve the same subproblem while each submodel remains visually interpretable.

For safety-related problems it is important to take into account that the commonly used strategies of extending binary classifiers to multi-class classifiers, which are illustrated in Fig. 5, may lead to regions with inconsistent decisions. In order to avoid an unintended labeling, the inconsistent decisions are solved according to a hierarchy of misclassification costs: for a given new data point $\mathbf{v}$ the class label is chosen which has the largest misclassification penalty.

4.1 Ensemble of Multi-Class Submodels

Using local multi-class models in a *Non-Hierarchical Ensemble Model* requires a *hierarchy of misclassification costs*, that is, it is assumed that there exists an *ordering* of the class labels, which allows statements like: *"class c_1 samples should never be misclassified, class c_2 samples might be misclassified only as class c_1 samples, class c_3 might be classified as class c_1 or c_2 samples, ..."*

$$penalty(c_1) > penalty(c_2) > penalty(c_3) > \ldots \qquad (4)$$

Such a hierarchy of misclassification costs leads to a confusion matrix as depicted in Table 2. This issue is closely related to ordinal classification problems. An SVM-based approach for ordinal classification can be found in [3].

Table 2 Confusion Matrix for multi-class submodels in a Non-Hierarchical Ensemble Model. The following hierarchy of misclassification costs is assumed: $penalty(c_1) > penalty(c_2) > penalty(c_3) > penalty(c_4) > penalty(...)$

	predicted class				
true class	c_1	c_2	c_3	c_4	...
c_1	$h_{1,1}$	0	0	0	...
c_2	$h_{2,1}$	$h_{2,2}$	0	0	...
c_3	$h_{3,1}$	$h_{3,2}$	$h_{3,3}$	0	...
c_4	$h_{4,1}$	$h_{4,2}$	$h_{4,3}$	$h_{4,4}$	...
...	...	...	...	...	...

Combining multi-class submodels of a Non-Hierarchical Ensemble Model becomes difficult because one can only rely on the prediction of the class $c_{\mathfrak{t}}$, which has the minimal misclassification cost – all other class label predictions might be false positives. Thus, it is necessary to include all samples that are not predicted as class $c_{\mathfrak{t}}$ in the training for the next submodel. This fact leads directly to the Hierarchical Separate-and-Conquer Ensemble approach, which is described in Sect. 4.2.

The extension of the DecisionTree-like Ensemble Model approach to solve a multi-class problem is straightforward – a novel subtree is generated for each class predicted by the multi-class submodel of the current node. The final classification decision is determined by the leaf node of the learned tree – similar to the standard decision tree approaches. To avoid inconsistent decisions it is also encouraged to use a hierarchy of misclassification costs in this approach.

4.2 Hierarchical Separate-and-Conquer Ensemble

This approach requires a hierarchy of the misclassification costs as already introduced for the Ensemble of Multi-Class Submodels approach. It is related to the commonly used one-against-rest approach. Instead of building all one-against-rest combinations of models, the class with the minimal classification costs is separated from all samples of the other classes via binary submodels. This approach is illustrated in Fig. 6. The procedure is the same as for the Non-Hierarchical Ensemble Model, which is described in Sect. 2.2. If the problem is solved for this class or there are no further improvements possible, all samples of this class are removed from the training data set and the procedure is repeated for the class which has now the smallest misclassification costs. This procedure is repeated until the data set of the next iteration has only a single class label. The resulting binary classifiers are evaluated according to the misclassification hierarchy, that is in the first step all submodels of the class with minimal misclassification costs are evaluated. If the novel sample cannot be assigned to the class with minimal misclassification costs, the procedure is repeated for the next class in the hierarchy of misclassification costs. If no submodel predicts the novel sample the sample is assigned to the class with maximal misclassification costs.

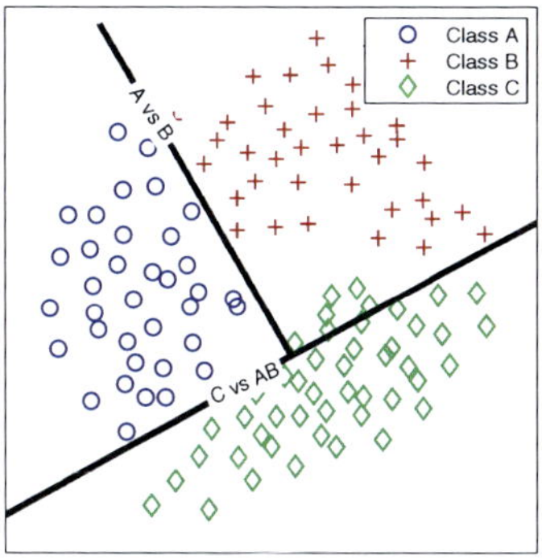

(a) Discriminant functions

	predicted class		
true class	A	B	C
A	36	0	0
B	0	38	0
C	0	6	41

(b) Confusion matrix

Fig. 6 Hierarchical Separate-and-Conquer Ensemble trained on the data from Fig. 5. The following hierarchy of misclassification costs is assumed: $penalty(A) > penalty(B) > penalty(C)$

4.3 One-versus-Rest Ensemble

This approach follows the one-against-rest multi-class classification approach. It is illustrated in Fig. 7. For every class $c_{\mathfrak{k}} \in \mathbb{K}$ versus $c_{\mathfrak{k}}^* = \mathbb{K} \setminus c_{\mathfrak{k}}$ a complete binary Non-Hierarchical Ensemble Model $\hat{f}_{c_{\mathfrak{k}}}(\mathbf{v})$ is trained. The class $c_{\mathfrak{k}}^*$ is chosen as the preferred class c_{pref} to avoid the misclassification of any sample belonging to $\mathbb{K} \setminus c_{\mathfrak{k}}$. For the sake of simplicity $c_{\mathfrak{k}}^*$ is encoded as -1. The resulting binary models can be combined by determining the maximum: $\hat{f}(\mathbf{v}) = \arg\max_{c_{\mathfrak{k}} \in \mathbb{K}} \hat{f}_{c_{\mathfrak{k}}}(\mathbf{v})$.

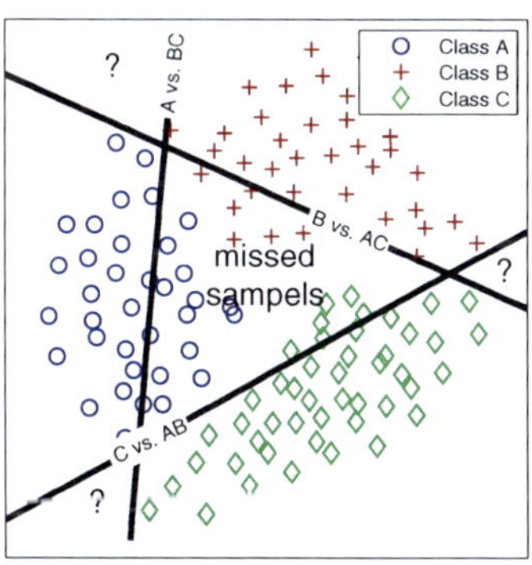

(a) Discriminant functions. Ambiguous regions are labeled with '?'

	predicted class			
true class	A	B	C	?
A	22	0	0	14
B	0	31	0	7
C	0	0	41	6

(b) Confusion matrix. The last column denotes missed samples

Fig. 7 One-versus-Rest Ensemble trained on the data from Fig. 5. Each model for class $c_{\mathfrak{k}}$ is trained with the objective to avoid the misclassification of the samples belonging to $c_{\mathfrak{k}}^* = \mathbb{K} \setminus c_{\mathfrak{k}}$

This approach is the easiest way to extend the binary submodeling approach to multi-class modeling, but it shows a lack of performance for overlapping data sets: it is possible that certain data points are assigned to the class $c_{\mathfrak{k}}^*$ by every submodel and some classes cannot be separated from the other classes due to overlapping of the classes in all projections. This approach still yields ambiguous decisions within the input space, as shown in Fig. 7. Such ambiguities can be resolved by following the hierarchy of misclassification costs.

4.4 An Illustrative Example (Cont'd)

We extend the example which is discussed in Sect. 2.3 to a four-class problem: the *Class 2* samples are drawn from $N((0.0,0.0,0.0)^T,0.2\mathbf{I})$, the *Class 3* samples are drawn from $N((0.5,0.5,0.5)^T,0.2\mathbf{I})$, the *Class 4* samples are drawn from $N((1.0,1.0,1.0)^T,0.2\mathbf{I})$, and the samples of *Class 1* are drawn from $N(\mathbf{e}_i+i/2,0.2\mathbf{I}), i=\{0,1,2\}$. For this multi-class problem the following hierarchy of misclassification costs is assumed:

$penalty(Class\ 4) > penalty(Class\ 3) > penalty(Class\ 2) > penalty(Class\ 1)$.

Ensemble of Multi-Class Submodels. The submodel of the root node of the Ensemble of Multi-Class Submodels approach is shown in Fig. 8(a). All predicted *Class 1* samples are *Class 1* samples, thus there is no further subtree-building needed for predicting *Class 1*. The second submodel, Fig. 8(b), is trained on all samples that are predicted as *Class 2* by the submodel of the root node. The same holds for the third and fourth submodel, Fig. 8(c)&(d), that are trained on the samples predicted as *Class 3* and *Class 4*, respectively. Further submodels cannot improve the overall performance of the global model. The final classification consists of four decision nodes and the maximal tree depth is two.

Hierarchical Separate-and-Conquer Ensemble. This approach solves the problem with four submodels, all shown in Fig. 9. The first submodel separates most of the *Class 1* samples from the samples of the other classes. The remaining *Class 1* samples are removed by the second submodel – further improvements in predicting *Class 1* are not possible. Thus, according to the hierarchy of misclassification costs, the third submodel separates the samples drawn from *Class 2* from the samples of *Class 3* and *Class 4*. Finally, the fourth submodel separates the *Class 3* samples from the *Class 4* samples.

One-versus-Rest Ensemble. This example shows the limitations of the One-versus-Rest Ensemble approach: it is not possible to build one-versus-rest models for *Class 2*, *Class 3*, and *Class 4* without misclassifying samples from *Class 1*. The only models returned by this approach are the same as shown in Fig. 9(a) and Fig. 9(b), that is, only *Class 1* samples can be predicted correctly, all other samples are predicted as 'don't know'.

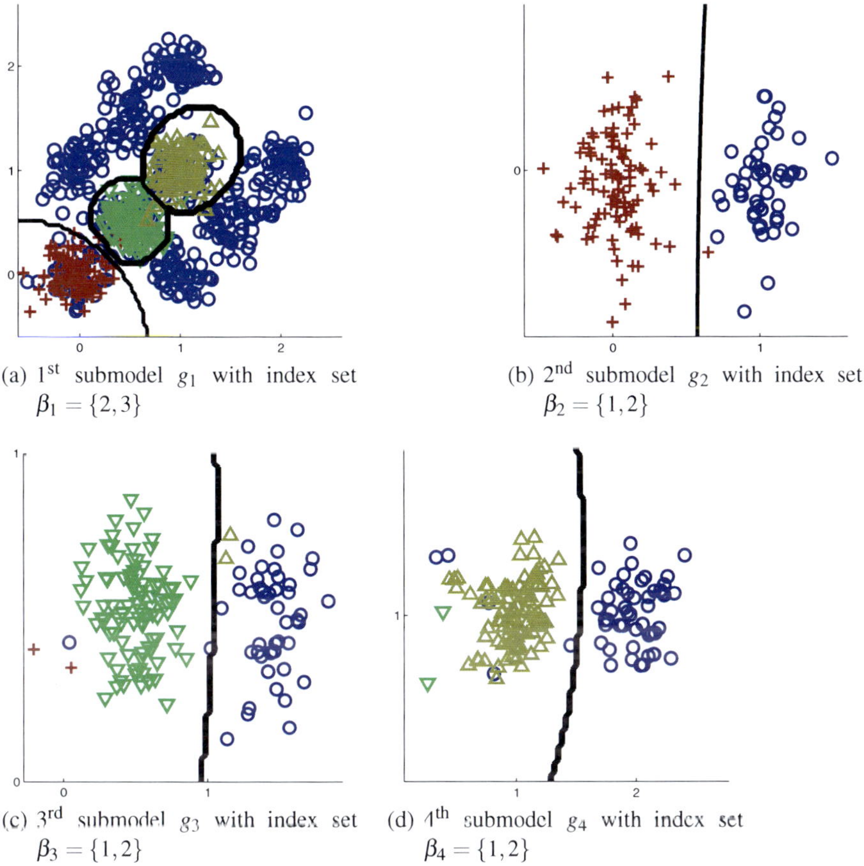

(a) 1st submodel g_1 with index set $\beta_1 = \{2,3\}$

(b) 2nd submodel g_2 with index set $\beta_2 = \{1,2\}$

(c) 3rd submodel g_3 with index set $\beta_3 = \{1,2\}$

(d) 4th submodel g_4 with index set $\beta_4 = \{1,2\}$

Fig. 8 Ensemble of Multi-Class Submodels approach and the Multi-Class *CUBES* data set: *Class 1* samples are shown as circles, *Class 2* samples are shown as crosses, *Class 3* samples are shown as downward-pointing triangles, and *Class 4* samples are shown as upward-pointing triangles. The decision borders of the submodels are drawn as solid lines

5 Experiments

Our ensemble methods are compared with a common support vector machine (SVM) implementation (libSVM from [4]) and a CART classification tree (*treefit* in MATLAB). The SVM and the classification tree are trained with a cost matrix in order to simulate the hierarchy of misclassification costs. Within Table 4, the Non-Hierarchical Ensemble Model is abbreviated as NHEM and the DecisionTree-like Ensemble Model is abbreviated as DTEM. The method in [18] is a variation of the Non-Hierarchical Ensemble Model that uses the Kolmogorov-Smirnoff test in order to select all the variables with different cumulative distributions conditional to the class labels as input of the submodels. The multi-class extensions of the ensemble models are abbreviated as HSCE (Hierarchical Separate-and-Conquer

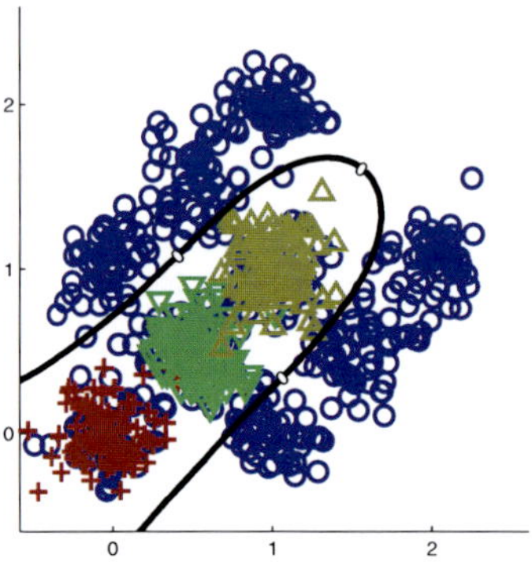

(a) 1st submodel g_1 with index set $\beta_1 = \{2,3\}$

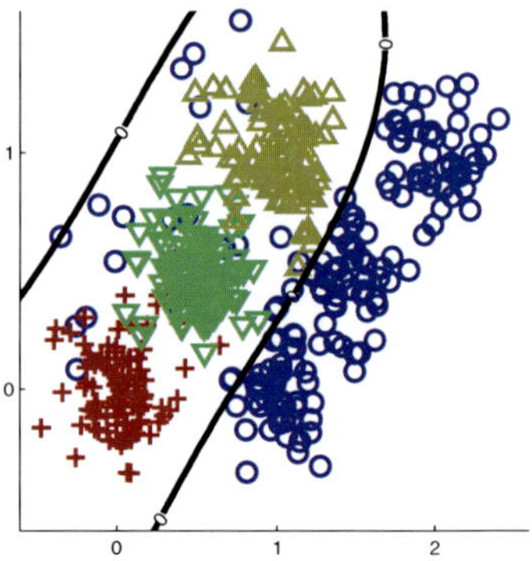

(b) 2nd submodel g_2 with index set $\beta_2 = \{1,3\}$

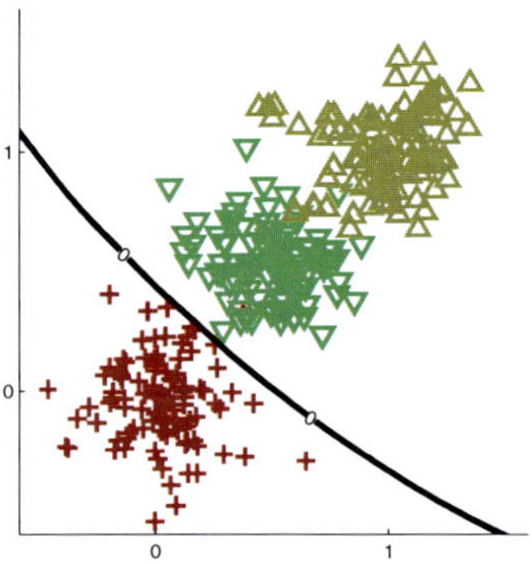

(c) 3rd submodel g_3 with index set $\beta_3 = \{1,2\}$

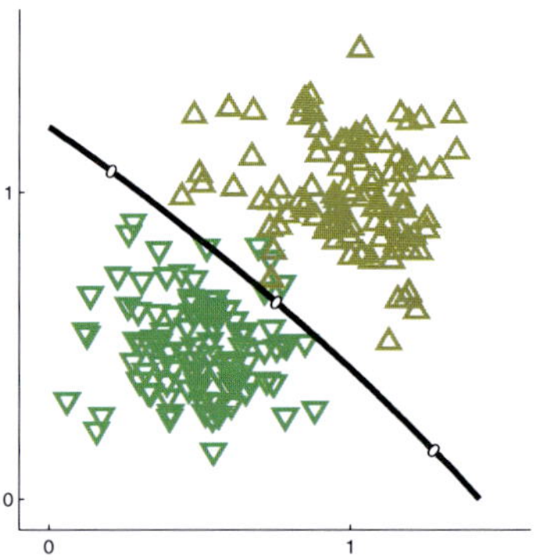

(d) 4th submodel g_4 with index set $\beta_4 = \{1,3\}$

Fig. 9 Hierarchical Separate-and-Conquer Ensemble and the Multi-Class *CUBES* data set: *Class 1* samples are shown as circles, *Class 2* samples are shown as crosses, *Class 3* samples are shown as downward-pointing triangles, and *Class 4* samples are shown as upward-pointing triangles. The decision borders of the submodels are drawn as solid lines

Ensemble), OvRE (One-versus-Rest Ensemble), and EMCS (Ensemble of Multi-Class Submodels) within Table 5. The submodels of our ensemble models are SVMs with Gaussian kernels. The parameter sets of the SVMs are chosen manually in order to obtain smooth decision surfaces within the submodels. The same parameter sets are used for the high-dimensional SVM. For feature selection, our ensemble methods perform an exhaustive search through all possible pairs of features.

For every classifier one can obtain a confusion matrix as depicted in Table 3.

Based on these confusion matrices, we are computing three evaluation measures in order to evaluate the performance of the learned solutions. First, the overall *predictive error* of the model is estimated by:

Table 3 Confusion matrix of a multi-class classifier (counts of data samples). The rows and columns are sorted according to the hierarchy of misclassification costs (cf. Sect. 4.1). The following ordering of the confusion matrix is assumed: $penalty(c_1) > penalty(c_2) > \ldots > penalty(c_k)$

	predicted class			
true class	c_1	c_2	$\cdots$	c_k
c_1	$h_{1,1}$	$h_{1,2}$	$\cdots$	$h_{1,k}$
c_2	$h_{2,1}$	$h_{2,2}$	$\cdots$	$h_{2,k}$
$\vdots$	$\vdots$	$\vdots$	$\ddots$	$\vdots$
c_k	$h_{k,1}$	$h_{k,2}$	$\cdots$	$h_{k,k}$

$$\widehat{err} = \frac{1}{m}\left(m - \sum_{i=1}^{k} h_{i,i}\right) ,$$

where m denotes the number of all data points.

The second evaluation measure, the *critical error*, concerns only the samples that violate the predefined hierarchy of misclassification costs:

$$\widehat{err}_{crit} = \frac{1}{m}\sum_{i=1}^{k}\sum_{j=i+1}^{k} h_{i,j} .$$

The critical error is the more important error measurement because it corresponds to a violation of the given domain knowledge.

As the third evaluation measure, the *one-point-estimate of the area under the ROC curve* (*AUC*), cf. [5], is used:

$$AUC = \max\left\{\frac{1}{k}, 1 - \frac{1}{k}\sum_{i=1}^{k}\sum_{j=1,i\neq j}^{k} n_{i,j}\right\} ,$$

where $n_{i,j} = h_{i,j}/\sum_{l=1}^{k} h_{i,l}$.

5.1 Binary Classification Problems

Hepatitis data set. The task of this data set is to predict whether a patient with hepatitis will die (*Class 1*) or survive (*Class 2*). *Class 2* is chosen as the default class, $c_{pref} = 2$. The original data set from [1] consists of 155 instances and 20 attributes. There are lots of missing values. We are ignoring all dimensions with more than eight missing values. The instances of the resulting data set that still have missing values are removed from the data set. Hence, the final data set used within our experiments consists of 143 instances and 14 attributes.

Respiratory data set (RDS). This data set[2] consists of 85 clinical records (each 17 attributes) for premature newborn children with two types of respiratory distress syndrome: *Hyaline Membrane Disease* (*Class 2*) and *non-HMD* (*Class 1*). The two classes require immediate and completely different treatments, therefore an accurate classification is crucial within the first few hours after delivery. The default class is set to $c_{pref} = 2$.

Wisconsin Breast Cancer data set (WBC). This database consists of 699 instances and nine attributes. There are 16 missing attribute values – samples with missing values are ignored within our experiments. The task is to determine whether a sample is *benign* (*Class 0*) or *malignant* (*Class 1*). *Class 1* is chosen as the default class $c_{pref} = 1$.

5.2 Multi-class Classification Problems

DERMATOLOGY data set. This example is a challenging problem in dermatology [10]. The task is to discriminate six differential diagnostics of erythemato-squamous diseases, namely: *Psoriasis* (*Class 1*), *Seboreic Dermatitis* (*Class 2*), *Lichen Planus* (*Class 3*), *Pityriasis Rosea* (*Class 4*), *Cronic Dermatitis* (*Class 5*), *Pityriasis Rubra Pilaris* (*Class 6*). All these diseases share the clinical features of erythema and scaling – with minor differences. The data set consists of 366 records and each record has 33 attributes. The age attribute of the original data set from the UCI Machine Learning Repository [1] is omitted here, because is has some missing values. The following hierarchy of misclassification costs is assumed: $penalty(Class\ 6) > penalty(Class\ 5) > penalty(Class\ 4) > penalty(Class\ 3) > penalty(Class\ 2) > penalty(Class\ 1)$.

LYMPH data set. This data set consists of 148 instances and 19 attributes. It concerns a four-class problem. The classes are: *Class 1* (*normal find*), *Class 2* (*metastases*), *Class 3* (*malign lymph*), and *Class 4* (*fibrosis*). The following hierarchy of misclassification costs is assumed: $penalty(Class\ 2) > penalty(Class\ 3) > penalty(Class\ 4) > penalty(Class\ 1)$.

NEWTHYROID data set. This problem concerns another typical medical data screening application. The classification task is to predict whether a patient's thyroid belongs to the class *euthyroidism* (*normal* = *Class 1*), *hyperthyroidism* (*hyper* = *Class 2*) or *hypothyroidism* (*hypo* = *Class 3*). The data set consists of 215 records and each record is described by five attributes. The following hierarchy of misclassification costs is assumed: $penalty(Class\ 3) > penalty(Class\ 2) > penalty(Class\ 1)$.

5.3 Comparison of the Ensemble Methods

Table 4 and Table 5 summarize the experiments performed on the benchmark data sets. Most of the data sets can be obtained from [1] – except for the *RESPIRATORY* data set.

[2] http://www.bangor.ac.uk/~mas00a/activities/real_data.htm

Table 4 10-fold cross-validation on binary classification problems. The error is averaged over 10 random fold initializations. *#M* denotes the number of models or the number of decision nodes within the decision tree. *#D* denotes the number of dimensions per (sub-) model or the dimensionality of the splits within the tree

Method	#M	#D	$\widehat{err}$ mean (std)	$\widehat{err}_{crit}$ mean (std)	AUC mean (std)
			HEPATITIS data set		
NHEM	2	2	16.71% (9.26)	2.43% (4.79)	0.660 (0.150)
SZEPANNEK	1	7	18.71% (8.84)	3.43% (5.12)	0.631 (0.136)
DTEM	5	2	19.43% (9.69)	6.29% (6.52)	*0.668* (0.149)
libSVM	1	14	21.00% (10.09)	7.43% (6.33)	0.646 (0.150)
treefit	12	1	21.21% (9.33)	7.86% (7.90)	0.646 (0.134)
			RESPIRATORY data set		
NHEM	4	2	10.13% (10.31)	4.88% (8.68)	0.898 (0.112)
SZEPANNEK	1	10	11.13% (10.79)	6.13% (8.24)	0.891 (0.115)
DTEM	2	2	13.00% (11.08)	7.25% (7.98)	0.873 (0.117)
libSVM	1	17	9.88% (10.10)	5.75% (8.41)	*0.899* (0.110)
treefit	5	1	19.00% (14.49)	6.00% (9.65)	0.813 (0.142)
			WISCONSIN BREAST CANCER data set		
NHEM	4	2	5.96% (6.89)	0.93% (1.27)	0.949 (0.051)
SZEPANNEK	1	9	4.85% (2.34)	2.26% (1.77)	0.947 (0.028)
DTEM	5	2	3.99% (2.26)	1.25% (1.44)	*0.961* (0.024)
libSVM	1	9	4.85% (2.34)	2.26% (1.77)	0.947 (0.028)
treefit	41	1	6.84% (3.43)	2.32% (1.80)	0.932 (0.034)

The evaluation measures are estimated by a 10-fold-cross-validation procedure [14] where the data set is divided into 10 roughly equal-sized subsets (folds). The models are trained on the data of nine folds and the remaining fold is used as an evaluation set. This is repeated for every single fold. Furthermore, we used 10 different fold initializations.

The binary classification ensembles show a good performance on the tested data sets. The predictive performance is competitive to the libSVM solutions while the interpretability of the ensemble models is much better than the interpretability of the higher-dimensional SVMs. The result on the multi-class data sets is quite similar. The One-versus-Rest Ensemble achieves a very good performance in terms of the critical error. On the other hand, the overall error of this approach is worse compared to all other methods. This poor performance is due to the large number of samples that are missed by the submodels and are always assigned to the "other" class. Nevertheless, samples that are labeled as "unrecognized" might be acceptable for some application problems. The Hierarchical Separate-and-Conquer Ensemble approach provides a good trade-off between the predictive performance and the interpretation of the models compared to the SVM solution that achieves the least overall error on all data sets but incorporates always the complete input space. The critical error of the Ensemble of Multi-Class Submodels approach is quite large in all experiments,

Table 5 10-fold cross-validation on multi-class classification problems. The error is averaged over 10 random fold initializations. Missed samples of the One-versus-Rest Ensemble are ignored when computing the *AUC*

Method	#M	#D	$\widehat{err}$ mean (std)	$\widehat{err}_{crit}$ mean (std)	AUC mean (std)	missed samples
			DERMATOLOGY data set			
HSCE	7	2	10.31% (5.27)	2.33% (2.15)	0.889 (0.056)	
SZEPANNEK	5	15	5.42% (3.69)	1.06% (1.80)	0.940 (0.044)	
EMCS	10	2	5.56% (3.68)	3.17% (2.90)	0.927 (0.058)	
OvRE	12	2	24.83% (6.96)	0.36% (0.94)	0.970 (0.045)	[22.6%]
libSVM	1	33	3.39% (2.84)	1.94% (2.25)	*0.963* (0.035)	
treefit	11	1	5.64% (3.65)	1.47% (1.95)	0.944 (0.044)	
			LYMPH data set			
HSCE	6	2	18.64% (9.52)	4.93% (5.89)	0.837 (0.131)	
SZEPANNEK	3	7	18.00% (10.53)	5.71% (6.18)	*0.838* (0.143)	
EMCS	9	2	17.43% (10.37)	6.79% (7.49)	0.814 (0.152)	
OvRE	16	2	36.29% (11.87)	4.00% (5.59)	0.876 (0.133)	[26.6%]
libSVM	1	18	22.07% (10.99)	4.50% (6.15)	0.768 (0.167)	
treefit	15	1	29.29% (11.86)	6.00% (6.94)	0.764 (0.139)	
			NEWTHYROID data set			
HSCE	3	2	4.19% (5.03)	1.10% (2.86)	*0.962* (0.063)	
SZEPANNEK	2	5	4.10% (4.64)	1.71% (3.80)	0.955 (0.076)	
EMCS	3	2	4.00% (4.53)	2.57% (3.79)	0.939 (0.087)	
OvRE	4	2	6.38% (5.63)	0.90% (2.41)	0.975 (0.062)	[4.7%]
libSVM	1	5	4.62% (4.56)	1.90% (3.71)	0.949 (0.075)	
treefit	6	1	4.29% (4.36)	1.29% (2.12)	0.957 (0.052)	

because the hierarchy of misclassification costs is ignored for this method within our experiments. On the other hand, this ensemble approach is the only variant that does not require a hierarchy of misclassification costs to build a multi-class model. While interpreting the decision boundaries of a high-dimensional SVM is infeasible, all ensemble approaches allow a visualization of the submodels and, thus, they facilitate the incorporation of domain knowledge via an interactive model selection process that increases the confidence about the learned solution.

5.4 Comparison of Different Feature Selection Methods

In the second experimental setting, we investigate the influence of the dimensionality of each submodel on the predictive error of the learned ensemble models. Therefore, the Hierarchical Separate-and-Conquer Ensemble is extended in order to use other feature selection methods: instead of the wrapper feature selection with the restriction to two-dimensional projections, the algorithm can use the Kolmogorov-Smirnoff test to determine all variables that have different conditional cumulative probability distributions given the class labels (cf. [18]) or it can use only those

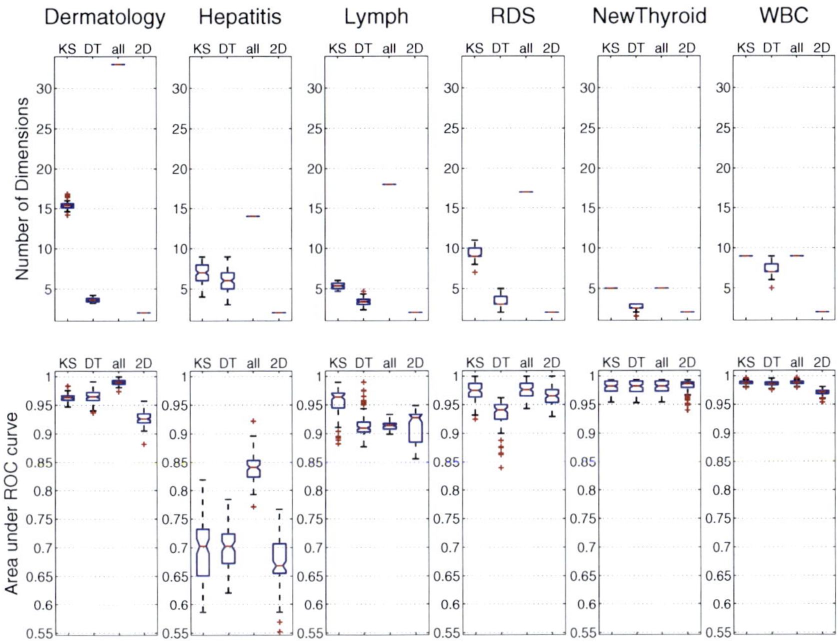

Fig. 10 Comparison of different feature selection methods on several benchmark data sets. KS – feature selection based on the Kolmogorov-Smirnoff test, DT - feature selection based on a decision tree, all – using all features, and 2D – only two-dimensional projections of the data set

variables that are used for the decision nodes within a CART model [2], which uses the Gini's diversity index as split criterion. In this study, we divided the data sets into four roughly equal-sized folds and we repeated the experiments for 100 different fold initializations. The results of this study are illustrated in Fig. 10. All results are obtained by applying the Non-Hierarchical Ensemble Model algorithm on the same data sets as in the previous study. For each data set the Non-Hierarchical Ensemble Model is trained with four different settings: (1) use only two-dimensional projections, (2) use Kolmogorov-Smirnoff test to determine all dimensions that have different conditional cumulative distributions per class label, (3) use all dimensions that are used by a decision tree learner, and (4) use all dimensions for learning the submodel. Since (2-4) can exploit a higher dimensionality, these variants are restricted to one submodel per class label.

There is no general trend which feature selection method performs best in all situations. The best predictive performance depends more on the structure of the data set than on the feature selection method itself. There is also a significant difference between the feature selection method based on the Kolmogorov-Smirnoff test and the decision tree-based feature selection method on the *LYMPH* and *RESPIRATORY* data sets. If one compares the average dimensionality of the submodels (which can be seen as an indicator of the interpretability of the models) for each different feature selection method the restriction to only two-dimensional submodels is competitive compared to the other approaches and provides an appropriate trade-off between interpretability and predictive performance.

6 Conclusions

In order to successfully apply machine learning approaches in the field of safety-related application problems it is crucial to provide interpretable and verifiable models. Unfortunately, it is infeasible to interpret high-dimensional models efficiently. Therefore, for safety-related problems, high-dimensional models are not to be applied. On the other hand, simple models which are easier to interpret show a lack of predictive performance. The framework proposed within this chapter provides a good trade-off between, on the one hand, the interpretation and verification of the learned (sub-)models, avoiding an unintended extrapolation behavior, and, on the other hand, the achievement of a high predictive accuracy. Each submodel can be visually interpreted and the ensemble of the submodels compensates for the limited predictive performance of each single submodel. In contrast to dimensionality reduction methods, which combine several dimensions of the input space, the submodels are trained on the original dimensions, allowing domain experts to evaluate the trained models directly. In our experiments, the benchmark data sets were successfully tested and provided competitive results.

References

1. Asuncion, A., Newman, D.J.: UCI Machine Learning Repository. University of California, Irvine, School of Information and Computer Sciences, `http://www.ics.uci.edu/~mlearn/MLRepository.html`
2. Breiman, L., Friedman, J.H., Stone, C.J., Olshen, R.A.: Classification and Regression Trees. CRC Press, Boca Raton (1984)
3. Cardoso, J.S., da Costa, P.J.F., Cardoso, M.J.: Modelling ordinal relations with SVMs: an application to objective aesthetic evaluation of breast cancer conservative treatment. Neural Networks 18(5–6), 808–817 (2005)
4. Chang, C.-C., Lin, C.-J.: LIBSVM: a library for support vector machines, `http://www.csie.ntu.edu.tw/~cjlin/libsvm`
5. Ferri, C., Hernández-Orallo, J., Salido, M.A.: Volume under the ROC surface for multi-class problems. In: Lavrač, N., Gamberger, D., Todorovski, L., Blockeel, H. (eds.) ECML 2003. LNCS (LNAI), vol. 2837, pp. 108–120. Springer, Heidelberg (2003)
6. Friedman, J.H.: Another approach to polychotomous classification. Department of Statistics, Stanford University (1996)
7. Friedman, J.H., Tukey, J.W.: A projection pursuit algorithm for exploratory data analysis. IEEE Trans. Comp. 23(9), 881–890 (1974)
8. Fürnkranz, J.: Separate-and-conquer rule learning. Artif. Intell. Review 13(1), 3–54 (1999)
9. Guyon, I., Elisseeff, A.: An introduction to feature extraction. In: Guyon, I., Gunn, S., Nikravesh, M., Zadeh, L. (eds.) Feature Extraction: Foundations and Applications, pp. 1–25. Springer, Heidelberg (2006)

10. Güvenir, H.A., Demiröz, G., Ilter, N.: Learning differential diagnosis of erythemato-squamous diseases using voting feature intervals. Artif. Intell. in Medicine 13(3), 147–165 (1998)
11. Hastie, T., Tibshirani, R.: Generalized Additive Models. Chapman & Hall, Boca Raton (1990)
12. Hsu, C.-W., Lin, C.-J.: A comparison of methods for multiclass support vector machines. IEEE Trans. Neural Networks 13(2), 415–425 (2002)
13. Huber, P.J.: Projection pursuit. The Annals of Statistics 13(2), 435–475 (1985)
14. Kohavi, R.: A Study of cross-validation and bootstrap for accuracy estimation and model selection. In: Proc. 14th Int. Joint Conf. Artif. Intell., Montréal, QB, pp. 1137–1143. Morgan Kaufmann, San Francisco (1995)
15. Kohavi, R., John, G.H.: Wrappers for feature subset selection. Artif. Intell. 97(1–2), 273–324 (1997)
16. Nusser, S., Otte, C., Hauptmann, W.: Learning binary classifiers for applications in safety-related domains. In: Mikut, R., Reischl, M. (eds.) Proc. 17th Workshop Computational Intell., Dortmund, Germany, pp. 139–151. Universitätsverlag Karlsruhe (2007)
17. Stone, C.J.: Additive regression and other nonparametric models. The Annals of Statistics 13(2), 689–705 (1985)
18. Szepannek, G., Weihs, C.: Local modelling in classification on different feature subspaces. In: Perner, P. (ed.) ICDM 2006. LNCS (LNAI), vol. 4065, pp. 226–238. Springer, Heidelberg (2006)
19. Tax, D.M.J., Duin, R.P.W.: Using two-class classifiers for multiclass classification. In: Proc. 16th Int. Conf. Patt. Recogn., Québec, QB, pp. 124–127 (2002)

Independent Data Model Selection for Ensemble Dispersion Forecasting

Angelo Ciaramella, Giulio Giunta, Angelo Riccio, and Stefano Galmarini

Abstract. This work aims at introducing an approach to analyze the independence between different data model in a multi-model ensemble context. The models belong to operational long-range transport and dispersion models, but they are also used for the real-time simulation of pollutant dispersion or the accidental release of radioactive nuclides in the atmosphere. In order to compare models, an approach based on the hierarchical agglomeration of distributions of predicted radionuclide concentrations is proposed. We use two different similarity measures: Negentropy information and Kullback-Leibler divergence. These approaches are used to analyze the data obtained during the ETEX-1 exercise, and we show how to exploit these approaches to select subsets of independent models whose performance is comparable to those from the whole ensemble.

Keywords: ensemble classifier, independence, data dimensionality, air pollutant dispersion.

1 Introduction

Standard meteorological/air quality practice, such as the prediction of the future state of the atmosphere, typically proceeds conditionally on one assumed model. The model is the result of the work of many area-expert scientists, e.g., meteorologists, computational scientists, statisticians, and others.

Angelo Ciaramella · Giulio Giunta · Angelo Riccio
Dept. of Applied Science, University of Naples "Parthenope", Isola C4,
Centro Direzionale I-80143, Napoli, Italy
e-mail: {angelo.ciaramella, giulio.giunta}@uniparthenope.it
angelo.riccio@uniparthenope.it

Stefano Galmarini
European Commission - DG Joint Research Centre, Institute for Environment and Sustainability, Ispra, Italy
e-mail: stefano.galmarini@jrc.it

O. Okun & G. Valentini (Eds.): Appli. of Supervised & Unsuper. Ensemble Meth., SCI 245, pp. 213–231.
springerlink.com
© Springer-Verlag Berlin Heidelberg 2009

Nowadays, several models are available for the forecast of variables of meteorological and/or air quality interest, but, even when using the same ancillary (e.g. initial and boundary) data, they could give different answers to the scientific question at hand. This is a source of uncertainty in drawing conclusions, and the typical approach, that is of conditioning on a single model deemed to be "the best", ignores this source of uncertainty and underestimates the possible effects of a false forecast.

Ensemble prediction aims at reducing this uncertainty by means of techniques designed to strategically sample the forecast pdf, e.g. the breeding of growing modes [20] or singular vectors [17] in the weather forecasting field. In recent years, a large literature has evolved toward the use of multi-model ensemble systems to improve weather, climate and air quality predictions, too [15]. This trend stems from many needs: spotting flood events; evaluating the effects of pollutant emissions; delivering more accurate seasonal and inter-annual weather predictions, etc.

The multimodel approach has been successfully applied to atmospheric dispersion predictions [8, 9, 10] where the uncertainty of weather forecast sums and mixes with that stemming from the description of the dispersion process. The methodology relies on the analysis of the forecasts of several models used operationally by national meteorological services and environmental protection agencies worldwide to forecast the evolution of accidental releases of harmful materials. The objectives are clear: after the release of hazardous material into the atmosphere, it is extremely important to support the decision-making process with any relevant information and to provide a comprehensive analysis of the uncertainties and the confidence that can be put into the dispersion forecast. Galmarini et al. [9] showed how the intrinsic differences among the models can become a useful asset to be exploited for the sake of a more educated support to decision making by means of the definition of ad-hoc parameters and treatments of model predictions. They proposed the so called 'Median Model', defined as a new set of model results constructed from the median of model predictions. The Median Model was shown to outperform the results of any single deterministic model in reproducing the concentration of atmospheric pollutants measured during the ETEX experiment [11].

Moreover, in [19] an approach for the statistical analysis of multi-model ensemble results is presented. The authors used a well-known statistical approach to multimodel data analysis, i.e. Bayesian Model Averaging (BMA), which is a standard method for combining predictive distributions from different sources. Moreover, similarities and differences between models were explored by means of correlation analysis.

However, we have to note that, if different models are used to simulate the same phenomenon, e.g. weather, climate or the dispersion of radioactive material, they probably will give similar responses. Potentially, model ensemble results may lead to erroneous interpretations, and this is more probable if models are strongly dependent. Models are certainly more or less dependent in the case of ensemble dispersion forecasting , since they often share similar initial/boundary data, numerical methods, parameterizations, and so on.

In this work, we use a statistical approach to analyze the independence between data model distributions and to select the models that have similar behavior.

Substantially, we use the agglomerative clustering approach to obtain a dendrogram that describes the relations between model data.

To compare models, we propose to use the entropy information approach. On the one hand, we consider as "distance" the Kullback-Leibler (KL) divergence [4, 16] . This divergence can be considered as a kind of distance between two probability densities, because it is always nonnegative, and zero if and only if the two distributions are equal.

On the other hand, Negentropy can be considered as a non-Gaussianity measure and, if we consider the residues between two distributions, we can estimate how these two distributions are different. We use approximations of Negentropy providing a very good compromise between the properties of the two classic non-Gaussian measures given by kurtosis and skewness [13].

In Sect. 2 the so called 'Median Model' approach to aggregate multiple predictions is briefly recalled. In Sect. 3 we introduce KL and Negentropy divergence and the agglomerative clustering. In Sect. 4 we show some results obtained from the application of these two techniques to data from the ETEX-1 experiment [11].

2 The 'Median Model' Approach

This work concerns the analysis of multi-model ensemble results from the ENSEMBLE system [3, 9, 10, 18]. ENSEMBLE (ensemble.jrc.ec.europa.eu) is a web based platform for the real-time exchange and ensemble treatment of operational atmospheric dispersion model predictions produced by different models world-wide. This project now aggregates the expertise of more than 25 research groups (atmospheric modelers, decision makers and technical advisors) and the results of almost 40 models. It is essentially based on the graphical representation and statistical treatment of multi-model dispersion and deposition forecast of radioactive and non-radioactive material in the atmosphere.

A series of activities, supported by the European Commission, IAEA (International Atomic Energy Agency) and WMO (World Meteorological Organization), underlined the importance to assess the reliability of model results, prior to their use as support to decision making. This was the aim of experimental campaigns, such as ATMES [14], ANATEX [6], CAPTEX [7] and the first and second European tracer experiments (ETEX-1 and ETEX-2) [11], as well as model inter-comparison exercises like RTMOD [2, 12].

The ETEX databases have been used in many inter-comparison studies. In Table 1 we show some results of the ENSEMBLE inter-comparison exercise applied to the ETEX data. The statistical treatment of multi-model ensemble results were based on the so-called 'Median Model', defined as the median value of all model results for each spatio-temporal location. Galmarini et al. [9, 10] already noted that the Median Model turned out to be superior to any single model in reproducing the measured cloud of the ETEX-1 experiment. Table 1 shows the root mean square error, correlation coefficient, FA2, FA5 and FOEX indexes of each model. FA2 and FA5 give the percentage of model results within a factor of 2 and 5, respectively, of the

Table 1 Root mean square error (RMSE), correlation coefficient (CC) and percentage of couple measured-modeled data within a factor of 2 (FA2) and 5 (FA5). The last column (FOEX) gives the percentage of over-prediction (> 0) or under-predictions (< 0). In the second last row, median model results, averaged over models from m01 to m16 are shown. In the last row, the median model results averaged over all models are shown

	RMSE	CC	FA2	FA5	FOEX
m01	4.76	0.17	14.25	37.65	77
m02	0.71	0.30	22.00	45.91	61
m03	6.04	0.22	19.13	42.04	55
m04	7.4e8	0.17	0.00	0.00	100
m05	2.05	0.27	13.02	32.72	71
m06	7.56	0.17	22.91	47.37	2
m07	0.93	0.26	19.98	42.91	36
m08	0.72	0.23	8.11	18.08	-42
m09	2.19	0.17	16.47	37.47	11
m10	1.81	0.41	15.11	35.32	17
m11	2.88	0.27	15.90	37.76	14
m12	2.27	0.26	21.00	42.43	34
m13	3.19	0.08	21.94	45.63	50
m14	3.06	0.13	12.34	28.35	56
m15	3.76	0.05	15.89	34.65	11
m16	8.53	0.08	21.97	44.39	36
m17	1.31	0.32	10.24	23.01	0
m18	2.89	0.20	17.61	37.82	-4
m19	1.47	0.27	21.81	46.12	76
m20	0.45	0.08	20.24	46.64	-8
m21	5.32	0.22	18.90	43.00	45
m22	1.79	0.24	27.96	54.76	21
m23	0.53	0.24	11.33	26.32	-28
m24	2.22	0.20	21.67	47.59	44
m25	3.27	0.24	22.93	46.64	50
m26	1.20	0.08	10.82	27.09	-7
MM 1-16	1.30	0.29	24.14	48.38	15
MM 1-26	1.15	0.30	26.43	50.99	13

corresponding measured value, while FOEX is the percentage of modeled concentration values that overestimate (positive) or underestimate (negative) the corresponding measurement.

Labels from m01 to m16 refer to the results from the sixteen models already reported in [9, 10]; from that time, several other models have been added to the ENSEMBLE system (labeled m17-m26 in Table 1). Table 1 also shows the values of statistical indexes for the Median Model, both using only models from m01 to m16 (label 'MM 1-16'), and all models from m01 to m26 (label 'MM 1-26'). Notably, the statistical performance of the Median Model outperforms that of any single model

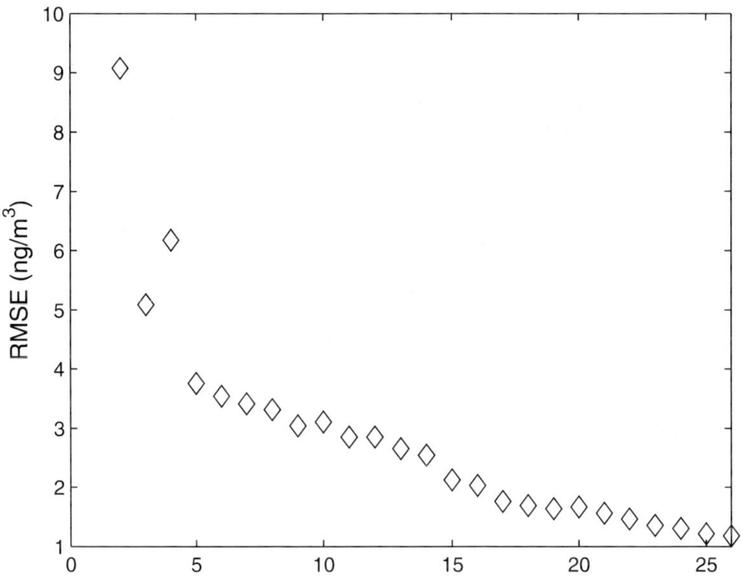

Fig. 1 RMSE errors of 'median models'. The kth median model results were calculated using data from $\{r_i\}_{i=1,\ldots,k}$; r_i denotes the models with the ith highest RMSE

in most cases; also, a noteworthy result is the additional improvement of the Median Model performance when all twenty-six models are included in the statistics.

However, we can look at the same data from a different perspective: sort models in descending order using the root mean square error as ordering criteria, and denote by $\{r_i\}_{i=1,\ldots,26}$ the permuted labels, so that r_1 indicates the model with the highest RMSE, r_2 the second highest, and so on. The median value can be recalculated at each spatio-temporal location using data from the first k models, i.e. from $\{r_j\}_{j=1,\ldots,k}$, with $k \in \{1,2,\ldots,26\}$. Of course there are twenty-six 'median models', now, depending on how many models are included in the statistics. Figure 1 shows the root mean square errors of these median models. As expected, the root mean square error decreases as the number of models included in the statistics increases; there is a drastic reduction after a few models, followed by a more slowly decreasing trend. This means that better predictive capabilities are obtained at the expense of a greater complexity (measured by the number of models), but just a few models are needed for a drastic reduction of the ensemble root mean square error.

This work is motivated by this consideration: select the 'best' subset of models without excessively sacrificing performance. In order to do that, we explored the possibility to select a subset, based on Negentropic or Kullback-Leibler divergence measures. These measures are well known information-theoretic criteria for measuring statistical independence and are widely used in statistics and signal processing. The idea behind the use of these criteria is based on the possibility to select a subset of model data that are really independent, thus avoiding wasteful procedures in collecting data from a large number of models.

3 Negentropy-Based Hierarchical Agglomeration

3.1 Kullback-Leibler Divergence

The KL divergence is defined between two discrete n-dimensional probability density functions $\mathbf{p} = [p_i \dots p_n]$ and $\mathbf{q} = [q, \dots q_n]$ as

$$KL(\mathbf{p}||\mathbf{q}) = \sum_{i=1}^{n} p_i \log\left(\frac{p_i}{q_i}\right) . \tag{1}$$

This is known as the relative entropy. It satisfies the Gibbs' inequality

$$KL(\mathbf{p}||\mathbf{q}) \geq 0 , \tag{2}$$

where equality holds only if $\mathbf{p} \equiv \mathbf{q}$. In general $KL(\mathbf{p}||\mathbf{q}) \neq KL(\mathbf{q}||\mathbf{p})$. In our experiments we use the symmetric version [4] that can be defined as

$$KL = \frac{KL(\mathbf{p}||\mathbf{q}) + KL(\mathbf{q}||\mathbf{p})}{2} . \tag{3}$$

The relative entropy is important in pattern recognition as well in information theory. It is also used in Independent Component Analysis to estimate the independence between distributions [13].

3.2 Negentropy Information

The definition of Negentropy J_N is given by

$$J_N(\mathbf{x}) = H(\mathbf{x}_{Gauss}) - H(\mathbf{x}) , \tag{4}$$

where $\mathbf{x}_{Gauss}$ is a Gaussian random vector of the same covariance matrix as $\mathbf{x}$ and $H(\cdot)$ is the differential entropy. Negentropy can also be interpreted as a measure of non-Gaussianity [13]. The classical method to approximate Negentropy relies on using higher-cumulants, through polynomial density expansion. However, such cumulant-based methods sometimes provide a rather poor approximation. A special approximation is obtained if one uses two functions G^1 and G^2, which are chosen so that G^1 is *odd* and G^2 is *even*. Such a system of two functions can measure the two most important features of non-Gaussian 1-D distributions. The odd function measures the asymmetry, and the even function measures the dimension of bimodality vs. peak at zero, closely related to sub- vs. super-gaussianity. Then the Negentropy approximation of equation (4) is:

$$J_N(\mathbf{x}) \propto k_1 E\{G^1(\mathbf{x})\}^2 + k_2(E\{G^2(\mathbf{x})\} - E\{G^2(\upsilon)\})^2 , \tag{5}$$

where υ is a Gaussian variable of zero mean and unit variance (i.e. standardized), the variable $\mathbf{x}$ is assumed to have also zero mean and unit variance and k_1 and k_2 are positive constants. We note that choosing the functions G^i that do not grow too fast, one obtains more robust estimators.

In this way approximations of Negentropy that give a very good compromise between the properties of the two classic non-Gaussianity measures given by kurtosis and skewness can be obtained [1, 5, 13]. They are conceptually simple, fast to compute, yet have appealing statistical properties, especially robustness.

3.3 Agglomerative Approach

We remark that our aim is to agglomerate by an unsupervised method the distributions obtained from the different models of the ensemble. Substantially the aim is to build a hierarchical tree (dendrogram) that permits to cluster models that have similar behavior. To obtain the dendrogram we calculate the dissimilarity matrix between the distributions of the models by using the Negentropy or the Kullback-Leibler divergence . The densities of the distributions are calculated with a simple non-parametric method. In particular, in our experiments we used a histogram approach.

Using this information we apply the agglomerative hierarchical clustering approach to obtain the dendrogram. In this case we use *complete linkage* or *furthest neighbor*

$$d(r,s) = \max(J(X_r, X_s)) \ , \tag{6}$$

where X_r and X_s are two distributions and $J(\cdot,\cdot)$ is one of the two entropy information.

4 Experimental Results

We apply this approach to the analysis of multi-model ensemble results. In [10], Galmarini et al. already showed how to exploit multi-model ensemble results to improve the forecast; the analysis of model results showed that while the single models produced a wide spectrum of time evolution of the concentration, the Median Model, on the contrary, provided a more accurate reproduction of the concentration trend and estimate of the cloud persistence at sampling locations.

In this work we show how the Negentropic or KL-based approaches allow to derive results that are comparable with the Median Model ones.

These approaches discriminate between data that are less dependent (in the statistical sense), so that 'redundant' information can be more easily discarded and equivalent performance can be achieved with considerably fewer models.

The ensemble analyzed in this work is an extended version of that originally presented in [10]. To summarize, we are looking at 26 simulations [19] of the ETEX-1 experiment [11]. The ETEX-1 experiment concerned the release of pseudo-radioactive material on 23 October 1994 at 16:00 UTC from Monterfil, southeast of Rennes (France). Briefly, a steady westerly flow of unstable air masses was present over central Europe. Such conditions persisted for the 90 h that followed the release

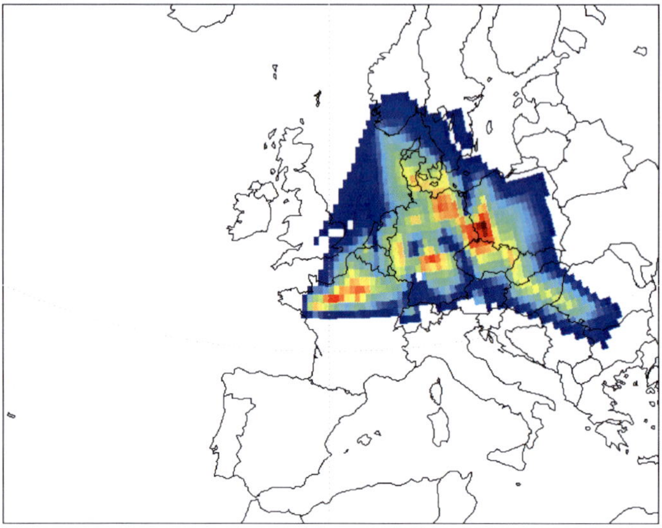

Fig. 2 ETEX-1 observations

with frequent precipitation events over the advection area and a slow movement toward the North Sea region. In Fig. 2 we show the integrated concentration after 78 hours from release.

Several independent groups worldwide tried to forecast these observations. Each simulation, and therefore each ensemble member, is produced with different atmospheric dispersion models and is based on weather fields generated by (most of the time) different Global Circulation Models (GCM). All the simulations relate to the same release conditions. For details on the groups involved in the exercise and the model characteristics, refer to [10] and [19].

In order to test the usefulness of these two information-theoretic approaches, we analyze two kind of dataset: the first dataset is made by the integrated surface concentrations from 26 models; the second dataset is made by the instantaneous surface concentrations from the same models, which are available every three hours up to three days from the start of release. The dendrogram was calculated for both dataset.

The dendrogram obtained by using the Negentropy information on the integrated concentrations after 78 hours is plotted in Fig. 3. We mark that the information related to models on the abscissa and the information related to model similarities obtained with the Negentropy on the ordinate. In Fig. 4 we plot the dendrogram obtained by using the KL divergence.

We observe that different clusters are obtained. In fact, in the Negentropy based dendrogram we can observe four mainly independent agglomerations (visualized by using different colors). These clusters are partially present in the KL dendrogram, though in this dendrogram we obtain, apparently, more than four agglomerations.

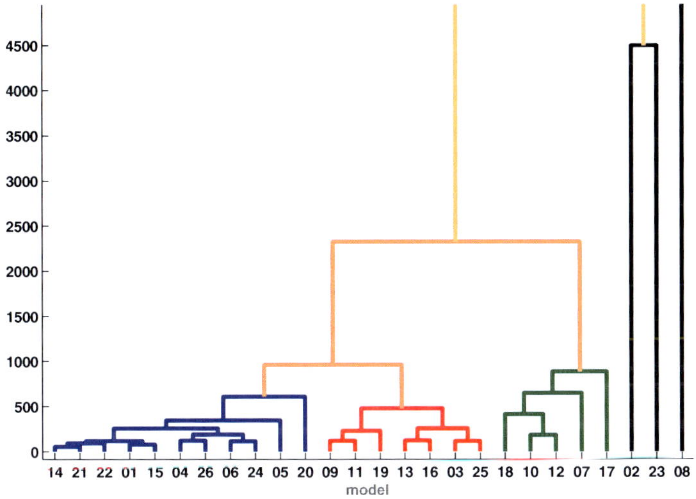

Fig. 3 Details of the dendrogram obtained by using the Negentropy information

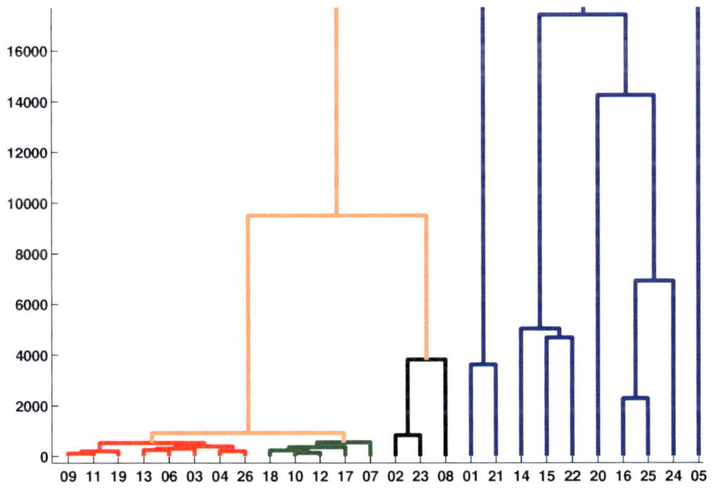

Fig. 4 Details of the dendrogram obtained by using the KL divergence

We can identify data that have similar behavior by analyzing the different clusters. For example, in Fig. 5 we show the data distributions in one of the clusters (blue cluster in the Negentropy dendrogram). The visualized models are $m14$, $m21$, $m01$, $m15$, $m06$ and $m24$, respectively.

In Fig. 6 we show the distributions in the red cluster of the Negentropy dendrogram. In this case the models are $m09$, $m11$, $m13$, and $m16$. We can stress that this cluster contains models that have a comparable distribution.

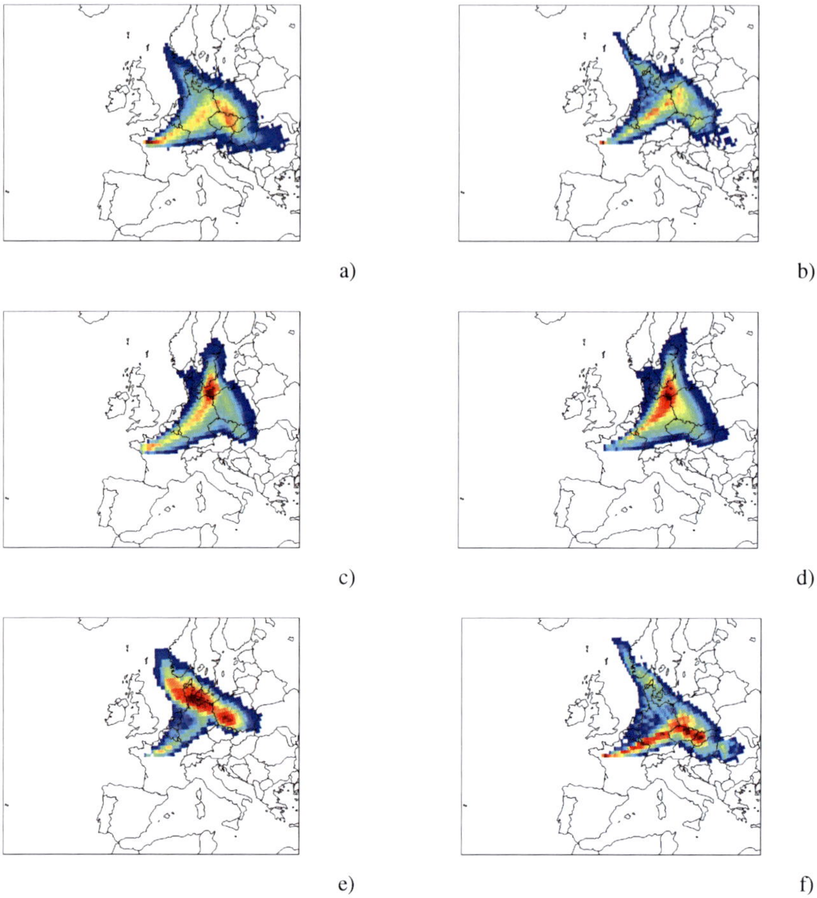

Fig. 5 Distributions after 78 hours of the models $m14$ (a), $m21$ (b), $m01$ (c), $m15$ (d), $m06$ (e), and $m24$ (f)

Moreover, models $m02$, $m23$ and $m08$ (Fig. 7) are far from the other clusters. They have similar distributions but they are more diffusive than the others (see Fig. 7). We, however, noticed that the same three models are agglomerated together in the KL dendrogram but they belong to another extended cluster.

We also noticed from the KL dendrogram that some models probably are erroneously assigned to some agglomerations. For example, model $m16$ is associated together with the model $m25$, but we can observe that its distribution is closer to that of model $m13$ than $m25$. In fact, in the Negentropy based dendrogram models $m13$ and $m16$ are agglomerated. Moreover model $m13$ in the KL dendrogram is agglomerated with model $m06$, but, as we can see in Figs. 8c and 8d, they have a rather different distribution. In Fig. 8 we plot the distributions of all these models.

In order to identify the group of models that more appropriately describe observations, we re-apply the clustering approaches also including the distribution of

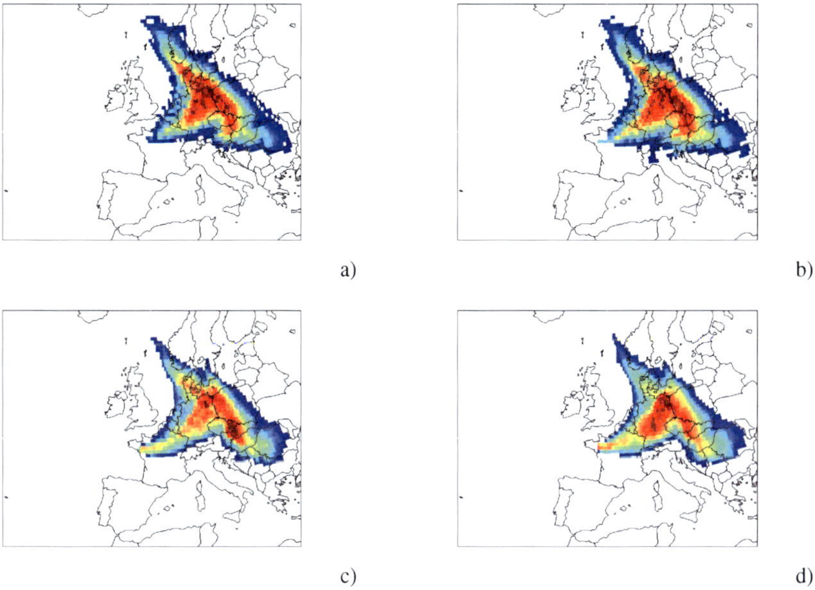

Fig. 6 Distributions after 78 hours of the models $m09$ (a), $m11$ (b), $m13$ (c), and $m16$ (d)

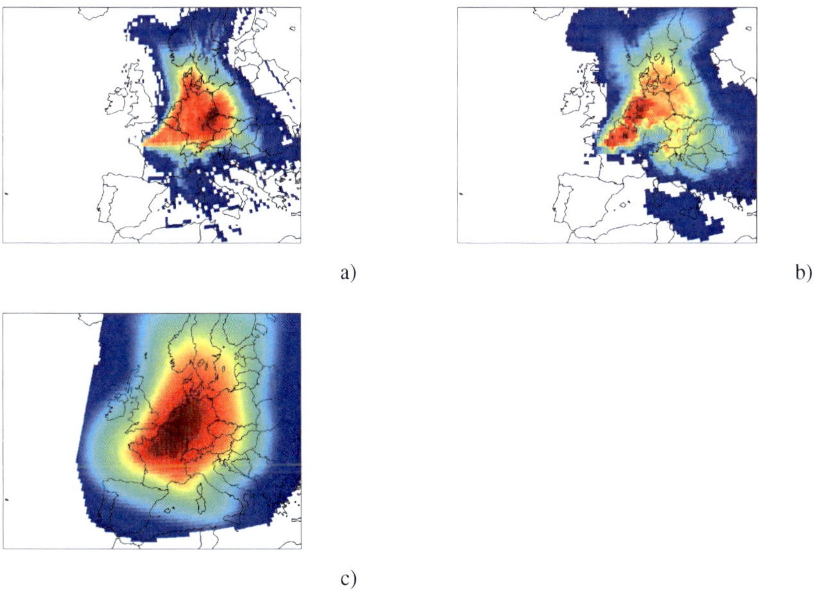

Fig. 7 Distributions after 78 hours of the models $m02$ (a), $m23$ (b), and $m08$ (c)

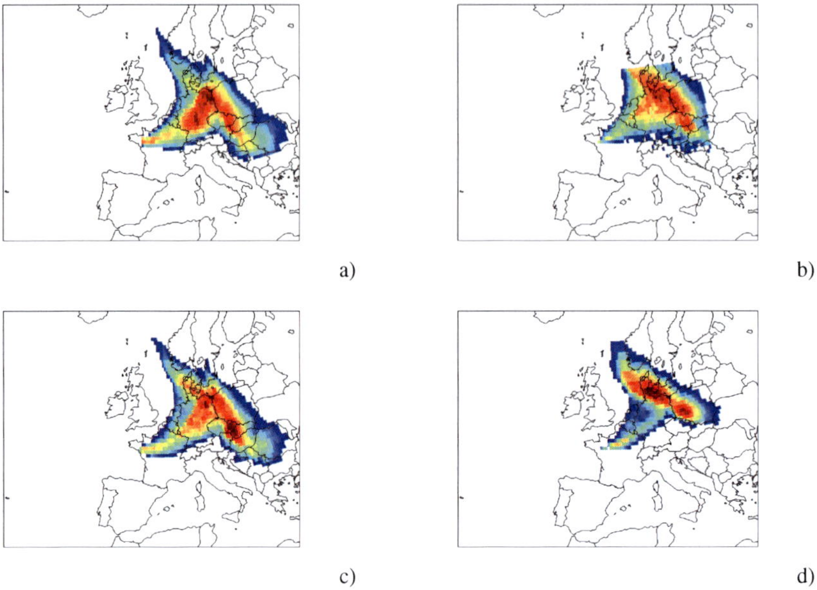

Fig. 8 Distributions after 78 hours of the models $m16$ (a), $m25$ (b), $m13$ (c), and $m06$ (d)

observed values. This distribution is named $m00$. In Fig. 9 we plot the two dendrograms. We note that in the Negentropy dendrogram the model $m00$ is agglomerated in the blue cluster, together with models $m01$ and $m15$. Instead, in the KL dendrogram it is associated only with model $m14$.

We also used Negentropy and KL divergence to analyze instantaneous concentrations. The interest in this kind of analysis is related to the possibility to assess the effectiveness of this clustering approach in reproducing the results already outlined in [9, 10], but using a fewer number of models.

In Fig. 10 we show the dendrograms applied to instantaneous concentrations derived by the Negentropy and KL algorithms. The different colors distinguish models for a given threshold dissimilarity value: models indicated with the same color are joined together at a dissimilarity value lower than the threshold. For both approaches we cut the dendrograms by selecting six clusters.

The repartition into 'independent' clusters can be exploited to calculate some useful statistics already used in the ENSEMBLE project, e.g., the APL (Above Percentile Level) index.

In [9], the $\mathrm{APL}_p(x,y,t)$ is defined as the pth percentile from all models at a specific time t and spatial location (x,y). The APL_p is defined as the two-dimensional surface obtained for all (x,y) locations of the domain at time t. The 'Median Model' is nothing else than the APL_{50}.

In [10] it was outlined that the Median Model effectively and systematically filters erroneous results from the ensemble distribution, so it is much more conservative to use the Median Model results than to rely on a single model, even if it is

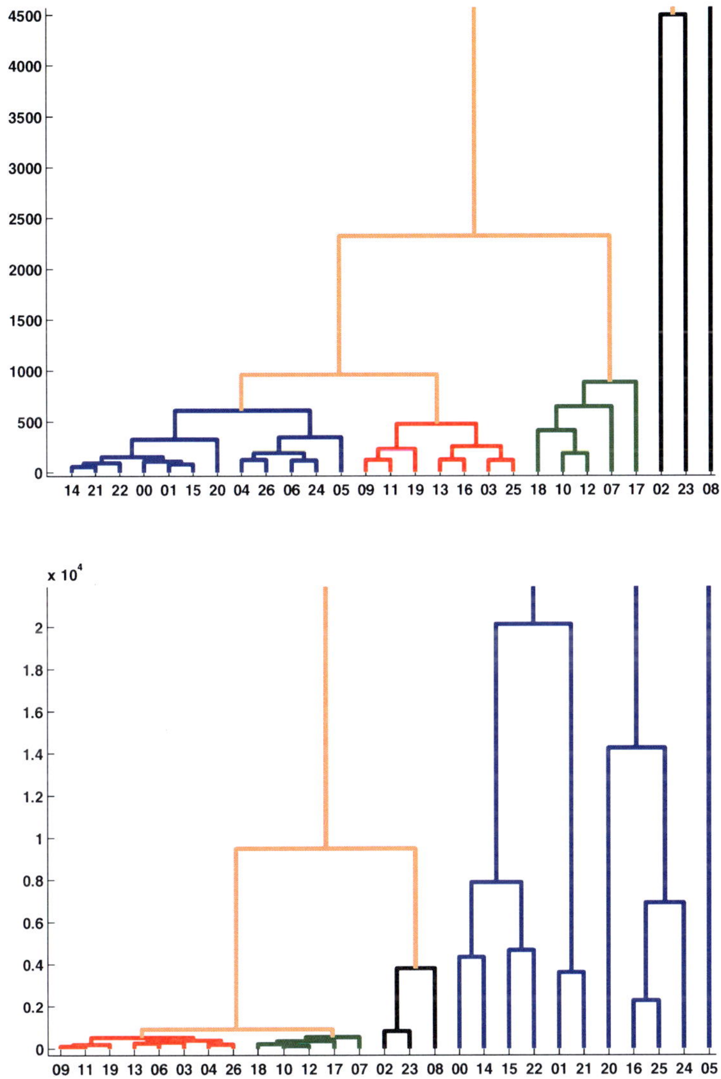

Fig. 9 Details of the dendrograms: (top) by using Negentropy ; (bottom) by using KL

deemed to be the 'best', especially for the organization of countermeasures and the assessment of consequences in the case of accidental radionuclide release.

In [10] the APL index was calculated using all available models, but this index can be straightforwardly estimated using a subset of models, too. In order to select a subset of models, we calculated the centroid of each cluster and, for each cluster, we selected the model closest (in the mean square sense) to the respective centroid.

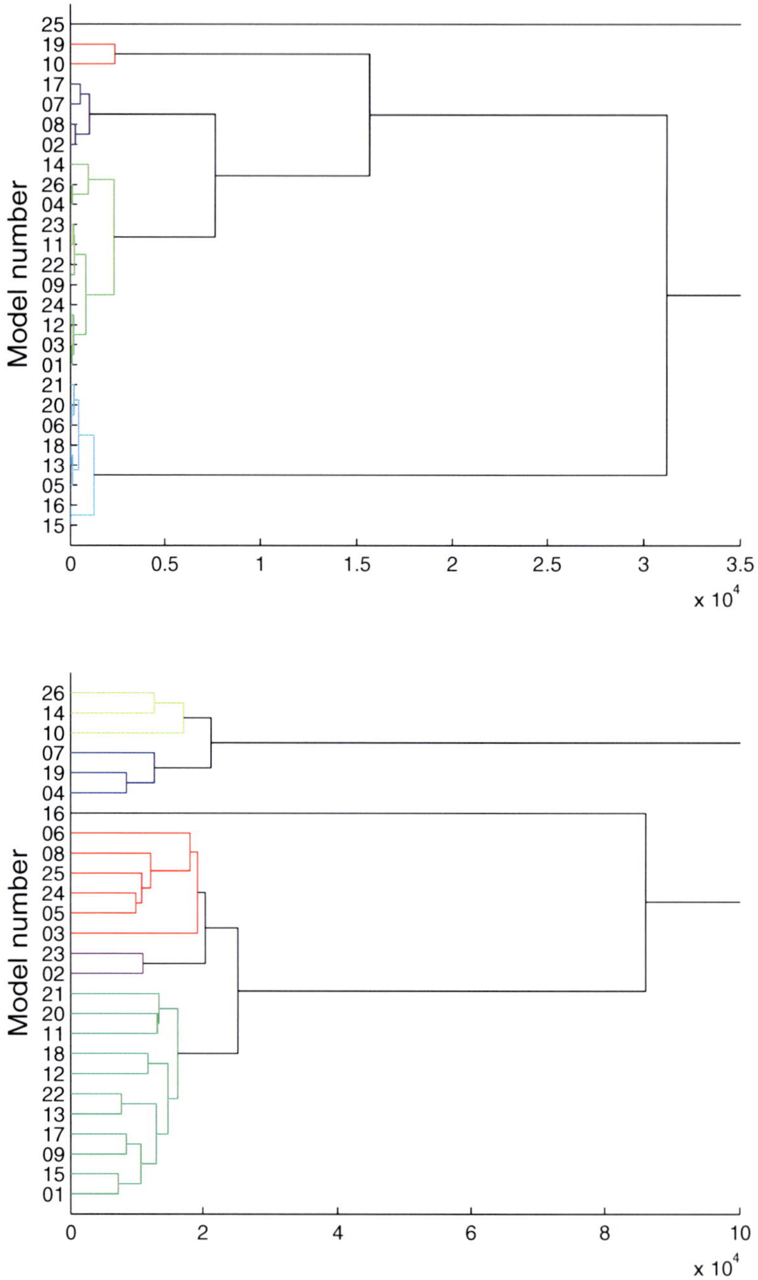

Fig. 10 Details of the dendrograms derived from the instantaneous concentrations: (top) by using Negentropy ; (bottom) by using KL

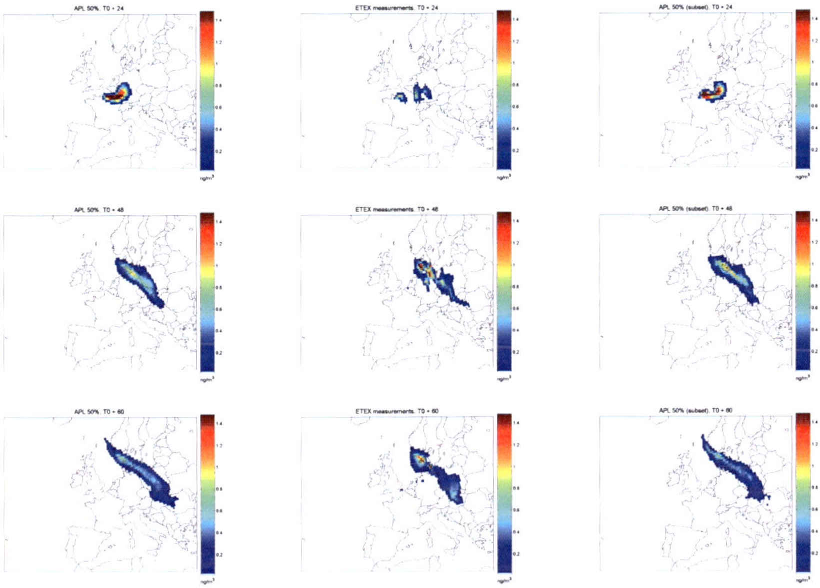

Fig. 11 APL_{50} from all models (adapted from [10]) (left column); observations (middle column); and APL_{50} from the six models selected from each cluster (right column) using Negentropy as dissimilarity measure at T0+24 (uppermost row), T0+48 (middle row) and T0+60 (lowermost row)

Figures 11 and 12 compare the observed values with APL_{50} and APL_{75}, both with the results obtained from all models and from the six models selected by using the Negentropy as dissimilarity measure. Results for the KL divergence do not differ significantly; hence, they are omitted. In Figs. 11 and 12 T0+XX stands for the prediction made XX hours after the starting of simulation.

First, it can be noted that the use of percentiles effectively filters erroneous results (e.g. extremely diffusive models, as $m08$ and $m23$ in Fig. 7) and compares favorably with observations. Figures 11 and 12 also show the APL surfaces at the same percentiles, calculated using six models, one from each cluster, as explained before. Though there are some differences concerning the absolute values, the area spanned by nonzero concentrations can be hardly distinguished, e.g., a small subset provide results similar to those from the whole ensemble.

It must be outlined that, in the absence of measurements for model validation, nobody can say *a priori* whether a model forecast will be reliable or not. In general, ensemble results represent a conservative choice than those produced by single models. In the case of a parsimonious choice of 'representative models', they must be carefully selected; the Negentropic or KL-divergence based procedures seem to preserve the main characteristics of ensemble results.

The empirical support to this statement is shown in Table 2, where the RMSE, FA2, FA5 and FOEX indexes are reported for the six models closest to the respective cluster centroid; as can be seen, the use of a single model is risky, since they usually

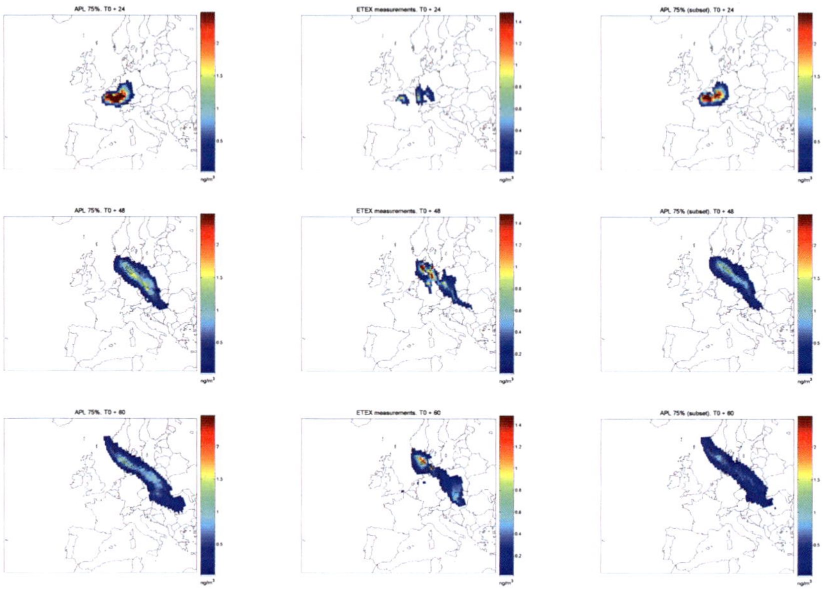

Fig. 12 APL_{75} from all models (adapted from [10]) (left column); observations (middle column); and APL_{75} from the six models selected from each cluster (right column) using Negentropy as dissimilarity measure at T0+24 (uppermost row), T0+48 (middle row) and T0+60 (lowermost row)

compare unfavorably with the Median Model (usually larger RMSE and lower FA2 and FA5 indexes), but the Median Model calculated over these six models (last row in Table 2) performs as well as, or even better than, the Median Model calculated over all 26 models (compare results from the last row with those from the Median Models in Table 1).

Table 2 Root mean square error (RMSE), correlation coefficient (CC), FA2, FA5 and FOEX indexes (look at Fig. 1 for their definition) for the six models from the six clusters identified using the Negentropy as dissimilarity measure In the last row, the same indexes calculated from the 'median model' results averaged over these six models ('MM' label)

	RMSE	CC	FA2	FA5	FOEX
m05	2.05	0.27	13.02	32.72	71
m07	0.93	0.26	19.98	42.91	36
m10	1.81	0.41	15.11	35.32	17
m19	1.47	0.27	21.81	46.12	76
m24	2.22	0.20	21.67	47.59	44
m25	3.27	0.24	22.93	46.64	50
MM	1.10	0.31	22.91	48.53	33

Table 3 Root mean square error (RMSE), correlation coefficient (CC), FA2, FA5 and FOEX indexes (look at caption 1 for their definition) calculated from the median of models within each cluster identified using the Negentropy as dissimilarity measure

	RMSE	CC	FA2	FA5	FOEX
M1	1.81	0.41	15.11	35.32	17
M2	1.47	0.27	21.81	46.12	76
M3	0.69	0.31	17.51	37.06	2
M4	1.56	0.27	24.58	50.15	14
M5	1.46	0.21	24.62	50.78	-4
M6	3.27	0.24	22.93	46.64	50

We think that the complementary between these models is not a fortuitous event, and this is highlighted by results in Table 3, where the same indexes are calculated from the median of models within each cluster. As can be seen, the average over subset of dependent, i.e., correlated, models does not improve significantly the performance of these ensembles (even though there are some clusters with a large number of models).

5 Conclusions

In this work an approach to analyze the independence between different data model describing atmospheric dispersion processes has been introduced.

The proposed approach is based on the Negentropy information or the Kullback-Leibler divergence . By using this information a hierarchical agglomeration can be obtained. The dendrogram describes the relations between data model distributions. The approach is used to analyze the data obtained during the ETEX-1 exercise.

Since model performance can be a case-dependent property, in the absence of measurements for model validation nobody can say *a-priori* whether a model forecast will be reliable or not. In general the results obtained by [10] with the Median Model seem to be more conservative than those produced by single models.

From the presented results, the Negentropic or KL-based approaches allow to derive results that are comparable with the Median Model ones. These approaches discriminate between data models that are less dependent (in the statistical sense), so that 'redundant' information can be more easily discarded and equivalent performance can be achieved with considerably fewer models.

References

1. Amato, R., Ciaramella, A., Deniskina, N., Del, C., di Bernardo, D., Donalek, C., Longo, G., Mangano, G., Miele, G., Raiconi, G., Staiano, A., Tagliaferri, R.: A multi-step approach to time series analysis and gene expression clusterings. To appear in Bioinformatics

2. Bellasio, R., Bianconi, R., Graziani, G., Mosca, S.: RTMOD: an Internet based system to analyse the predictions of long-range atmospheric dispersion models. Computers and Geosciences 25(7), 819–833 (1999)
3. Bianconi, R., Galmarini, S., Bellasio, S.: A WWW-based decision support system for the management of accidental releases of radionuclides in the atmosphere. Environ. Model. Software 19(4), 401–411 (2004)
4. Ciaramella, A., Tagliaferri, R.: Amplitude and permutation indeterminacies in frequency domain convolved ICA. In: Proc. IEEE Int. Joint Conf. Neural Networks, Portland, OR, pp. 708–713. IEEE Comp. Soc., Los Alamitos (2003)
5. Ciaramella, A., Napolitano, F., Miele, G., Raiconi, G., Staiano, A., Tagliaferri, R.: Clustering and visualization approaches for human cell cycle gene expression data analysis. To appear in the special issue, Approximate Reasoning and Machine Learning for Bioinformatics of Int. J. Approx. Reason.
6. Clark, T.L., Cohn, R.D., Seilkop, S.K., Draxler, R.R., Heffter, J.L.: Comparison of modelled and measured tracer gas concentrations during the Across North America Tracer Experiment (ANATEX). In: Proc. 17th Int. Tech. Meeting of NATO-CCMS on Air Pollution Model. and its Appl., Cambridge, UK (1988)
7. Ferber, G.J., Heffter, J.L., Draxler, R.R., Lagomarsino, R.J., Thomas, F.L., Dietz, R.N., Benkovitz, C.M.: Cross-Appalachian tracer experiment (CAPTEX-83). Final Report, NOAA Tech. Memo ERL ARL-142, Air Resources Laboratory, Silver Spring, MD (1986)
8. Galmarini, S., Bianconi, R., Bellasio, R., Graziani, G.: Forecasting consequences of accidental releases from ensemble dispersion modelling. J. Environ. Radioactiv. 57(3), 203–219 (2001)
9. Galmarini, S., Bianconi, R., Klug, W., Mikkelsen, T., Addis, R., Andronopoulos, S., Astrup, P., Baklanov, A., Bartniki, J., Bartzis, J.C., Bellasio, R., Bompay, F., Buckley, R., Bouzom, M., Champion, H., D'Amours, R., Davakis, E., Eleveld, H., Geertsema, G.T., Glaab, H., Kollax, M., Ilvonen, M., Manning, A., Pechinger, U., Persson, C., Polreich, E., Potemski, S., Prodanova, M., Saltbones, J., Slaper, H., Sofief, M.A., Syrakov, D., Sorensen, J.H., Van der Auwera, L., Valkama, I., Zelazny, R.: Ensemble dispersion forecasting – Part I: concept, approach and indicators. Atmos. Environ. 38(28), 4607–4617 (2004)
10. Galmarini, S., Bianconi, R., Addis, R., Andronopoulos, S., Astrup, P., Bartzis, J.C., Bellasio, R., Buckley, R., Champion, H., Chino, M., D'Amours, R., Davakis, E., Eleveld, H., Glaab, H., Manning, A., Mikkelsen, T., Pechinger, U., Polreich, E., Prodanova, M., Slaper, H., Syrakov, D., Terada, H., Van der Auwera, L.: Ensemble dispersion forecasting – Part II: application and evaluation. Atmos. Environ. 38(28), 4619–4632 (2004)
11. Girardi, F., Graziani, G., van Veltzen, D., Galmarini, S., Mosca, S., Bianconi, R., Bellasio, R., Klug, W.: The ETEX project. EUR Report 181-43 EN. Office for official publications of the European Communities, Luxembourg (1998)
12. Graziani, G., Galmarini, S., Mikkelsen, T.: RTMOD: real-time model evaluation, JRC Report TN.I.00.11 (1999)
13. Hyvärinen, A., Karhunen, J., Oja, E.: Independent Component Analysis. John Wiley & Sons, Hoboken (2001)
14. Klug, W., Graziani, G., Grippa, G., Pierce, D., Tassone, C. (eds.): Evaluation of long range atmospheric transport models using environmental radioactivity data from the Chernobyl accident. EUR Report 14147 EN. Office for Official Publications of the European Communities, Luxembourg (1992)

15. Krishnamurti, T.N., Kishtawal, C.M., Zhang, Z., LaRow, T., Bachiochi, D., Williford, E., Gadgil, S., Surendran, S.: Multimodel ensemble forecasts for weather and seasonal climate. J. Climate 13(23), 4196–4216 (2000)
16. MacKay, D.J.C.: Information Theory, Inference and Learning Algorithms. Cambridge University Press, Cambridge (2003)
17. Molteni, F., Buizza, R., Palmer, T.N., Petroliagis, T.: The ECMWF ensemble system: methodology and validation. Quart. J. Royal Meteor. Soc. 122(529), 73–119 (1996)
18. Potempski, S., Galmarini, S., Addis, R., Astrup, P., Bader, S., Bellasio, R., Bianconi, R., Bonnardot, F., Buckley, R., D'Amours, R., van Dijk, A., Geertsema, A., Jones, A., Kaufmann, P., Pechinger, U., Perssonm, C., Polreich, E., Prodanova, M., Robertsonm, L., Sørensen, J., Syrakov, D.: Multi-model ensemble analysis of the ETEX-2 experiment. Atmos. Environ. 42(31), 7250–7265 (2008)
19. Riccio, A., Giunta, G., Galmarini, S.: Seeking for the rational basis of the median model: the optimal combination of multi-model ensemble results. Atmos. Chem. Phys. 7(24), 6085–6098 (2007)
20. Toth, Z., Kalnay, E.: Ensemble forecasting at the NMC: the generation of perturbations. Bull. Amer. Meteor. Soc. 74(12), 2317–2330 (1993)

Integrating Liknon Feature Selection and Committee Training

Erinija Pranckeviciene

Abstract. We address a practical application of feature selection and training a committee of classifiers in high dimensional classification problems. Embedded Liknon feature selection method is integrated into the training of a committee of classifiers via external K-fold crossvalidation with an inner loop. In problems, characterized by nonlinear class separation, the Liknon-selected feature profiles, optimal for linear class separation, also identify the relevant feature subspaces. The capabilities of the proposed approach are illustrated in a benchmark of NIPS2003 feature selection challenge.

Keywords: feature selection, Liknon, committee of classifiers, multiple classifier system, NIPS2003 feature selection challenge, high data dimensionality, feature filtering, linear programming, support vector machine.

1 Introduction

Ensemble methods are used in many real life classification problems [18]. A great variety of methods exist for building multiple classifier systems [14] . We focus on the ensemble of classifiers trained on different subsets of features that are selected in disjoint partitions of the data [23]. Such ensemble exploits a variety of the local data representations [4, 5], reduces the error variance [26] and is useful in problems, characterized by high data dimensionality and small sample size.

Integrated feature selection and ensemble training were applied in handwritten character recognition [8] and classification of biomedical data [19, 21]. An ensemble, built by random subspace method [13] in feature space, was used for accumulation of interpretable feature profiles of biomedical spectra [2]. Usually size

Erinija Pranckeviciene
Department of Human and Medical Genetics, Faculty of Medicine, Vilnius University, Santariskiu 2, LT08661 Vilnius, Lithuania
e-mail: erinija.pranckeviciene@mf.vu.lt

O. Okun & G. Valentini (Eds.): Appli. of Supervised & Unsuper. Ensemble Meth., SCI 245, pp. 233–250.
springerlink.com © Springer-Verlag Berlin Heidelberg 2009

of the optimal feature subset is not known. Finding the optimal subset of the original features by the state-of-the-art methods, such as floating forward feature selection (FFFS) [25], may be very time consuming in high dimensional data. Small sample size constrains the size of a validation set. Without proper validation, feature selection is prone to overfitting and selection bias [1]. Integration of feature selection and committee training helps in overcoming the mentioned problems. We explore Liknon feature selection for the training of the committee of classifiers. This approach appeared among the top-ranked methods in Agnostic Learning versus Prior Knowledge (ALvsPK) challenge [12, 22].

Embedded feature selection method Liknon, also known as Linear Programming Support Vector Machine [3, 7] , produces a small subset of features, optimal for linear class separation. In Liknon, a number of features in the selected subset is bounded by the number of the available training samples. In disjoint data partition into K-folds, K linear discriminants are obtained by solving Liknon. Large weights of the discriminant identify feature subsets of various sizes, in which a "linear part" of the relevance of the features for class separation is signified. Such feature subsets are used for training a committee of linear or nonlinear classifiers.

Our study is organized as follows. Section 2 explains the method, the data partitioning and the Liknon feature selection. A practical example is presented by classifying an artificial "Banana" dataset. Section 3 overviews a benchmark of the NIPS2003 feature selection (NIPS2003 FS) challenge [10, 11] in terms of the characteristics of the feature subsets and the achieved classification performance of the various feature selection methods. In Sect. 4, the performance of the investigated method is analyzed by comparing it with the methods in the benchmark. Section 5 concludes the study.

2 Computational Paradigm and Parameters

Training a committee via feature selection includes several independent steps: choice of a crossvalidation scheme, feature selection method and selection of a base classifier.

1. **Choice of a crossvalidation scheme.** Training an ensemble requires some data partitioning scheme. Relevant approaches include Bagging and Boosting. We use an external K-fold crossvalidation, usually 5 or 10, with an inner loop. In the outer loop the data is subdivided into Fold Training and Fold Test sets. Further, in the inner loop, the Fold Training set is randomly partitioned M times into two: a balanced *Training* set and the remaining *Monitoring* set. In every split, the *Training* set is used for feature selection and the *Monitoring* set for assessment of the selected features. Monitoring helps in alleviation of the overfitting and feature selection bias. Further validation can be performed on the independent *Validation/Test* sets, if such datasets are available. The proposed procedure is presented in the top panel of Fig. 1.

2. **Feature selection.** In Liknon feature selection, a sequence of linear discriminants of increasing complexity defines the feature subsets of the increasing size. Such a sequence is obtained by solving Liknon *NM* times on the *Training* set with the different, increasing *NM* values of a regularization parameter. The performance of each discriminant in the sequence is assessed on the *Monitoring* set. A balanced error rate (*BER*) [11] is the optimality criterion. *BER* is computed from the confusion matrix $\begin{pmatrix} tp & fp \\ fn & tn \end{pmatrix}$ of a classifier in the following way: $BER = \frac{1}{2}\left(\frac{fp}{(fp+tp)} + \frac{fn}{(fn+tn)}\right)$, where tp is the true positive, fp is the false positive, fn is the false negative and tn is the true negative. In case of the unbalanced classes, the *BER* accounts for a contribution to the classification error of the majority and the minority class equally. The optimal feature subset is determined by the Liknon discriminant with minimum *Monitoring BER*. The determination of the optimal Liknon discriminant is illustrated in the bottom panel of Fig. 1. The feature subsets, determined by the optimal discriminants in *M* random splits, differ. The *M* optimal subsets identified in a single fold can be united into a profile of features.
3. **Choice of a base classifier.** The type of a base classifier is determined either by the problem-oriented knowledge, or selected from a pool of candidates. In the latter case, several different classifiers are trained on the *K* feature profiles, obtained in folds. The type of the classification rule, best suited to the data, is selected by comparing the overall performances of the classifiers. The classifier with the best performance across all the *M* Fold Training splits is selected as base. The *K* individual base classifiers, trained in each fold, are combined into the committee. We use a majority voting as a combination method.

The *K* folds, *M* splits, and testing of the *NM* values of the regularization parameter require the $K \times M \times NM$ of Liknon optimizations. The classification of the "Banana" dataset [6] will be used to illustrate the outlined approach in the subsequent sections. Details of the Liknon feature selection method are presented in the following section.

2.1 Liknon-Based Feature Selection

Liknon, originally introduced by [7], was found useful in applications of profiling of gene expression microarrays and face recognition, in which the data dimensionality exceeds the available number of samples by orders of magnitude [3, 21]. Liknon implements a linear rule for two-class classification:

$$y_s = sign(\mathbf{x}_s \mathbf{w}^T + w_0) \ , \tag{1}$$

where $\mathbf{x}_s = [x_s^1, \ldots, x_s^D]$ are *D*-dimensional samples, $\mathbf{y} = [y_1, \ldots, y_N]$ is a vector of class labels, assuming values $+1$ for the positive class and -1 for the negative class, $s = 1, \ldots, N$ indicates the sample index, $N = N_1 + N_2$ is the total number of samples

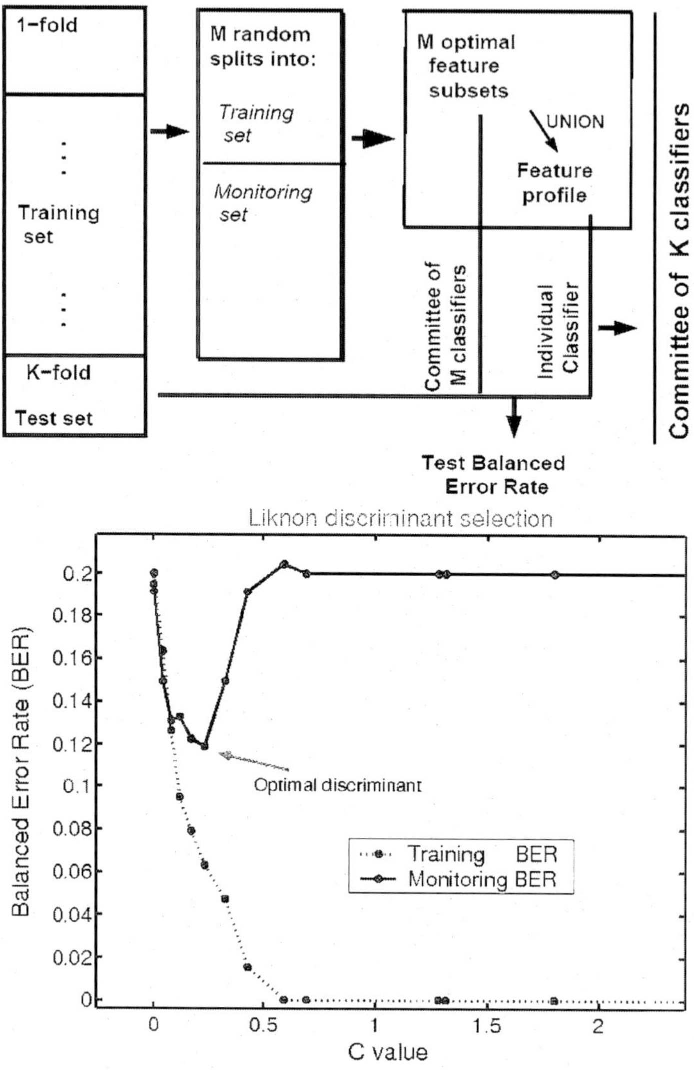

Fig. 1 Scheme of training the committee via feature selection. Top panel: Independent parts of the procedure. The data is partitioned into K-folds, with an inner loop in each, in which M random splits into the *Training* and *Monitoring* sets are performed. Optimal Liknon features are selected in the splits. The M feature subsets are either combined into a single feature profile, or used in an ensemble, which can be considered as a special type of the classifier. A type of a base classifier is selected from several candidates. The base classifier has best overall performance in M splits and K folds. Bottom panel: Selection of the optimal Liknon discriminant in the single split

in the two classes. The weight vector $\mathbf{w}$ is obtained by solving the optimization problem:

$$(\mathbf{w}^*, \ldots, \xi_1^*, \ldots, \xi_N^*) = \underset{(\mathbf{w}, \xi_1, \ldots, \xi_N)}{\arg\min} \left(\|\mathbf{w}\|_1 + C \sum_{s=1}^{N} \xi_s \right) \ ,$$
$$\text{s.t.:}$$
$$y_s \left(\mathbf{x}_s \mathbf{w}^T + w_0 \right) + \xi_s \geq 1 \ ,$$
$$\xi_s \geq 0 \ ,$$
$$s = 1, \ldots, N \ . \tag{2}$$

The ξ_s are slack variables. The * denotes the optimal solution. The L_1 norm is $\|\mathbf{w}\|_1 = \sum_{j=1}^{D} |w_j|$. The solution $\mathbf{w}^*$ is sparse. Because of sparsity, it is used for feature selection. Features that are important for classification are identified by large weights w_j^* of the vector $\mathbf{w}^*$.

The solution $\mathbf{w}^*$ is obtained by means of the Linear Programming (LP) optimization method. Because of the properties of the LP method, the number of selected features is bounded by the number of samples in the training set. If feature to sample ratio is high, then the Liknon solution will have small number of non-zero weights w_j^*. Let x_{s1}^f and x_{s2}^f denote the feature f values for the data samples $s1$ and $s2$ of the opposite classes. The f denotes the feature index, where $f = [1, \ldots, D]$. The parameter C controls the sparseness of the solution. It controls selection of the individual features by bounding the absolute difference between the classes:

$$|a^f| = |\sum_{s1-1}^{N_1} x_{s1}^f - \sum_{s2-1}^{N_2} x_{s2}^f| \ , \quad |a^f| \ > \ \frac{1}{C} \ . \tag{3}$$

A group of the individual features, in which the difference (3) between the classes is greater than $\frac{1}{C}$, is included into the Liknon feature subset. Descending ordering of the class differences (3) of the individual features $|a_{\max}^f| \geq \ldots \geq |a^f{}_{min}|$ determines the order of the C computation. A sequence of the C values is computed by setting the parameter equal to the inverse of every member of the sequence, $\frac{1}{|a_{\max}^f|} \leq \ldots \leq \frac{1}{|a_{\min}^f|}$ as $C = \frac{1}{|a^f|}, f = [1, \ldots, D]$. This sequence of the C values is used to compute a sequence of Liknon discriminants. In case of a high data dimensionality, the number of the C values has to be reduced, in order to avoid a computational burden. The reduction is performed by sampling $|a^f|$ from the interval $[|a^f{}_{max}|, |a^f{}_{min}|]$. The underlying idea is thoroughly explained in [24].

The difference a^f represents the dot product of the feature values and the class labels [21]. The absolute value of a^f can be used for univariate feature filtering. First, the features are ranked by the decreasing value of $|a^f|$, and some percentage of low-$|a^f|$-value features is discarded. Liknon feature selection is performed on the rest. In general, the ranking of the individual features by the absolute difference between the two classes (3) is different from the ranking by the distances between the class means[1].

2.2 *Banana Example: Classification in Nonlinear Class Separation*

In the artificial, noise-augmented Banana dataset, the classes are nonlinearly separated. The dimensionality of the dataset is $D = 100$, a number of samples in both classes is $N_1 + N_2 = 210 + 190$. Features 29 and 30 separate the two classes nonlinearly. In the remaining 98 features, both classes are distributed normally $N(0,1)$. The distributions of the class data in features, containing structure and no structure, are presented in the right and left top panels of Fig. 2. The *Validation* set was generated independently, by using the same "Banana" data model, implemented in PRTools [6], and the same Matlab script as for generating the Training data. We perform Liknon feature/classification model selection in 5-fold crossvalidation ($K = 5$) with a single ($M = 1$) random split in the inner loop. The number of the C values in this example is $NM = 8$. Table 1 summarizes the sizes of the data partitions.

Table 1 Sizes of the data partitions in the Banana example

Fold Training	Fold Test	Training	Monitoring	Validation
$152+168$	$38+42$	$76+76$	$76+92$	$1981+2019$

The absolute values of the weights of the optimal discriminants $\mathbf{w}^*$ selected in each fold, identify the relevant features. They are shown as a heat map in the right bottom panel of Fig. 2. The color-coded values of the discriminant weights are interpreted as the indicators of the feature importance. The ground truth features 29 and 30 have large weights in all folds. The classification performances of linear Fisher and nonlinear NN3 classifiers trained using the optimal feature subsets in folds, are presented in Table 2. The BERs attained in Fold Test and the independent *Validation* sets are presented graphically in the left bottom panel of Fig. 2. Overall, the classification performance of the NN3 rule is superior in this example. NN3 achieved the best performance in Fold 3, in which the smallest and cleanest feature subset was identified.

In the Banana example the Liknon-identified feature subsets include the ground truth features, adequately representing a character of class separation in various "projections" of feature and sample spaces. The classification performances of the different classifiers are constrained by a capability of the classifiers to handle a complexity of the classification problem. Liknon generates the multivariate feature

[1] For example, percentages of disagreeing ranks of the features, ordered by $|a^f|$, and by the difference between the means, of the NIPS2003 FS training datasets are: ARCENE - 66.90%; DEXTER - 24.61%; DOROTHEA - 59.48%; GISETTE - 3.28%; MADELON - 5.20%.

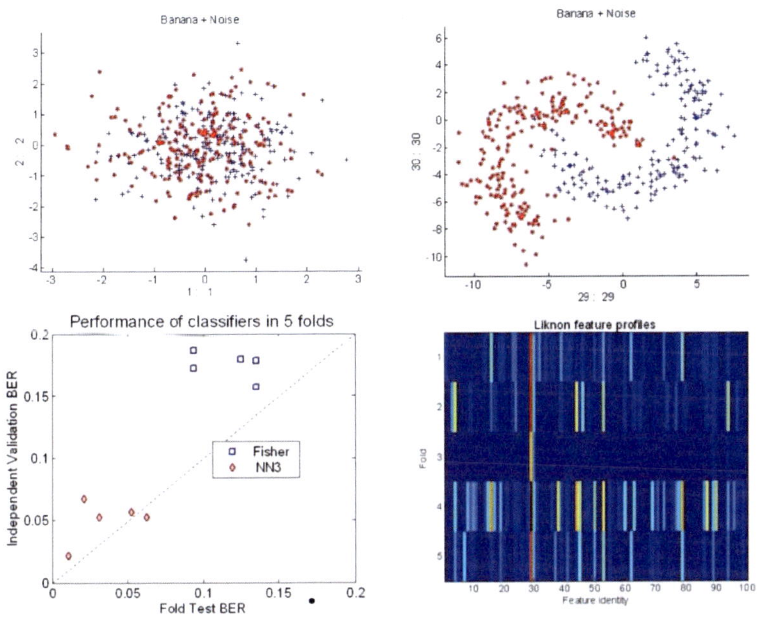

Fig. 2 Feature/classification model selection in Banana example. The top panel shows the class distributions in relevant and noisy features. Liknon feature profiles, identified in folds, are presented in the bottom right panel. The performances of the individual Fisher and NN3 classifiers, trained on the Liknon feature profiles in folds, are shown in the bottom left panel

Table 2 Fold Test BER of Fisher and NN3 classifiers in the Banana example

Fold	1	2	3	4	5
NN3	0.0521	0.0625	0.0104	0.0208	0.0313
Fisher	0.1354	0.1250	0.1354	0.0938	0.0938

subsets, optimal for linear separation. If the classes are nonlinearly separated, then the Liknon discriminant defines the linear boundary with some margin, outside of which classes are linearly separated. Liknon features account for that "linear part" of the class separation. By comparing the performances of the nonlinear versus the linear classifier, trained using the selected feature subsets, we gain an insight into the character of the class separation. The small balanced error rate and an agreement between the values of the BER, attained on Fold Test and *Validation* sets, point to the classifier, which suits the classification problem better. Performance of the NN3 committee, trained on Liknon feature profiles in more complex cases of class separation, is investigated further by using the datasets of the benchmark of NIPS2003 FS challenge.

3 The Benchmark of NIPS2003 Feature Selection Challenge

The benchmark of NIPS2003 feature selection challenge is a platform for testing feature selection methods. The datasets of the NIPS2003 FS benchmark represent two class high-dimensional classification problems. The original data is augmented with probes - the fake features [10, 11]. Ground truth about the feature identity (useful or probe) enables an assessment of the feature selection method. In the benchmark the methods are assessed on-line, by submitting the labels of the *Test* set, assigned by the classifier, and the list of features to the benchmark website [17]. Three complex, nonlinear datasets were used in our study: ARCENE - mass spectra data, GISETTE - handwritten digit recognition data, and MADELON - the artificially generated data. The dimensionality, the sizes of the data sets, percentage of probes and a number of feature selection methods, which performed feature selection in the benchmark to date, are presented in Table 3.

Table 3 Characteristics of the NIPS2003 FS benchmark datasets

Dataset	ARCENE	GISETTE	MADELON
Dimensionality	10000	5000	500
Total probes%	30	50	96
Training	44 + 56	3000 + 3000	1000 + 1000
Validation	44 + 56	500 + 500	300 + 300
Test	310 + 390	3250 + 3250	900 + 900
Number of feature selection methods in the benchmark to date	440	241	320

3.1 Performance, Size and Purity of the Feature Subsets in the Benchmark

Feature selection and training of a classifier are intrinsically combined. Different rules may improve their classification performance with the different features. A classifier trained on a probe-free feature subset of the optimal size is expected to have improved performance. For the three datasets, distributions of the BER of the benchmark methods with respect to the size and purity of the feature subsets, show an association of the better classification performance with the purer feature subsets. The size of a multivariate feature subset is measured by a percentage of the total features, and the purity is a percentage of the useful (alternative to probes) features in the feature subset. We summarize the observed trends[2] in Figs. 3, 4 and 5. Continuous size and purity of a feature subset, were mapped into the categories, representing the intervals of a fixed percentage as $< 10\%$, $10\% - 20\%$, and so on, till 100%. The *Test* BER was mapped into the performance categories. The cut-points

[2] The exploratory analysis was carried out by SPSS(v.16) package.

of the categories were set equal to the 25%, 50% and 75% percentiles, ensuring the approximately equal number of cases in each category. The top panels of the Figs. 3, 4 and 5 show how the benchmark methods are distributed with respect to the size and purity of the feature subsets. The performance categories, defined by the attained *Test* BER values, are indicated by different colors. The names of the methods are displayed in an optimized way. The bottom panels represent the box-whisker plots of the attained *Test* BER within the categories of the subset size, clustered by the purity. The box plots summarize the characteristics of the *Test* BER. The black line represents a median, the first and third quartiles bound the box, the whiskers indicate the minimum and maximum values, which are not statistically outlying. The outliers are denoted by circles and extremes are denoted by asterisks. The number of cases in the category is shown in a text-box over each box-whisker plot.

In ARCENE, the quarter of the methods, achieving the best performance, attained the *Test* BER below 0.22 and close to 0.11. A majority of those methods selected/used the feature subsets of the sizes of up to 30%, in which more than 80% were the useful features, see Fig. 3. The best attainable classification performance associates with the purer subsets of the smaller size. The best method on ARCENE dataset is New-Bayes-nn+v which is: *"Bayesian neural network with two hidden layers (20 and 8 units), applied to a set of features selected by examining the ARD hyperparameters found in New-Bayes-lr-sel and New-Bayes-nn-sel. An ARD prior was used here as well to allow some of the features to have more influence than others. The model was fitted to both the training and validation data"*[17]. The number of features in this method is 100, all useful. The best performing methods used pure subsets of the size of up to 20%.

In GISETTE, the half of the best performing methods attained the *Test* BER below 0.0258, close to 0.007. The feature subsets of the better performing methods had sizes from 10% to 40%, in which from 80% to 100% were the useful features (see Fig. 4). The box-whisker plots of the *Test* BER in the size categories show a trend towards the association of the purity of the feature subset with an improving classification performance. However, best performance is attained in a group of 21 methods, using small subsets, containing only up to 60% of the useful features. This group comprises recently developed methods, setting a new performance threshold for the benchmark [11].

In MADELON, the half of the best methods attained the *Test* BER below 0.1156 and operated on the pure and small feature subsets, containing only 4% of the total number of features (see Fig. 5). The best performance was achieved on the subsets of the size of up to 4%, out of which more than 95% were useful features. Note that MADELON dataset contains 96% of probes.

The observed benchmark results on the three datasets, except for several methods, exhibit a trend of association between the purity of the feature subsets and improved classification performance. A considerable improvement can be gained by combining a pure subset with an appropriate classification method.

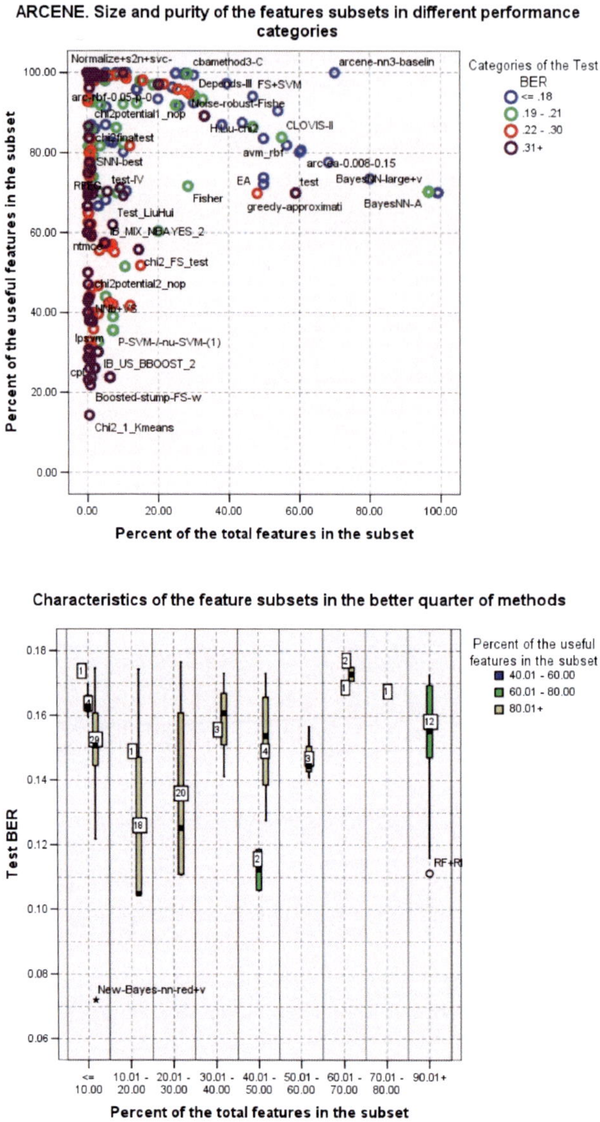

Fig. 3 ARCENE. Top panel: The purity and size of the feature subsets of the benchmark methods in the different categories of the *Test* BER. Bottom panel: The *Test* BER in the different categories of the feature subsets of the best quarter of the methods

4 NN3 Committee and Liknon Feature Profiles in the Benchmark

In this section, we set the NN3 committee , trained on the Liknon feature profiles, against only those benchmark methods, which did not use *Validation* set for

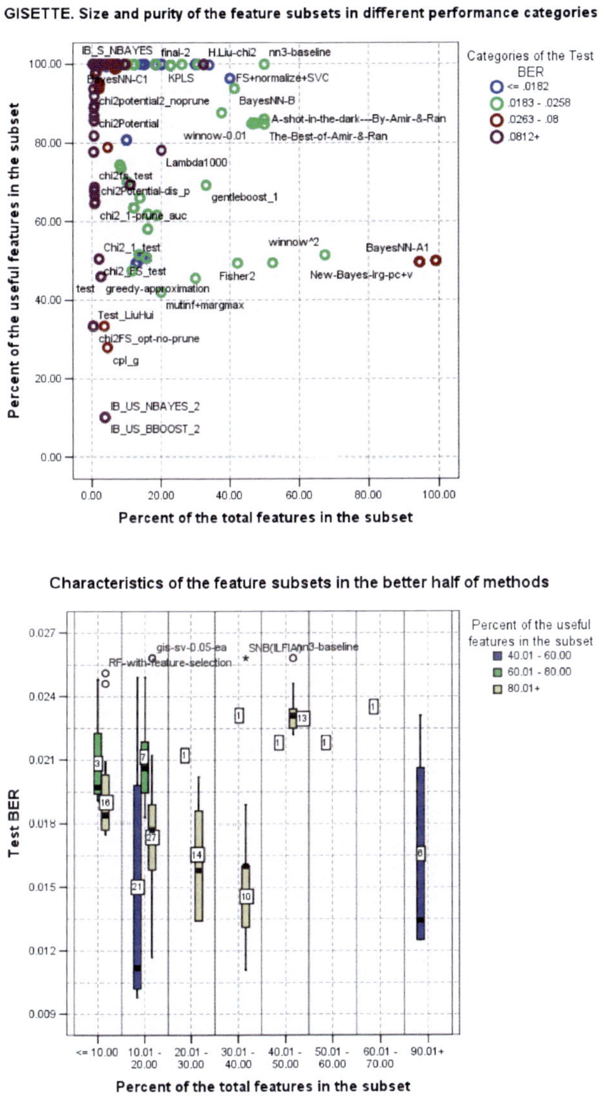

Fig. 4 GISETTE. Top panel: The purity and size of the feature subsets of the benchmark methods in the different categories of the *Test* BER. Bottom panel: The *Test* BER in the different categories of the feature subsets of the best half of the methods

training, since *Validation* set was not used for feature selection and training in our study. Table 4 summarizes a number of the original challenge entries, their mean and median *Test* BER, the *Test* BERs of NN3 baseline and NN3 committee , a percentage of features and useful features (size and purity) in the feature subsets, used by the committee .

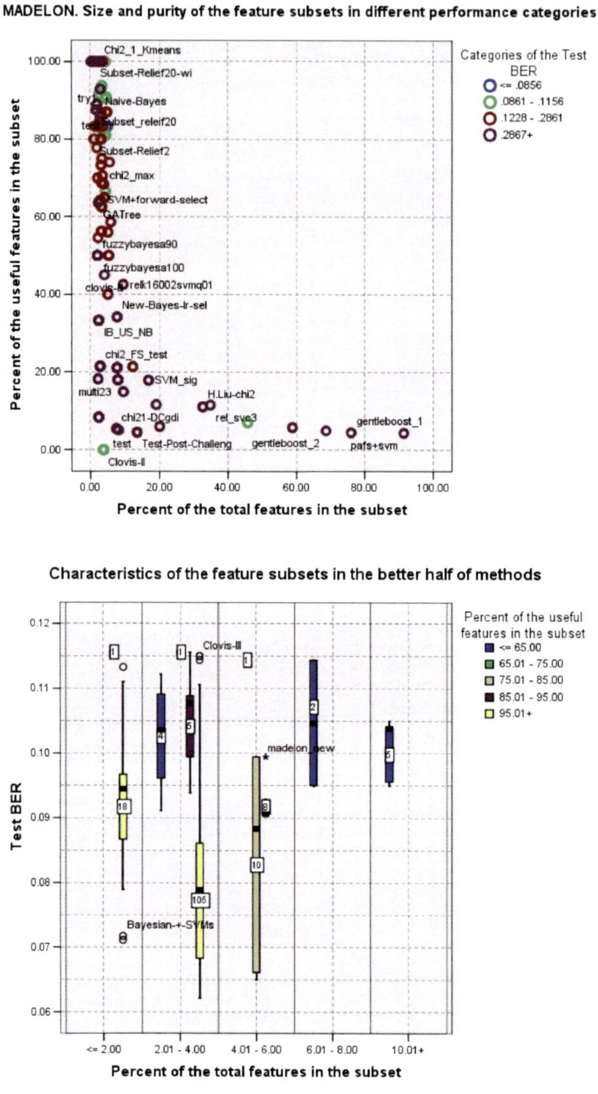

Fig. 5 MADELON. Top-panel: The purity and size of the feature subsets of the benchmark methods in the different categories of the *Test* BER. Bottom-panel: The *Test* BER in the different categories of the feature subsets of the best half of the methods

The computational parameters of building the NN3 committee on Liknon feature profiles are presented in Table 5. The committee comprises the K expert NN3 classifiers, trained on Liknon feature profiles. The decisions of the experts are combined by voting. NN3, trained using all training samples on all useful features, provides a baseline performance of the rule, without the feature selection. Normally, in real

Table 4 Summary of the performance of the methods compared to NN3 committee

Dataset	ARCENE	GISETTE	MADELON
Number of the original entries	180	94	108
		Test BER	
Mean	0.2543	0.065	0.212
Median	0.2209	0.0268	0.1542
NN3 baseline	0.1749	0.0258	0.0994
NN3 committee	0.1877	0.0292	0.1144
		Size and purity	
Features #	345 (3.45%)	357 (7.14%)	21 (4.20%)
Useful #	282 (81.74%)	354 (99.16%)	14 (66.67%)

Table 5 Data partitions and computational parameters for NN3 committee training

Dataset	ARCENE	GISETTE	MADELON
Discarded features %	85	95	98
Training	30 + 30	150 + 150	300 + 300
Monitoring	5 + 15	2250 + 2250	500 + 500
Fold Test	9 + 11	600 + 600	200 + 200
M #splits	5	5	5
K #experts	5	5	5
NM # C values	8	8	8

life, such baseline can not be computed, because we do not have information about the feature identity.

The NN3 committee belongs to the group of the best performing original challenge methods. The Liknon feature profiles contain less than 10% of the total features and more than 65% of the useful features. Liknon-identified features are optimal for linear class separation. The classes in ARCENE, GISETTE and MADELON are separated in a complex, nonlinear way. In these difficult problems, after the Liknon-based univariate filtering (discussed in Sect. 2.1) small and pure feature subsets were identified. It is an encouraging result for possible applications of Liknon feature selection in the cases of nonlinear class separation. The positions of the NN3 committee and baseline among the other, better performing original methods in the NIPS2003 FS challenge, are presented in Figs. 6, 7 and 8. For ARCENE and GISETTE datasets, the improved classification performance is associated with the larger subset sizes than those of Liknon-identified subsets. The limit for a size of the Liknon feature subset is set by the initial filtering and a partitioning of the data into K folds, determining the number of samples for training and monitoring. Although the NN3 committee appeared in the group of the best performing methods, the NN3 rule might not be the most appropriate classifier for the types of the

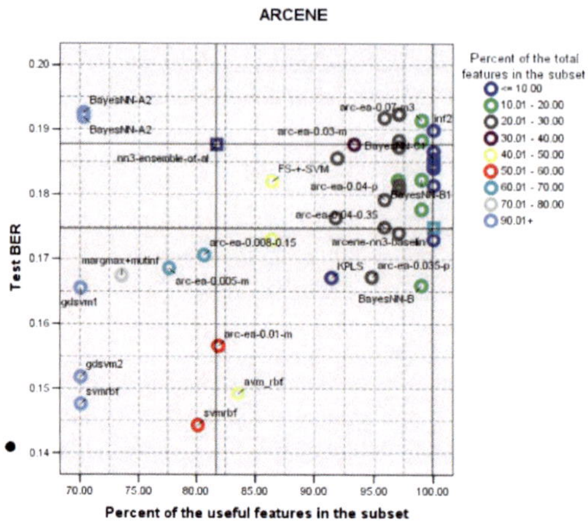

Fig. 6 ARCENE. The NN3 committee and baseline among the best performing original methods of the NIPS2003 FS challenge. The two methods are indicated by the square marker

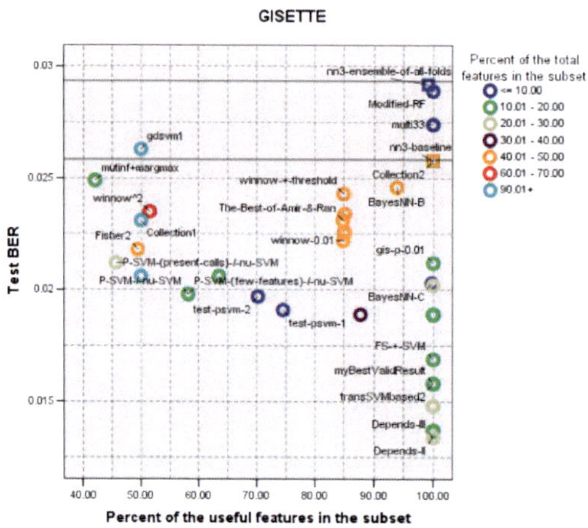

Fig. 7 GISETTE. The NN3 committee and baseline among the best performing original methods of the NIPS2003 FS challenge.The two methods are indicated by the square marker

benchmark data. The *Test* BER of the NN3 baseline is outperformed by Bayesian Neural Network and kernel methods.

The NN3 classifiers trained on the feature subsets selected in the disjoint partition into K-folds, made a diverse committee . The performance of the committee is better than the majority of the individual classifiers. The characteristic of the individual classifiers for ARCENE and MADELON datasets are presented in Tables 6 and 7.

A variance of Fold Test BER and *Validation* BER arises due to the different training sets in folds and different feature profiles. Smaller values of the *Validation* and Fold Test BERs may be interpreted as indicators of more relevant and purer feature

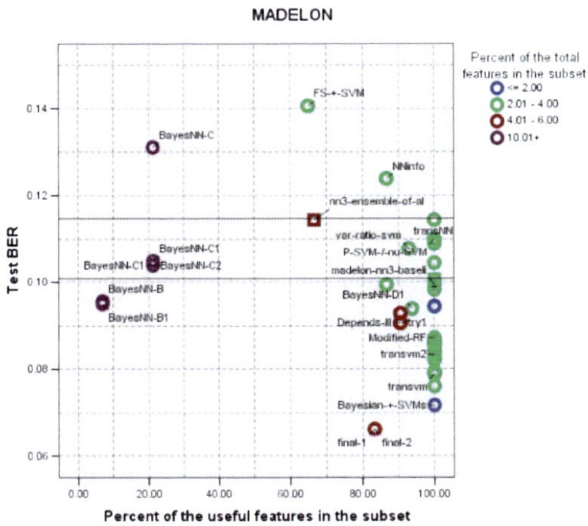

Fig. 8 MADELON. The NN3 committee and baseline among the best performing original methods of the NIPS2003 FS challenge.The two methods are indicated by the square marker

Table 6 ARCENE. Performance of the individual NN3 experts in folds and the committee

Fold	*Validation* BER	*Test* BER	Fold Test BER	features/probes
1	0.2597	0.2166	0.0556	103/20 (19.42%)
2	0.1599	0.2015	0.1667	99/14 (14.14%)
3	0.2484	0.2335	0.0	91/15 (16.48%)
4	0.3052	0.2214	0.1111	103/20 (19.42%)
5	0.2378	0.2650	0.1111	95/36 (37.89%)
Committee	0.1810	0.1877	-	345/63 (18.26%)

Table 7 MADELON. Performance of the individual NN3 experts in folds and the committee

Fold	*Validation* BER	*Test* BER	Fold Test BER	features/probes
1	0.2133	0.1911	0.1875	15/4 (26.67%)
2	0.1550	0.1406	0.1400	10/0 (0%)
3	0.2267	0.2139	0.1525	6/0 (0%)
4	0.1200	0.1089	0.1400	12/1 (8.33%)
5	0.1217	0.1156	0.1500	15/2 (13.33%)
Committee	0.1183	0.1144	-	21/7 (33.33%)

subsets (MADELON). One should be cautious when interpreting the estimates computed from small samples (ARCENE) because of high variance of such estimates. Several methodologies for a determination of "enough samples" for training [16] and testing [9] were suggested in the specific applications.

5 Conclusion

Our study practically addressed the integration of Liknon feature selection into the committee training for high-dimensional classification problems. As a computational paradigm for building the committee, we suggested K-fold external cross-validation with the inner loop, in which the feature selection is performed. Generally, other feature selection methods can be used within this framework. In the present study we focused on the embedded Liknon feature selection method, which finds the subset optimal for linear class separation, that can be also used in the cases of nonlinear class separation. We showed this in the controlled way by classifying the "Banana" dataset. The most appropriate base classifier was selected by comparing several different classifiers trained on Liknon feature profiles in all folds. In the cases of the complex nonlinear class separation, the performance of the NN3 committee trained on Liknon feature profiles, compared favorably with the performance of the best part of the original NIPS2003 FS challenge methods on three benchmark datasets.

In our study we have chosen a simple nonlinear classification rule as a base classifier. Indeed, a neural net or a support vector machine with nonlinear kernel could constitute a committee . Such complex classifiers would involve an additional tuning of the parameters, hence additional validation. We avoided more validation and used a simple model. In feature selection, the performances of the individual experts can be taken into account in assessing a purity (relevance) of the feature subset, provided there are enough samples for training and testing. The benchmark results showed the trend of the association between the improved classification performance and the purity of the feature subsets.

Usually, in the assessment of the feature selection/classification method by its performance, it is difficult to foresee whether the features are relevant or not, or if the subset size is optimal or not, or if a method is capable of handling a complexity of the data or not. Efforts are being made towards development of a methodology [15] for choosing an appropriate classification rule, also in problems characterized by a deficit of the training data [20].

We presented a brief overview of the performance of a variety of feature selection methods in relation to the size and purity of the feature subsets. We believe that a reader, interested in feature selection, gained some insights about the attainable performances in the benchmark of the NIPS2003 FS challenge. The details of the original challenge methods are very well explained in the book on feature selection and extraction [10].

References

1. Ambroise, C., McLachlan, G.J.: Selection bias in gene extraction on the basis of microarray gene-expression data. Proc. Natl. Academy Sci. 99(10), 6562–6566 (2002)
2. Bamgbade, A., Somorjai, R., Dolenko, B., Pranckeviciene, E., Nikulin, A.E., Baumgartner, R.: Evidence accumulation to identify discriminatory signatures in biomedical spectra. In: Miksch, S., Hunter, J., Keravnou, E.T. (eds.) Proc. 10th Conf. Artif. Intell. in Medicine, pp. 463–467. Springer, Heidelberg (2005)

3. Bhattacharyya, C., Grate, L.R., Rizki, A., Radisky, D., Molina, F.J., Jordan, M.I., Bissell, M.J., Mian, I.S.: Simultaneous relevant feature identification and classification in high-dimensional spaces: application to molecular profiling data. Signal Proc. 83(4), 729–743 (2003)
4. Chawla, N.V., Moore, T.E., Hall, L.O., Bowyer, K.W., Kegelmeyer, W.P.: Distributed learning with bagging-like performance. Pattern Recogn. Lett. 24(1-3), 455–471 (2003)
5. Cunningham, P., Carney, J.: Diversity versus quality in classification ensembles based on feature selection. In: Lopez de Mantaras, R., Plaza, E. (eds.) ECML 2000. LNCS (LNAI), vol. 1810, pp. 109–116. Springer, Heidelberg (2000)
6. Duin, R.P.W., Juszczak, P., Paclik, P., Pekalska, E., De Ridder, D., Tax, D.M.J.: PRTools4: A Matlab toolbox for pattern recognition (2004), `http://www.prtools.org`
7. Fung, G., Mangasarian, O.: A feature selection Newton method for support vector machine classification. Computational Optimization and Appl. 28(2), 185–202 (2004)
8. Gunter, S., Bunke, H.: Creation of classifier ensembles for handwritten word recognition using feature selection algorithms. In: Proc 8th Int. Workshop Frontiers in Handwritten Recognition, Niagara-on-the-Lake, OT, Canada, pp. 183–188. IEEE Comp. Soc., Los Alamitos (2002)
9. Guyon, I., Makhoul, J., Schwartz, R., Vapnik, V.: What size test set gives good error rate estimates? IEEE Trans. Pattern Analysis Mach. Intell. 20(1), 52–64 (1998)
10. Guyon, I., Gunn, S., Nikravesh, M., Zadeh, L. (eds.): Feature Extraction: Foundations and Applications. Springer, Heidelberg (2006)
11. Guyon, I., Li, J., Mader, T., Pletsher, P.A., Schneider, G., Uhr, M.: Competitive baseline methods set new standards for the NIPS 2003 feature selection benchmark. Pattern Recogn. Lett. 28(12), 1438–1444 (2007)
12. Guyon, I., Safari, A., Dror, G., Cawley, G.: Agnostic learning vs. prior knowledge challenge. In: Proc 20th Int. Joint Conf. Neural Networks, Orlando, FL, pp. 829–834. IEEE Comp. Soc., Los Alamitos (2007)
13. Ho, T.K.: The random subspace method for constructing decision forests. IEEE Trans. Pattern Analysis Mach. Intell. 20(8), 832–844 (1998)
14. Kuncheva, L.: Combining Pattern Classifiers. John Wiley & Sons, Hoboken (2004)
15. Mansilla, E.B., Ho, T.K.: On classifier domains of competence. In: Proc. 17th Int. Conf. Pattern Recogn., Cambridge, UK, pp. 136–139. IEEE Comp. Soc., Los Alamitos (2004)
16. Mukherjee, S., Tamayo, P., Rogers, S., Rifkin, R., Engle, A., Campbell, C., Golub, T.R., Mesirov, J.P.: Estimating dataset size requirements for classifying DNA microarray data. J. Computational Biology 10(2), 119–142 (2003)
17. NIPS2003 Feature Selection challenge (2003), `http://www.nipsfsc.ecs.soton.ac.uk`
18. Polikar, R.: Ensemble based systems in decision making. IEEE Circuits and Syst. Magazine 6(3), 21–45 (2006)
19. Pranckeviciene, E., Baumgartner, R., Somorjai, R.: Using domain knowledge in the random subspace method. Application to the classification of magnetic resonance spectra. In: Oza, N.C., Polikar, R., Kittler, J., Roli, F. (eds.) MCS 2005. LNCS, vol. 3541, pp. 336–345. Springer, Heidelberg (2005)
20. Pranckeviciene, E., Ho, T.K., Somorjai, R.: Class separability in spaces reduced by feature selection. In: Proc. 18th Int. Conf. Pattern Recogn., Hong-Kong, pp. 254–257. IEEE Comp. Soc., Los Alamitos (2006)
21. Pranckeviciene, E., Somorjai, R.: Liknon feature selection for microarrays. In: Masulli, F., Mitra, S., Passi, G. (eds.) WILF 2007. LNCS (LNAI), vol. 4578, pp. 580–587. Springer, Heidelberg (2007)

22. Pranckeviciene, E., Somorjai, R., Tran, M.N.: Feature/model selection by the Linear Programming SVM combined with state-of-art classifiers: what can we learn about the data. In: Proc. 20th Int. Joint Conf. Neural Networks, Orlando, FL, pp. 1627–1632. IEEE Comp. Soc., Los Alamitos (2007)
23. Pranckeviciene, E.: Integrating feature selection and committee training. In: Okun, O., Valentini, G. (eds.) Proc. 2nd Workshop Supervised and Unsupervised Ensemble Methods and Their Appl., Patras, Greece, pp. 69–72 (2008)
24. Pranckeviciene, E., Somorjai, R.: Liknon feature selection: behind the scenes. In: Guyon, I., Cawley, G., Dror, G., Safari, A. (eds.) Hands-on Pattern Recognition: Challenges in Data Representation, Model Selection, and Performance Prediction (2009)
25. Pudil, P., Novovicova, J., Kittler, J.: Floating search methods in feature selection. Pattern Recogn. Lett. 12(3), 1119–1125 (1994)
26. Valentini, G., Dietterich, T.G.: Bias-variance analysis of support vector machines for the development of SVM-based ensemble methods. J. Mach. Learn. Res. 1, 1–48 (2000)

Evaluating Hybrid Ensembles for Intelligent Decision Support for Intensive Care

Pedro Gago and Manuel Filipe Santos

Abstract. The huge amount of data available in an Intensive Care Unit (ICU) makes ICUs an attractive field for data analysis. However, effective decision support systems operating in such an environment should not only be accurate but also as autonomous as possible, being capable of maintaining good performance levels without human intervention. Moreover, the complexity of an ICU setting is such that available data only manages to cover a limited part of the feature space. Such characteristics led us to investigate the development of ensemble update techniques capable of improving the discriminative power of the ensemble. Our chosen technique is inspired by the Dynamic Weighted Majority algorithm, an algorithm initially developed for the concept drift problem. In this paper we will show that in the problem we are addressing, simple weight updates do not improve results, whereas an ensemble, where we allow not only weight updates, but also the creation and eliminations of models, significantly increases classification performance.

Keywords: classifier ensemble, data dimensionality.

1 Introduction

Since the 1960's computer applications whose purpose was that of supporting the decision making process have been designed [26]. Even though the first computer applications in business environments were intended to make easier operational activities like order processing, billing or inventory control, the need arose for tools that could ease the tasks related to decision support [1].

Pedro Gago
ESTG, Instituto Politécnico de Leiria, Portugal
e-mail: pgago@estg.ipleiria.pt

Manuel Filipe Santos
DSI, Universidade do Minho, Portugal
e-mail: mfs@dsi.uminho.pt

O. Okun & G. Valentini (Eds.): Appli. of Supervised & Unsuper. Ensemble Meth., SCI 245, pp. 251–265.
springerlink.com © Springer-Verlag Berlin Heidelberg 2009

In the medical area several expert systems were built and deployed [2, 11, 20]. However, the failure rate was high as the effort required to update the knowledge base was excessive and the scope of the expert systems was very limited.

Researchers started shifting their attention to the automation of the knowledge acquisition process by using methods from several areas of expertise (e.g. machine learning, statistics). Knowledge Discovery from Databases (KDD) [5] is well suited for this task. In fact, given that there is enough data, KDD techniques make knowledge acquisition easier, thus simplifying the task of building decision support tools. However, despite KDD being a semi-automatic process, the predictive models still need to be re-evaluated on a regular basis to detect any loss of predictive accuracy. In fact, model performance is known to degrade over time as the world does not remain in a stationary state [9] (e.g. in the medical area new drugs and therapeutic procedures are constantly being developed). Whenever performance drops below acceptable values it is necessary to repeat the KDD process or, at the very least, retrain the models using the latest data. Thus, an adaptive Decision Support System (DSS) must include mechanisms to detect the degradation in performance and to act accordingly in order to maintain the needed performance levels [18]. Moreover, it is now well established that prediction accuracy can usually be improved by using ensembles of prediction models instead of a single model [4, 12]. Despite ensemble performance being usually better than that of a single model the quality of ensemble predictions also degrades with the passing of time [9].

In this paper we present several experiments aiming at predicting the final outcome for patients staying in an Intensive Care Unit (ICU). Our final goal is that of building a DSS connected to the hospital's computer network allowing for real-time prediction and continuous performance assessment. The prediction is the result of an ensemble of classifiers, composed of both neural networks and decision trees as it has been shown that the different model types contribute to a lower number of coincident failures, thus increasing the ensemble performance [31]. Whenever necessary the system automatically alters the ensemble, either by changing the models weights, by deleting poor performing models or by adding new models.

In Sect. 2 we present some previous work in this area that led to the development of the INTCare System (that is described in Sect. 3). In Sect. 4 we present an overview the overall problem we are trying to solve and also describe the data used in the experiments presented in this paper. Section 5 contains a brief overview of related work. In Sect. 6 we describe the experimental setting and in Sect. 7 we present the experimental results. Finally, Sect. 8 includes the discussion of the results and in Sect. 9 we present conclusions and pointers to future work.

2 Previous Work

Artificial Intelligence techniques have been used in order to enhance Decision Making Support Systems (DMSS). One can find examples of Artificial Neural Networks (ANN) and Genetic Algorithms [8] , Decision Trees (DT) [27], Fuzzy Logic [16] and Data Mining (DM) [18] embedded in decision support systems. Likewise, agents are becoming commonplace as they provide a useful abstraction that facilitates systems conceptualization. An example of use of agents in the medical area is the AMPLIA system [29]. In an ICU setting the Guardian system [13] monitors and makes diagnoses of intensive-care patients. Another system, the Rihad ICU Program [3] was used to predict death.

In our previous work, several DM models were applied to ICU data [22, 23, 24]. These studies used off-line learning, where all data was stored and then accessed repeatedly by the DM algorithms. The clinical data was collected during the EURICUS II research programme [19], which involved a massive study in 42 ICUs from 9 countries during a period of 10 months, from 1998 to 1999. The database included thousands of daily records related to bedside measurements of critically ill patients, including features such as: the case mix - an information that remains unchanged during the patient's stay in the ICU (e.g. age or admission origin); the intermediate outcomes - being triggered from four monitored biometrics (e.g. the systolic blood pressure or urine output); and the patient's state - based on daily organ failure scores (e.g. SOFA index [30]) and the final outcome (death/no death). Intermediate outcomes are represented by the number of Events and Critical Events per day - the number of daily occurrences of values out of the established limits for four physiologic variables that are monitored continuously. These four variables are the Heart Rate, the Systolic Blood Pressure, the Oxygen Saturation and the Urine Output. A group of clinical specialists determined the intervals considered normal for each of these parameters [24]. This vast amount of information was modeled by:

- A Learning Classifier System to predict the length of stay (short or prolonged) and the outcome (death/no death) of patients from the HGSA ICU, attaining accuracy of 72% [23];
- A clustering framework to organ dysfunction diagnosis [22], where patients were segmented into clusters (using a Self-Organizing Map) and then each cluster was processed with the C.5 DT algorithm (the classification accuracies of the clusters ranged from 74% to 98%); and
- An ANN approach [24], where easily acquired daily inputs (the intermediate outcomes) were fed into Multilayer Perceptrons in order to predict the failure of six organic systems (e.g. liver or respiratory systems), achieving the overall accuracy of 70%.

The outputs of the organ failure prediction models are measured by the SOFA (Sequential Organ Failure Assessment) score, which provides values within the range 0, 1, 2 (normal function) and 3, 4 (failure). As for mortality assessment, the value one indicates in-hospital death while zero represents no in-hospital death. Intermediate values can be read as probabilities of occurrence. The results obtained proved the usefulness of the DM approach but, at the same time, suggested the need for more work in order to gain:

- real time data collection from bed-side monitors;
- real time outcome and intermediate outcomes prediction;
- scenario evaluation;
- increased predictive accuracy over time;

- explanation capabilities;
- simulation of therapeutic procedures consequences;
- control of effectors (e.g. infusion pumps).

To achieve these objectives the INTCare project was started.

3 INTCare System

INTCare is an agent based system, composed by several semi-autonomous agents in charge for the functionalities inherent to the system. Conceptually, it can be viewed as set of four subsystems (Fig. 1): Data Entry, Knowledge Management, Inference and Interface. Formally, the INTCare system is defined as a tuple $\Xi = \langle C_{INTCare}, \Delta_{INTCare}, a_{pp}, a_{cde}, a_{dm}, a_{pf}, a_{mi}, a_{dr}, a_{pd}, a_{sc}, a_{int}, a_{ic}\rangle$, where:

- $C_{INTCare}$ is the context and corresponds to a logical theory, represented as a triple $\langle Lg, Ax, \Delta\rangle$, where Lg stands for an extension to the language of programming logic, Ax is a set of axioms over Lg, and Δ is a set of inference rules;
- $\Delta_{INTCare}$ is the set of bridge rules defining the interaction among the systems' components (the agents);
- $a_{pp}, \ldots, a_{ic}$ are the system agents.

This formalism corresponds to a logical framework, suitable to specify agent-oriented systems based on the notion of context logic, and some properties of object-oriented design such abstraction, encapsulation, modularity and hierarchy [21]. In this work, the agents are represented as logical theories with a specific context (different agents may involve different contexts). Several agents (i.e. contexts) can be put together and be able to reason about the behavior of the entire system as a (heterogeneous) logical theory. A set of special rules called bridge-rules is applied to provide the interface among agents and systems of agents. These rules describe the agents' reactions to events occurring in their environment. The agents include a set of event types, and a set of time points. Next, the overall system is described making use of this formalism, explaining it in some detail.

3.1 Description of the Agents

The INTCare agents follow the subordinated architecture [32]; they are social computational entities, semi-autonomous, reactive, with internal-state, and pre-defined goals incited by the system creators, whose activity contributes to the goal of the overall system. In a technical perspective the INTCare agents are high granularity objects aggregating a great number of capacities.

The social model is static, pre-defined, and incorporated into the system by the developers. The intelligent behavior, the accuracy, the robustness, the flexibility and efficiency of this kind of system emerges from the agents and their interaction. The overall behavior is the combined result of its purpose and its interaction with the environment [28]. All agent interactions are defined by using events. The next lines describe the agents' functionalities and the associated events:

- Clinical Data Entry (a_{cde}) agent is responsible for the capture of clinical data from the medical and nursing staff;
- Pre-Processing (a_{pp}) agent is responsible for the correct linking of all the values in order to create a valid (even if limited in scope) medical record for the patient. It proceeds with the copy of the values entered by the medical and nursing staff (or recorded via the bedside monitors), examines them and derives new fields, if necessary (such as Critical Events);
- Data Mining (a_{dm}) agent belongs to the Knowledge Management sub-system and is responsible for the retrieval of the relevant data ($read_{dw}$) in order to make possible the application of AI algorithms to train new models (train), whenever requested by the Performance (a_{pf}) agent, storing them into the Knowledge Base ($update_{models}$). After training, the models are stored in the Knowledge Base;
- The Performance (a_{pf}) agent pro actively scans the the Data Warehouse for updates that allow statistics collection (e.g. discharge data that may or may not confirm a prediction made) so that it is able to calculate a set of assessment parameters maintained in the Performance Database. The evaluation metrics include classification accuracy, sensitivity and specificity values [24]. These statistics are updated every time new relevant information is collected. Whenever the collected statistics show that the performance has fallen bellow a predefined threshold (a configuration parameter), some action must be taken by the data mining agent in order to try to improve the performance of the system ($message_{dm}$);
- Model Initialization (a_{mi}) agent populates the Knowledge Base with the models obtained from off-line training. This agent is currently used only when first starting INTCare. This agent may be asked to perform a conversion before loading the models into the Knowledge Base;
- Data Retrieval (a_{dr}) agent is an information agent, whose only objective is that of retrieving, from the Data Warehouse, the information requested by the interface agent (get_{data}) and returning it ($send_{data}$);
- Prediction (a_{pd}) agent answers user questions by applying the adequate models (get_{models}, predict) contained in the Knowledge Base to the data stored in the Data Warehouse. Next, results are sent back to the calling agent ($send_{data}$) and if the calling agent is the interface agent, the prediction made is recorded into the Data Warehouse ($update_{dw}$);
- Scenario Evaluation (a_{sc}) agent makes it possible to the doctor to create and evaluate what-if scenarios. After receiving the data from the interface agent, the scenario evaluation agent requests a forecast from the prediction agent ($re_{prediction}$), the scenario is then stored in the Scenarios Database ($store_{scenario}$) and the result is sent back to the interface agent ($send_{result}$);
- Interface (a_{int}) agent allows (web-based) interaction with the system by providing an easy way for doctors to request prognostics and evaluate scenarios. Whenever new data is needed, this agent messages the data retrieval agent ($message_{dr}$).

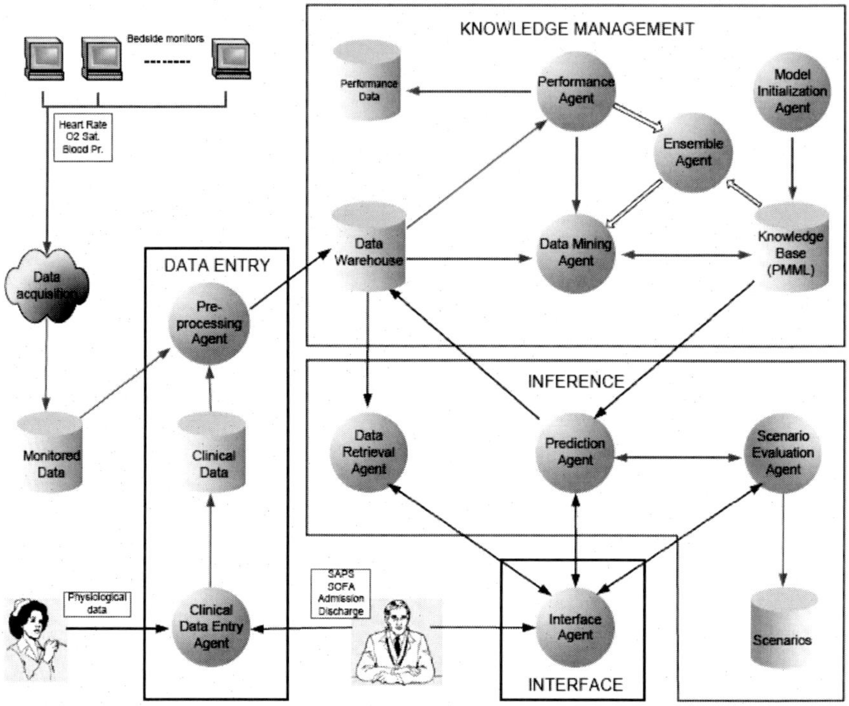

Fig. 1 INTCare's architecture

4 Problem Description

As described earlier we are currently developing an intelligent Decision Support System called INTCare [6]. Operating in an ICU setting, INTCare uses data available in the first 24 hours, after ICU admission in order to predict the patient's outcome (the patient status at the time of hospital discharge: dead or alive) and also to predict organ failure for six organ or systems (cardiovascular, respiratory, hepatic, renal, central nervous system and hematologic). Initially, the models included in INTCare were obtained via batch off-line training even though the INTCare's architecture allows for integration with the Hospital's Electronic Patient Records. Such integration will allow INTCare to be semi-autonomous as it will be possible to automatically collect data both for making predictions and to the evaluation of its predictive performance. This autonomous behavior demands the inclusion of adjustment mechanisms into INTCare, making it capable of maintaining an acceptable performance as time passes.

In this paper we present the results of several experiments on the use of this information to build a predictive model that maintains an interesting predictive performance in the ICU. In particular, we are going to address the problem of creating a system capable of predicting hospital mortality for ICU patients using data collected during the first 24 hours after ICU admission. Moreover, such a system must be able to function without human intervention, i.e. it must automatically adjust to new data. We are interested in comparing the performance of a static ensemble system with that of several dynamic ensemble systems. It is our belief that the dynamic systems better suit the problem at hand as the data available is never enough to cover the diversity of patients and the related clinical phenomena that occur in an ICU and thus the training set does not contain enough examples to completely describe the concept being investigated.

4.1 Data Description

The available data is composed of data collected during the first day of ICU stay for approximately 13000 patients. Outcome information was added to each record, indicating the status of the patient at the time of hospital discharge (dead or alive).

Four variables contain the information that remains unchanged during the patient's stay, including the site where the patient came from, the type of admission, the patient's age and the Simplified Acute Physiology Score (SAPS II) score [15] (SAPS II is a severity of disease classification system). The remaining variables (except for the outcome) contain values collected during the patient's first day in the ICU. The remaining attributes were derived from the information available on the intermediate outcomes, which are defined from four monitored characteristics: the systolic blood pressure, the heart rate, the oxygen saturation and the urine output (UR). The information regarding these monitored biometrics was condensed into 12 variables (3 for each characteristic) indicating the existence and duration of what was defined as relevant clinical events. Information regarding the definition of events and critical events may be found in [25]. Finally, the last attribute denotes the patient's final outcome (status at the time of hospital discharge).

5 Related Work

The weight update procedure we use to supervise the ensemble is inspired by both the Weighted Majority algorithm [17] and the Dynamic Weighted Majority (DWM) [10]. Even though DWM is intended to track concept drift we found the general idea appealing and decided to investigate the use of a similar algorithm in a more stable task of predicting the outcome of patients in a ICU.

Even if inspired by DWM, our implementation is different in several details. One of the most relevant is that we do not use incremental learning algorithms. After being included in our ensemble the models (or experts) are not modified in any way, only their weights are changed as determined by the algorithm. Unlike DWM, the creation and elimination of experts is not directly dependent on any individual prediction made. Rather it is the result of the overall prediction results over the entire batch of records being processed.

6 Experimental Setting

Two different ensemble evolution strategies were evaluated and compared to a "traditional" ensemble (Configuration A). In the first one (Configuration B) the model weights are changed after the evaluation of each batch of records. The weights update procedure is also included in the second configuration (Configuration C). Also, in this configuration new models are created using the records in each batch and the poor performing models (those with negative weights) are removed from the ensemble.

In order to investigate the effect of evaluating batches of records of different sizes we tested batches of 10, 20, 50, 100 and 200 records. The rationale behind this is that bigger batches may reduce the responsiveness of the system and thus lower its tendency for over training.

The same initial ensemble is used for each configuration. This ensemble was created using a process similar to the Random Subspace Method [7]. In our approach all the available training records ($\frac{1}{3}$ of the records, random selection) are used when creating a new model but each attribute has only a 50% chance of being selected. Our initial ensemble is composed of 50 models, 25 of which are neural networks , with the other 25 being decision trees . The algorithms used for model creation were those present in the Weka tool [33]. We used j48 for decision trees and the multilayer perceptron for neural networks . Both algorithms were run with the default parameter values. In our base ensemble, all the models are assigned the same initial weight ($\frac{1}{50}$).

In configurations B and C, in order to evaluate the effects of allowing changes in the model's weights we decided to adjust the weights after each prediction. The models that had made a correct prediction had their weights increased. Those who failed had their weights reduced. We started by doing this after each prediction, but then investigated whether evaluating batches of records before making the weights updates would lead to better results. We tested updating the weights after each batch of 10, 20, 50, 100 and 200 records. After each batch of predictions (of size N) we had the number of correct predictions (C). First we calculated the fraction of the increment for each weight:

$$P = \frac{C - \frac{N}{2}}{\frac{N}{2}} , \tag{1}$$

where P is positive if the model is correct in more than half of the records in the batch being considered. It is zero for those cases where exactly half of the answers are correct and it is negative for the remaining cases. If all the predictions are correct, the value for P is 1, if all are wrong the value is -1. The weight update is then the result of Eq. (2), where w_i stands for the weight of model i (by dividing P by $\frac{1}{10*number\,of\,models}$ we assure that the weight changes are not too abrupt (the initial weights are set to $\frac{1}{number\,of\,models}$)).

$$w_i = w_i + P\frac{1}{10 * number\,of\,models} . \tag{2}$$

In configuration C new models were added to the ensemble after the evaluation of each batch of records. Two new models are created: one decision tree and one neural network. Both are trained on that batch of records using the same method described above for the initial training of the ensemble. The correspondent weight is equal to the average of the weights of the other models in the ensemble. Next, we present the algorithm (as in DWM, we used the term "expert" instead of "model"). The function *Increment* computes the value of the increment as described in Eq. (1) and *UpdateWeights* updates the experts weights as indicated in Eq. (2). *DeleteNeg* removes from the ensemble all experts with negative weights. Moreover, *Create-DecTree* creates a decision tree and *Create-NNetwork* creates a neural network. Both functions use the records from the last evaluated batch. Finally, the function *Normalize* normalizes the experts' weights (so that the sum of the weights is 1). The system outputs its prediction (*answer*) for the record under evaluation.

Weight Updates Algorithm

$\{\mathbf{x}, y\}$: training data
$\{e, w\}_m^1$: set of experts and their weights
p : number of records in each batch
n : total number of records
δ_i : fraction of weight increment for expert i

num $= 0$
for i = 1,...,n
    answer $\leftarrow$ 0
    for j = 1, ..., m
        predicted $\leftarrow$ Classify(e_j, x_i)
        answer $\leftarrow$ answer + predicted * w_j
        if (predicted = y_i) num $_j \leftarrow$ num $_j$ + 1; **end if**
    end for
    output answer
    if (i mod $p = 0$)
        $\delta \leftarrow$ Increment(num, p)
        $\mathbf{w} \leftarrow$ UpdateWeights($\mathbf{w}, \delta$)
        $\{e, w\} \leftarrow$ DeleteNeg($\{e, w\}$)
        $m \leftarrow m + 1$
        $e_m \leftarrow$ Create-DecTree(precords)
        $w_m \leftarrow \bar{w}$
        $m \leftarrow m + 1$
        $e_m \leftarrow$ Create-NNetwork(precords)
        $w_m \leftarrow \bar{w}$
        $\mathbf{w} \leftarrow$ Normalize($\mathbf{w}$)
    end if
end for

To evaluate the results we used the average of the values of the area under the Receiver Operating Characteristic curve (AUC ROC) obtained after 30 runs of each experiment. The ROC curves are often used in the medical area to evaluate computational models for decision support, diagnosis and prognosis [14, 34]. A model presenting an AUC of 1 has perfect discriminative power (perfect predictive ability) while a value of 0.5 corresponds to random guessing.

7 Results

We divided the available data into two mutually exclusive datasets. Models were created using the first dataset. Those models were then evaluated using the second dataset. Several experiments were conducted with different parameter settings with regard to the frequency of weights updates and the possibility of creating new models.

We started by evaluating the performance of the static ensemble (no changes are made to the ensemble during the evaluation of the records). With no changes in the ensemble composition the AUC ROC was 78.80% ±0.93 %.

Next we tested configurations B and C by investigating the effect of using different intervals for evaluation before the weights were changed. Different configurations include weight changes after every record was evaluated or after batches of 10, 20, 50, 100 or 200 records. In Table 1 we present the AUC under the ROC curve for each of the configurations considered, allowing us to see the ensembles with better discrimination capabilities (better able to distinguish between the patients with outcome 0 or 1).

Table 1 Results for the ensemble evolution (% of AUC ROC). B - weight updates. C - weight updates and model creation and elimination

Config.	B	C
1	78.79±0.93	–
10	78.72±0.93	85.05±0.07
20	78.61±0.93	84.83±0.10
50	78.44±0.92	84.58±0.15
100	78.03±0.92	84.41±0.21
200	77.45±0.91	82.94±0.33

These results seem to indicate that it is better to update the models weights immediately after the evaluation of each record. In order to further clarify this point we decided to evaluate the ROCs in a segmented manner. Considering batches of 200 records and computing the AUC ROC for each of the batches we got the results shown in Fig. 2. The graph shows an example of the evolution of the partial AUC ROC values for one of the experiment runs.

We can clearly see that after the first weight update, the performance of the ensemble using configuration C is significantly better than that of the static ensemble (configuration A). However, even if the overall trend is positive, there are some batches of records where the AUC ROC value drops. That may be explained by the

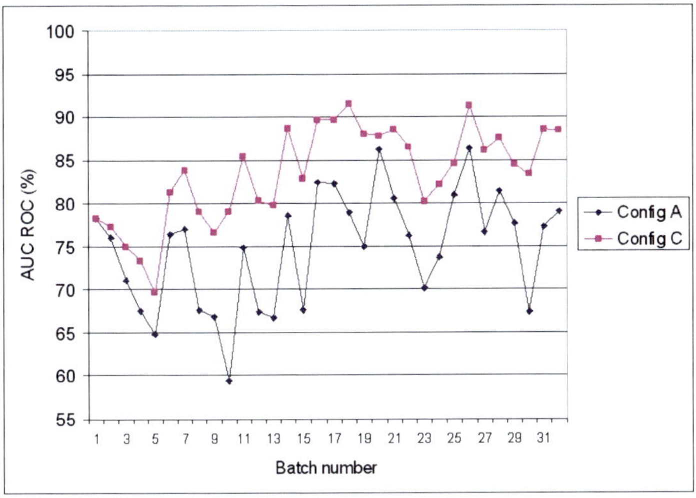

Fig. 2 Evolution of AUC ROC values over time

variables we used and by the composition of the batches. Indeed, there are some medical conditions that cannot be detected by analyzing the available variables (e.g. a patient with head trauma often has normal values for the four characteristics included in this work or a patient that rejects some therapeutic procedures due to his religious beliefs). Batches with a higher number of such records are likely to lead to lower AUC ROC values. The increase in discriminative power may be better perceived in Fig. 3. Here we show the increase in AUC ROC after each batch of 200 records is processed.

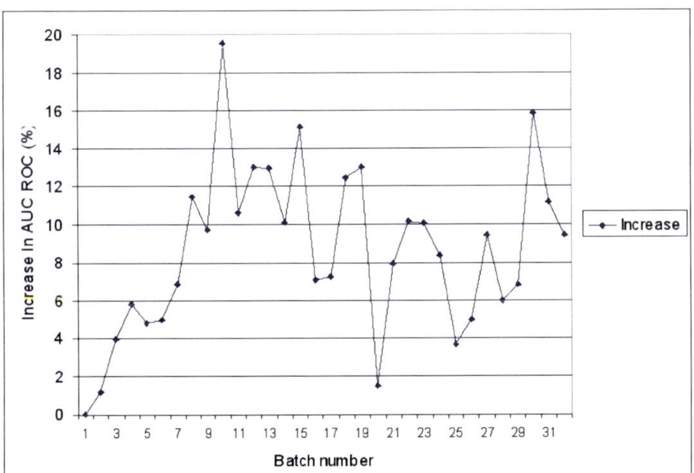

Fig. 3 AUC ROC increase

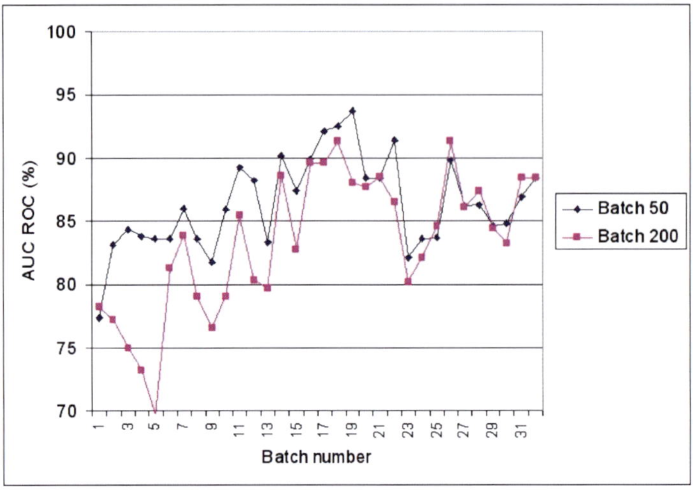

Fig. 4 AUC ROC variation with weights updated after 50 or after 200 records

It is clear that there is a real gain in using the evolution algorithm (configuration C) as opposed to a merely static ensemble (configuration A). Indeed, after the first two batches the increase in AUC ROC is always greater than 4% except for one batch.

Finally, Fig. 4 shows the differences in performance for configuration C when different batch sizes are considered. In order to allow a clear understanding we plotted only those values for batches of 50 and batches of 200 records.

8 Discussion

Ensemble methods are known to improve results when compared to those from single classifiers. Our tests show that allowing for dynamic adjustments ensemble (both in terms of models' weights and in terms of number of models) leads to overall better discriminative power of our ensemble classifier. Moreover, the intervals at which such changes are made seem to be an issue worthy of further study as different values were obtained with different intervals. Finally, the issue of whether or not to change the ensemble composition seems to point to a solution where the creation and elimination of models is encouraged.

The size of the batches used is still an open issue as smaller batch sizes increase the number of models in the ensemble and may prove impractical in a real setting. Indeed, Fig. 4 seems to suggest that in the short term smaller batches are recommended. However, if we examine more closely the right side of that figure (after allowing the ensemble to evolve over several batches of records) it seems that the performance of the second ensemble is converging to (if not surpassing) that of the first ensemble. This suggests investigating update procedures with batches of increasing size.

9 Conclusion

Future decision supports systems must be capable of adapting to changes in their environment [18, 28]. In the medical area this will allow for an easier integration of these tools in everyday use as its reliability tends to increase. In this paper we presented the results of a set of experiments in building the adaptive module of a decision support system (INTCare). Considering the available data, both in the present time and in the foreseeable future we tested dynamic hybrid ensemble architectures allowing for unassisted operation while maintaining acceptable performance. We concluded that for our problem one should allow for dynamic ensembles with the addition of new models after each batch of records is examined and the elimination from the ensemble of those models that have negative weights.

Future work includes extending the architecture to include organ failure prediction. Other necessary developments include the need to incorporate all the available data as it is being registered. For patients staying several days in the ICU the prediction models must take into account not only data from the initial 24 hours but also all the data stored since then. Moreover, relevant clinical information is possibly hidden in the sequence in which clinical adverse events occur. As the INTCare system has access to data collected via bedside monitoring, this seems to be an interesting path to be explored.

Acknowledgements. We would like to thank the referees for their comments which helped improve this paper. Part of this work was supported by the Fundação para a Ciência e Tecnologia (grant SFRH BD 28840 2006). Financial support for this study was received from the FCT project PTDC/EIA/72819/2006 and from the Algoritmi research center.

References

1. Arnott, D., Pervan, G.: A critical analysis of decision support systems research. J. Inf. Tech. 20(2), 67–87 (2005)
2. Buchanan, B., Shortliffe, E.: Rule-based expert systems: the MYCIN experiments of the Stanford Heuristic Programming Project. Addison-Wesley, Reading (1984)
3. Chang, R., Jacobs, S., Lee, B.: Predicting outcome among intensive care unit patients using computerised trend analysis of daily apache ii scores corrected for organ system failure. Intensive Care Medicine 14(5), 558–566 (1988)
4. Dietterich, T.G.: Ensemble methods in machine learning. In: Kittler, J., Roli, F. (eds.) MCS 2000. LNCS, vol. 1857, pp. 1–15. Springer, Heidelberg (2000)
5. Fayyad, U.M., Piatetsky-Shapiro, G., Smyth, P.: From data mining to knowledge discovery: an overview. In: Advances in Knowledge Discovery and Data Mining, pp. 1–34. American Association for Artificial Intelligence, Menlo Park (1996)
6. Gago, P., Santos, M.F., Silva, Á., Cortez, P., Neves, J., Gomes, L.: Intcare: a knowledge discovery based intelligent decision support system for intensive care medicine. J. Decision Syst. 14(3), 241–259 (2005)
7. Ho, T.-K.: The random subspace method for constructing decision forests. IEEE Trans. Pattern Analysis Mach. Intell. 20(8), 832–844 (1998)

8. Kim, Y., Street, W.N.: An intelligent system for customer targeting: a data mining approach. Decision Support Syst. 37(2), 215–228 (2004)
9. Klinkenberg, R., Ruping, S.: Concept drift and the importance of examples. In: Franke, J., Nakhaeizadeh, G., Renz, I. (eds.) Text Mining - Theoretical Aspects and Applications, pp. 55–77. Physica-Verlag, Heidelberg (2003)
10. Kolter, J.Z., Maloof, M.A.: Dynamic weighted majority: A new ensemble method for tracking concept drift. In: Proc. 3rd IEEE Int. Conf. Data Mining, Melbourne, FL, pp. 123–130. IEEE Comp. Soc., Los Alamitos (2003)
11. Kulikowski, C.A., Weis, S.M.: Representation of expert knowledge for consultation: the CASNET and EXPERT projects. In: Artificial Intelligence in Medicine, pp. 21–56. Westview Press, Boulder (1982)
12. Kuncheva, L.I.: Combining Pattern Classifiers: Methods and Algorithms. Wiley Interscience, Hoboken (2004)
13. Larssan, J.E., Hayes-Roth, B.: Guardian: an intelligent autonomous agent for medical monitoring and diagnosis. Intell. Syst. and their Appl. 13(1), 58–64 (1998)
14. Lasko, T.A., Bhagwat, J.G., Zou, K.H., Ohno-Machado, L.: The use of receiver operating characteristic curves in biomedical informatics. J. Biomedical Informatics 38(5), 404–415 (2005)
15. Le Gall, J.R., Lemeshow, S., Saulnier, F.: A new simplified acute physiology score (saps ii) based on a European/North American multicenter study. J Amer. Med. Assoc. 270(24), 2957–2963 (1993)
16. Lin, C., Hsieh, P.-J.: A fuzzy decision support system for strategic portfolio management. Decision Support Syst. 38(3), 383–398 (2004)
17. Littlestone, N., Warmuth, M.: The weighted majority algorithm. Information and Computation 108(2), 212–261 (1994)
18. Michalewicz, Z., Schmidt, M., Michalewicz, M., Chiriac, C.: Adaptive Business Intelligence. Springer, Heidelberg (2006)
19. Miranda, D., Nat, R., Nijk, A., Schaufeling, W., Lapichino, G.: Nursing activities score. Critical Care Medicine 31(2), 374–382 (2003)
20. Pople, H.: Evolution of an expert system: from Internist to Caduceus. In: de Lotto, I., Stefanelli, M. (eds.) Artificial Intelligence in Medicine, pp. 179–208. Elsevier Science Publisher, Amsterdam (1985)
21. Santos, M.F.: Sistemas de Classificação em Ambientes Distribuidos. PhD thesis, Universidade do Minho (1999)
22. Santos, M.F., Pereira, J., Silva, Á.: A cluster framework for data mining models – an application to intensive medicine. In: Chen, C.-S., Filipe, J., Seruca, I., Cordeiro, J. (eds.) Proc. 7th Int. Conf. Enterprise Information Systems, Miami, FL, pp. 163–168 (2005)
23. Santos, M.F., Neves, J., Abelha, A., Silva, Á., Rua, F.: Augmented data mining over clinical databases: using learning classifier systems. In: Proc. 4th Int. Conf. Enterprise Information Systems, Ciudad Real, Spain, pp. 512–516 (2002)
24. Silva, Á., Cortez, P., Santos, M.F., Gomes, L., Neves, J.: Multiple organ failure diagnosis using adverse events and neural networks. In: Seruca, I., Cordeiro, J., Hammoudi, S., Filipe, J. (eds.) Enterprise Information Systems VI, pp. 127–134. Springer, Dordrecht (2006)
25. Silva, Á., Cortez, P., Santos, M.F., Gomes, L., Neves, J.: Mortality assessment in intensive care units via adverse events using artificial neural networks. Artif. Intell. in Medicine 36(3), 223–234 (2006)
26. Shim, J.P., Warkentin, M., Courtney, J.F., Power, D.J., Sharda, R., Carlsson, C.: Past, present, and future of decision support technology. Decision Support Syst. 33(2), 111–126 (2002)

27. Tsang, E., Yung, P.: Eddie-automation: a decision support tool for financial forecasting. Decision Support Syst. 37(4), 559–565 (2004)
28. Vahidov, R., Kersten, G.E.: Decision station: situating decision support systems. Decision Support Syst. 38(22), 283–303 (2004)
29. Vicari, R.M., Flores, C.D., Silvestre, A.M., Seixas, L.J., Ladeira, M., Coelho, H.: A multi-agent intelligent environment for medical knowledge. Artif. Intell. in Medicine 27(3), 335–366 (2003)
30. Vincent, J., Moreno, R., Takala, J., Willatts, S., De Mendonça, A., Bruining, H., Reinhart, C.K., Suter, P.M., Thijs, L.G.: The sofa (sepsis-related organ failure assessment) score to describe organ dysfunction/failure. Intensive Care Medicine 22(7), 707–710 (1996)
31. Wang, W., Partridge, D., Etherington, J.: Hybrid ensembles and coincident-failure diversity. In: Proc. Int. Joint Conf. Neural Networks, Washington, DC, pp. 2376–2381. IEEE Comp. Soc., Los Alamitos (2001)
32. Weiss, G.: Multiagent Systems A Modern Approach to Distributed Artificial Intelligence. MIT Press, Cambridge (1999)
33. Witten, I.H., Frank, E.: Data Mining: Practical Machine Learning Tools and Techniques. Morgan Kaufmann, San Francisco (2005)
34. Zweig, M.H., Campbell, G.: Receiver-operating characteristic (roc) plots: a fundamental evaluation tool in clinical medicine. Clinical Chemistry 39(4), 561–577 (1993)

Index